STELLAR ASTROPHYSICAL FLUID DYNAMICS

In all phases of the life of a star, hydrodynamical processes play a major role. This volume gives a comprehensive overview of the current state of knowledge in stellar astrophysical fluid dynamics, and marks the 60th birthday of Douglas Gough, Professor of Theoretical Astrophysics at the University of Cambridge and leading contributor to stellar astrophysical fluid dynamics. Topics include properties of pulsating stars, helioseismology, convection and mixing in stellar interiors, dynamics of stellar rotation, planet formation, and the generation of stellar and planetary magnetic fields. Each chapter is written by leading experts in the field, and the book provides an overview that is central to any attempt to understand the properties of stars and their evolution. With extensive references to the technical literature, this is a valuable text for researchers and graduate students in stellar astrophysics.

MICHAEL THOMPSON is Professor of Physics at the Imperial College of Science, Technology and Medicine, London.

JØRGEN CHRISTENSEN-DALSGAARD is Professor of Helio- and Asteroseismology in the Department of Physics and Astronomy, University of Aarhus, Denmark.

STELLAR ASTROPHYSICAL FLUID DYNAMICS

Edited by

MICHAEL J. THOMPSON
Imperial College London

JØRGEN CHRISTENSEN-DALSGAARD
University of Aarhus

CAMBRIDGE
UNIVERSITY PRESS

CAMBRIDGE UNIVERSITY PRESS
Cambridge, New York, Melbourne, Madrid, Cape Town, Singapore, São Paulo

Cambridge University Press
The Edinburgh Building, Cambridge CB2 8RU, UK

Published in the United States of America by Cambridge University Press, New York

www.cambridge.org
Information on this title: www.cambridge.org/9780521818094

© Cambridge University Press 2003

First published 2003
This digitally printed version 2008

A catalogue record for this publication is available from the British Library

Library of Congress Cataloguing in Publication data

Thompson, Michael, 1959–
Stellar astrophysical fluid dynamics / Michael Thompson, Jorgen Christensen-Dalsgaard.
p.   cm.
Includes bibliographical references and index.
ISBN 0 521 81809 5
1. Fluid dynamics.   2. Astrophysics.   I. Christensen-Dalsgaard, Jørgen, 1950–   II. Title.
QB466.F58 T46   2003
523.8–dc21   2002031585

ISBN 978-0-521-81809-4 hardback
ISBN 978-0-521-05020-3 paperback

Contents

Preface

This volume, "Stellar Astrophysical Fluid Dynamics", arises from a meeting held 25-29 June 2001 to celebrate the sixtieth birthday earlier that year of Douglas Gough. Douglas has been and continues to be an inspiring and enthusiastic teacher and colleague to many, as well as a highly original and influential researcher in astrophysical fluid dynamics. Many colleagues and former research students (the categories are far from mutually exclusive) came together to celebrate, of course, but also for scientific discussions of the highest quality. The meeting fully lived up to its title of "New Developments in Astrophysical Fluid Dynamics", and although the title of the present volume has been specialised a little to emphasise the dominant stellar aspect, the full breadth of the meeting's science is retained.

The choice of venue at the Chateau de Mons, an armagnac-producing chateau in the Gers region of south-west France, was inspired and highly appropriate given Douglas's love of the region and its spirit. The food, wine and armagnac blended with the science, celebration and personal interactions to make a truly memorable week. One particular high spot occurred during a banquet after the first day of the meeting when Douglas was initiated as a Mousquetaire d'Armagnac, a brotherhood dedicated to promoting the enjoyment of armagnac throughout the World. Another was Mike McIntyre's performance at the piano of an original composition of his based on Rosanne and Douglas's telephone number ("DR, ou je te veux, encore"); and shall we ever forget Sylvie and Gerard Vauclair's persuasively feline rendition of Rossini's Cat Duet?

Of course, the meeting could not have happened without the hard work of many people. Although Douglas was unaware until the day, the plans were formed over nearly five years, in places as diverse as the Serendipity Cafe in Boulder, Colorado, and the IAU general assembly in Kyoto. We are grateful for guidance in the scientific planning of the meeting from Nigel Weiss, Wojciech Dziembowski and Juri Toomre. Rosanne Gough, Kate Thompson and Karen Christensen-Dalsgaard were involved from beginning to end particularly in planning the social side of the week. The whole thing could not have happened without our co-organiser Sylvie Vauclair, who from the first time we discussed plans with her in Kyoto put a huge amount of effort into making the meeting the success it was. As well as playing a full part in the scientific planning, Sylvie found and arranged the venue, made all local arrangements for excursions, music, banquets, handled the finances, successfully sought support for the meeting from numerous sources, and carried the brunt of all local liaison. Sylvie, thank you.

We gratefully acknowledge and thank the following organisations for financial or other support of the meeting:

- Conseil Régional Midi-Pyrénées
- Conseil Général du Gers
- Université Paul Sabatier (Toulouse)
- Lycée Bossuet de Condom
- Ferme des Etoiles (Mauroux)
- Theoretical Astrophysics Center (Denmark)

We also thank Gerard Vauclair for his help in the planning and execution of the meeting, Birte Christensen-Dalsgaard for the photograph used for the frontispiece to this book (Douglas wearing the sash and regalia of a Mousquetaire d'Armagnac), and Simon Mitton for his encouragement in bringing this volume to fruition. Last but by no means least we thank the staff of the Chateau de Mons for looking after us so well, and special thanks to their director, M. Michel Pourquet, who calmly assured us that things would happen (and they did) and made our stay at the chateau so pleasant and successful.

M.J.T., J.C.-D.

Professor Douglas Gough – June 2001, Condom, France

1

A selective overview

JØRGEN CHRISTENSEN-DALSGAARD

Teoretisk Astrofysik Center, Danmarks Grundforskningsfond, und
Institut for Fysik og Astronomi, Aarhus Universitet, DK-8000 Aarhus C, Denmark

MICHAEL J. THOMPSON

Space & Atmospheric Physics, The Blackett Laboratory, Imperial College,
London SW7 2BZ, UK

1.1 Introduction

Although sometimes ignored, there is no doubt that hydrodynamical processes play a central role in virtually all areas of astrophysics. If they are neglected in the analyses of observations and the modelling, the results for any object must become questionable; the same is therefore true of the understanding of basic astrophysical phenomena and processes that result from such investigations.

Investigations of astrophysical fluid dynamics are hampered by both theoretical and observational problems. On the theoretical side it is evident that the systems being studied are so complex that realistic analytical investigations are not possible. Furthermore, the range of scale, extending in the case of stars from the stellar radius to scales of order 100 m or less, entirely prevents a complete numerical solution. Observationally, the difficulty is to find data that are sensitive to the relevant processes, without being overwhelmed by other, similarly uncertain, effects. Progress in this field therefore requires a combination of physical intuition combined with analysis of simple model systems, possibly also experiments analogous to astrophysical systems, detailed numerical simulations to the extent that they are feasible, together with a judicious choice of observations and development and application of analysis techniques that can isolate the relevant features. Douglas Gough has excelled in all these areas.

In this brief introduction we make no pretense of reviewing the whole vast field of hydrodynamical processes in astrophysics, or even in stars. The excellent contributions to the rest to the book will do a far better job than we can here of discussing the current state and outstanding issues of many aspects of the subject. Nor do we try to review all the many contributions that Douglas has thus far made to the subject. We must content ourselves with a highly subjective selection of a few of the major themes of Douglas's work to date, on this the occasion of his sixtieth birthday, providing some (though again by no means comprehensive) context of associated work in those areas.

At this point two investigations somewhat outside even the broad general range of Douglas's research deserve to be mentioned. One (Gough & Lynden-Bell 1968) was a simple, but ingenious, experiment to study effects of turbulence in a rotating fluid and involving a rather unusual application of Alka-Seltzer tablets. The second (Bastin & Gough 1969), published a few months before the first manned lunar landing, was a computation of the thermal and radiative properties of the lunar surface, as determined by its scales of roughness, and a comparison with the observed thermal properties of the Moon. The resulting inferences of the properties of roughness can surely be characterized as a selenological inverse problem.

1.2 On taking mixing-length theory seriously

In stellar astrophysics, the most obvious hydrodynamical problem concerns convection, the effects of which are directly observable on the solar surface in the granulation. The conditions under which instability arises, *viz.* a density gradient that decreases too slowly with distance from the centre of the star such that an adiabatically rising element of gas finds itself lighter than the surroundings, are well understood, although, as discussed by Gough & Tayler (1966) additional effects such as magnetic fields may substantially complicate the stability analysis. The subsequent development of the instability, on the other hand, and the resulting energy transport and hence the relation between the temperature gradient and the flow of energy in the star, is very uncertain. The 'classic' treatment of convection in stellar modelling is through the so-called mixing-length theory, whereby convection is described by the motion of convective elements over a certain characteristic length, often taken to be a multiple of the local pressure scale height, after which the element is dissolved, delivering its excess heat to the surroundings (*e.g.* Böhm-Vitense 1958).

Analyses of convection often explicitly or implicitly make the *Boussinesque approximation*, where density variations are neglected except in the buoyancy term, the fluid being otherwise treated as incompressible. However, unlike most laboratory experiments, stellar convection typically takes place over regions of very substantial variation in density and hence inherently involves compressibility. The resulting presence of sound waves is a major complication from a numerical point of view, requiring much shorter time steps than those needed to resolve the convective motions. To bypass such complications, Gough (1969) developed the *anelastic approximation* through a formal scale analysis of the fully compressible equations; in this approximation, sound waves are precluded, resulting in a set of equations that are appropriate for the treatment of convection or, *e.g.*, internal gravity waves. As discussed in this volume by Toomre and Elliott the anelastic approximation is still commonly used in large-scale simulations of solar convection.

The visual appearance of laboratory or solar convection in the form of a more or less regular flow pattern suggests a possibly manageable type of numerical computation, whereby the horizontal properties of convection are modelled in terms of an expansion in planforms, perhaps limited to a single term, whereas the vertical behaviour is computed in detail and with substantial numerical resolution. Although still highly simplified, one may hope that such a description can provide physical insight into the behaviour of convection under more realistic circumstances. The basic properties of such *modal descriptions* were elucidated by Gough, Spiegel & Toomre (1975), who included a discussion of the asymptotic limits of weak and strong convective instability, for the case of simple laboratory situations. Numerical solutions for this case were presented by Toomre, Gough & Spiegel (1977). Latour *et al.* (1976) extended the formalism for application to stellar modelling, in the anelastic approximation. This was used by Toomre *et al.* (1976) in the study of the near-surface convection zones in A-type stars, again restricting the description to a single horizontal planform; the results suggested substantial overshoot between the separate hydrogen and helium convection zones found in such stars, suggesting that the intervening region may be mixed, with significant effects on the surface abundances of these stars.†

Even simplified numerical calculations of convection are too complex and time consuming to be included in the general modelling of stars or their pulsations. Thus simpler prescriptions, such as mixing-length theory, are unavoidable. Although mixing-length theory was originally formulated in a heuristic fashion, Spiegel (1963) and Gough & Spiegel (1977) pointed out that it could be derived in terms of a more precise physical picture. This involves a probabilistic description of the creation, motion and destruction of convective elements, depending on the degree of linear instability of the convective motion, and the assumed aspect ratio of the convective elements. For a static star the result, with proper choices of parameters, yielded the

---

† Interestingly, modern large-scale simulations often use expansions in horizontal planforms, such as in the case of the anelastic spherical-harmonic code discussed by Toomre (this volume). In such simulations, obviously, a very large number of horizontal modes are included.

same result as the normal mixing-length description. However, as developed by Gough (1965), and in more detail by Gough (1977a), the physical description lent itself to generalization to convection in a pulsating star, providing expressions for the perturbations to the convective flux and the turbulent pressure. Baker & Gough (1979) applied the resulting expressions to the modelling of RR Lyrae variables; they showed that convective effects caused a return to stability at sufficiently low effective temperature, providing a natural explanation for the red edge of the instability strip. Spiegel (1963) and Gough (1977b) noted that the treatment of mixing-length theory could be generalized to account for non-local effects, involving both the motion of a given convective eddy over a finite distance of varying stellar conditions and the average over convective eddies yielding the energy transport at a given location. Balmforth & Gough (1990) extended the non-local treatment to pulsating stars. This was applied by Balmforth (1992) to demonstrate that solar acoustic modes are likely stable, the dominant stabilizing effect being the perturbation to the turbulent pressure. Also, Houdek (2000) showed that this could account for the red edge of the δ Scuti and Cepheid instability strip.

1.3   The solar spoon

A strong indication that normal models of stellar evolution might be inadequate resulted from the initial attempts to detect neutrinos from the Sun: the observed upper limit (Davis, Harmer & Hoffman, 1968) was substantially below the model predictions (*e.g.* Bahcall, Bahcall & Shaviv, 1968). Given the simplifications made in modelling the Sun, this was perhaps not surprising. In particular, it was pointed out by, for example, Ezer & Cameron (1968) that mixing of the solar core might reduce the neutrino flux. Such mixing could result from hitherto neglected instabilities of the solar interior. Dilke & Gough (1972) found, based on a simple model of solar oscillations, that g modes might be destabilized by nuclear reactions. They speculated that such instability might trigger sudden mixing of the solar core, through the onset of convective instability. After such a mixing episode the flux of solar neutrinos would be temporarily depressed, over a few million years, and the suggestion was therefore that we are currently in such a period. Dilke & Gough also pointed out that the accompanying relatively rapid fluctuations in the solar luminosity might have acted as a trigger for the series of ice ages that is currently affecting the climate of the Earth, thus making more likely the required special nature of the present epoch. More detailed calculations by Christensen-Dalsgaard, Dilke & Gough (1974), Boury *et al.*

(1975) and Shibahashi, Osaki & Unno (1975) confirmed the reality of the instability, resulting from the build-up of the gradient of the ^{3}He abundance, the first instability setting in after about 200 Myr of evolution. However, the subsequent non-linear development into convective mixing has not been demonstrated. Dziembowski (1983) showed that resonant coupling would limit the amplitudes at a level far below what might be expected to initiate mixing. Furthermore, analyses by Merryfield, Toomre & Gough (1990, 1991), albeit for a simplified physical system, found a tendency for non-linear development to lead to sustained finite-amplitude oscillations rather than direct convective mixing. Even so, it should be kept in mind that solar models are subject to instabilities which have not been properly taken into account in solar modelling.

The closely related issue of the effects of a sustained flow on the solar ^{3}He profile and neutrino flux is addressed by Jordinson (this volume).

As discussed by Shibahashi (this volume), inferences from helioseismology place strong constraints on solar structure; the inferred helioseismic model is in fact very close to the 'standard solar models', including also the predicted neutrino flux. In particular, the excellent agreement between normal, unmixed, models and helioseismic inferences of the solar core indicates that no substantial mixing has taken place, in contrast to the proposal by Dilke & Gough (1972). Although helioseismology does not provide direct information about the solar core temperature upon which the neutrino production predominantly depends, these results nevertheless make it plausible that the cause for the neutrino discrepancy lies in the properties of the neutrino, rather than in errors in solar models. Specifically, if neutrinos have finite mass, the electron neutrinos generated in the nuclear reactions in the Sun may in part be converted to muon or tau neutrinos before reaching the detectors. Strong indications of this process were obtained by measurements from the Sudbury Neutrino Observatory (SNO) which allowed the total neutrino capture rate to be compared with the capture rate of electron neutrinos (Ahmad *et al.* 2001). Direct measurements of all types of neutrinos by Ahmad *et al.* (2002) have very recently provided dramatic confirmation of the transformation between the different types of neutrinos; the total neutrino flux was found to be fully consistent with the predictions of standard solar models, confirming the indications from helioseismology that such models are in fact good representations of solar structure, as reflected in the neutrino production.

It may be somewhat disappointing to Douglas that the conclusion of the efforts, to which he has contributed substantially, over more than three decades to understand the origin of the solar neutrino problem does not point towards novel aspects of stellar interior physics. However, the prospects now of using the predicted solar neutrino production, constrained by the results of helioseismology, to investigate the properties of the neutrinos are perhaps even more exiting.

1.4 Deep roots of solar cycles

The solar 11-year sunspot cycle, and the 22-year magnetic cycle, are clear indications of large-scale dynamical processes in the Sun. These are normally assumed to result from 'dynamo processes', involving a coupling between the solar differential rotation, convection and the magnetic field. However, although models exist which reproduce aspects of the solar cycle, these cannot be regarded as sufficiently definitive that other mechanisms, perhaps involving magnetic oscillations of the deep solar interior, can be ruled out. Indeed, Dicke (1970, 1978) proposed that the periodicity might arise from a regular oscillator in the deep solar interior, modulated by the convection zone to give rise to the apparently somewhat irregular periodicity observed in the solar cycle; he noted that this would be reflected in the long-term phase stability of the cycle, in contrast to the dynamo models where a more erratic phase behaviour might be expected. Gough (1978a, 1981a) analysed examples of the two models, as well as the available data on sunspot maxima and minima; the results suggested that the solar behaviour was intermediate between the models, with no firm conclusion possible.

The variations in solar irradiance accompanying the solar cycle are well established (*e.g.* Willson & Hudson, 1991; Pap & Fröhlich, 1999); there have also been studies, although somewhat conflicting, concerning possible variations in the solar radius (*e.g.* Gavryusev *et al.* 1994; Noël 1997; Brown & Christensen-Dalsgaard 1998). As discussed by Gough (1981a), the changes in solar structure associated with the solar cycle depend on the physical location of the dominant mechanisms causing these changes, thus providing the possibility of obtaining information about the physical nature of the cycle. In particular, he noted that the ratio between the luminosity variations, assumed to be reflected in the irradiance, and the radius variations would depend on the location of the physical mechanism responsible for the solar cycle. Gough analysed the ratio $W = \Delta \ln R / \Delta \ln L$, R being the surface radius and L the luminosity, showing that it increases with increasing depth of the perturbation that produces the variations (see also Däppen 1983). While the irradiance variations, corresponding to an increase of about 0.1 % at sunspot maximum relative to sunspot minimum, has been measured precisely from space, results on the radius variations are somewhat contradictory. Major difficulties are the correction for changes in solar atmospheric structure and, for the visual observations which span the most extended period, corrections for observational bias. It has been pointed out (*e.g.* Parkinson, Morrison & Stephenson 1980) that the width of the path of totality at a total solar eclipse provides an accurate measure of the solar radius, given that the lunar diameter and the ephemeris are known with great accuracy. To apply this idea Gough & Parkinson (1983) carried out determinations of the edges of the eclipse path during the 1983 solar eclipse in Indonesia, by distributing teams of observers to span the northern and southern limits of totality. Unfortunately, some confusion amongst the observers led to inconsistencies in the results which made the desired precise determination impossible.

1.5 Helioseismology: oscillations as a diagnostic of the solar interior

A major development in observational stellar astrophysics took place in 1975 with the first announcements of observations of coherent solar oscillations. Oscillatory signals on the solar surface had been detected substantially earlier, the first detailed results having been obtained by Leighton, Noyes & Simon (1962). These observations showed localized oscillations with periods near five minutes, but with apparently limited spatial and temporal coherence. Thus the oscillations were generally regarded as an atmospheric phenomenon, although a more global nature was suggested by Ulrich (1970), Leibacher & Stein (1971) and Wolff (1972), based on somewhat more detailed observations by Frazier (1968). The global nature of the oscillations was definitely established by the observations by Deubner (1975) and Rhodes, Ulrich & Simon (1977): by obtaining fairly extensive data as a function of both position and time they were able to construct two-dimensional power spectra, as functions of horizontal wave number and frequency, and demonstrate that power was concentrated in ridges corresponding to the modal nature of the oscillations, as earlier predicted by the theoretical analyses. Furthermore, Hill, Stebbins & Brown (1976) announced the detection of oscillations, with periods between 13 and 50 minutes in the apparent solar diameter. Also, Brookes, Isaak & van der Raay (1976) and Severny, Kotov & Tsap (1976) independently presented evidence for an oscillation with a period of very close to 160 minutes in solar Doppler observations.

It was immediately obvious that frequencies of global solar oscillation would be extremely powerful probes of the solar interior. This led to expectations that a 'heliological inverse problem' might be formulated, analogous to the inverse problems used in geophysics to infer the structure of the Earth's interior (Christensen-Dalsgaard & Gough 1976). An initial comparison was also made by Christensen-Dalsgaard & Gough between the frequencies reported from the observations by Hill *et al.* (1976) and a solar model; some significance was attached, a little prematurely, by the junior author to the apparently reasonable agreement between observations and model. Indeed, initially, much interest centred on the observations by Hill *et al.* which, since they involved a full solar diameter, appeared to be truly global; in contrast, the modes detected by Deubner and by Rhodes *et al.* were concentrated in a relatively superficial part of the Sun. Thus several additional comparisons were made between the observed frequencies and those of solar models (*e.g.* Scuflaire *et al.* 1975; Iben & Mahaffy 1976; Rouse 1977). Even so, possible alternative explanations unrelated to global solar oscillations were also sought; an interesting hydrodynamical possibility was seiches in supergranules, similar to those observed *e.g.* in a glass of beer (Gough, Pringle & Spiegel, 1976). The longer-period oscillation discovered by Brookes *et al.* and Severny *et al.* was potentially of even greater interest: if it were truly a global solar oscillation it would have to be a g mode, as demonstrated by Christensen-Dalsgaard, Cooper & Gough (1983), and hence it would be sensitive to the deep solar interior. However, the rather peculiar nature of the mode, particularly the long period and the absence of neighbouring modes in the expected dense spectrum of g modes, prompted alternative explanations; in particular, the close proximity of the period to 1/9 of a day led to suspicions that the observations might be related to phenomena in the Earth's atmosphere.

Although of a more superficial nature, the original high-degree five-minute oscillations are of very substantial diagnostic potential for the solar interior. They are concentrated in the upper parts of the convection zone. However, since the convection zone is essentially adiabatically stratified, apart from a very thin upper layer, its structure is fully determined by specifying the (constant) value of the specific entropy. As a result, the constraints from the five-minute oscillations on the upper parts of the convection zone essentially allow extrapolation to the rest of the convection zone. Consequently, Gough (1977c) and Ulrich & Rhodes (1977) were able, on the basis of the early data, to infer that the convection zone was substantially deeper than obtained from the then current solar models. Also, the modes are directly sensitive to the equation of state in the ionization zones of hydrogen and helium, providing a first helioseismic test of the physics of the solar interior (Berthomieu *et al.* 1980; Lubow *et al.* 1980). Gough (1984a) emphasized the great significance of using such analyses to test the highly complex thermodynamic properties of matter under solar conditions, as well as to determine the present solar envelope helium abundance. As discussed by Däppen in this volume, such investigations of the 'microphysics' of the solar interior have been very successful, allowing tests of subtle aspects of the equation of state in the solar convection zone.

The oscillations are also sensitive to motions in the solar interior. Deubner, Ulrich & Rhodes (1979) noted that the frequencies of the five-minute oscillations were shifted by the sub-surface rotational velocity. More general rotation laws were considered by Gough (1981b), and independently by Hansen, Cox & van Horn (1977) and Cuypers (1980), thereby laying the foundation for the use of observed rotational frequency splittings to investigate solar internal rotation, by extending earlier work by Ledoux (1951) who had considered uniform rotation.

It is interesting that much of the early enthusiasm for helioseismology was inspired by the diameter observations by Hill *et al.* and the detection of the 100-minute oscillation. Later observations, including the detailed data obtained for periods below 15 minutes and the extensive attempts to detect long-period oscillations, make it overwhelmingly likely that the early results were in fact unrelated to global solar effects. However, they served an important role as inspirations for other observational efforts which have led to the dramatic success of helioseismology based on the five-minute oscillations, also evident in the present volume.

1.6 Inverting helioseismic data

A key issue in the applications of helioseismology is evidently the ability to carry out *inversion, i.e.,* to infer from the observed frequencies localized information about the properties of the solar interior. Similar problems have a long history of study in geophysics.† It was realized by Gough (1978b) that the method proposed by Backus & Gilbert (1968) was well suited to obtain inferences of, for example, solar rotation in the form localized averages characterized by well-defined *averaging kernels*. Establishing a procedure that has been used extensively since, Gough tested the method on artificial data similar to the observations presented by Deubner *et al.* (1979). Gough (1982) analyzed rotational splittings inferred from observations of the solar diameter by Bos & Hill (1983), to constrain the solar internal rotation and thence the gravitational quadrupole moment J_2 of the Sun, which is of obvious importance to tests based on planetary motion of gravitational

---

† A discussion of the information transfer between geo- and helioseismology was provided by Gough (1996a).

theories. In addition to the Backus-Gilbert technique he applied a least-squares polynomial fit as well as a minimization of the integral of the square of the radial gradient of angular velocity, subject to the constraints of the observed splittings being satisfied exactly. It is interesting to note that the latter technique resembles the regularized least-squares fitting technique now commonly used in helioseismic analyses.

The solar quadrupole moment affects the gravitational field outside the Sun and hence the motion of planets. This effect must be taken into account in tests of general relativity based on, for example, the precession of the orbit of Mercury. Given the value of J_2 obtained by Gough (1982), radar observations of planetary motion were indeed consistent with general relativity. Although, as already mentioned, the observations of oscillations in the solar diameter were questionable, more recent inferences of J_2 based on helioseismic inversions of frequencies in the five-minute region have confirmed this conclusion (*e.g.* Pijpers 1998; Roxburgh 2001).

Douglas, with his student A. Cooper, also developed inversion techniques for the inference of solar internal structure. Since the dependence of the oscillation frequencies on, *e.g.*, the sound speed and density of the solar interior is strongly nonlinear, this is normally carried out by linearization around a reference model, relating the frequency differences between observations and model to the differences in structure between the Sun and the model. By means of tests involving structure and frequency differences between simplified solar models, they investigated the ability of the Backus-Gilbert technique to construct localized information about the density difference between the Sun and the model. The success depended greatly on the assumed mode set (Cooper 1981; Gough 1984b): with just low-degree p modes only broad averages in the outer parts of the model could be obtained, whereas with a set including also low-degree g modes good resolution was possible throughout the model. These results were of substantial significance in the evaluation of the DISCO mission to observe solar oscillations, proposed to ESA (*e.g.* Bonnet *et al.* 1981; Balogh *et al.* 1981): it was estimated that this mission would indeed be able to observe low-order, low-degree g modes. While DISCO was in the end not selected, and the detection of g modes, even from the much more powerful instruments on the SOHO spacecraft, remains elusive (*e.g.* Appourchaux *et al.* 2000), the detailed p-mode data have dramatically realized the potential for structure inversion (Shibahashi, this volume).

Approximate analyses of stellar oscillation frequencies have played a major role in the development of helio- and asteroseismology. An early example was the derivation by Gough, Ostriker & Stobie (1965) of approximate expres-

sions for the periods of radially pulsating stellar atmospheres. A powerful
technique for helioseismic inversion followed from the realization by Duvall
(1982) that the observed frequencies satisfied a simple relation with radial
order and degree. Gough (1984c) demonstrated that this followed naturally
from the asymptotic properties of high-order acoustic modes, relating the
observed behaviour to the sound-speed profile in the Sun. He also showed
that as a result the solar internal equatorial rotation rate could be simply
inferred from the observed frequencies. Furthermore, Douglas developed
a similar procedure allowing the determination of the sound speed of the
solar interior without recourse to a solar model (Christensen-Dalsgaard *et
al.* 1985). A similar asymptotic inversion technique was developed in geophysics (*e.g.* Brodskiĭ & Levshin 1977). This type of analysis, extended to
include higher-order asymptotic effects, remains a very useful technique for
inversion, and furthermore has provided valuable insight into the relations
between the properties of the solar interior and the frequencies (see also
Gough 1986, 1993).

1.7 On the detection of subphotospheric convective velocities and temperature fluctuations

Global-mode analysis of the sun's oscillations is not well suited to detecting subsurface structures – such as convective cells – that vary in longitude
as well as latitude and depth. To first order the global modes sense only
the longitudinally averaged, north-south symmetric average of the internal
structure. In recent years a battery of so-called local helioseismic techniques
have been developed that use analyses of the wavefield in a local region of
the solar surface to probe such subsurface variation. One such technique
is ring analysis (Hill 1988, Morrow 1988). The method is being used to
map subsurface flows and their temporal variation (see Toomre, this volume) and wavespeed variations due to magnetic and thermal perturbations.
This is achieved by inferring from the surface observations the local dispersion relation of the p-mode waves under patches of the visible disk. The
methodology was proposed and laid out in a paper by Gough & Toomre
(1983), wherein the possibility of using local perturbations of the dispersion relation of p modes to detect subphotospheric convective velocities and
thermal fluctuations was put forward and kernels relating the perturbations
to thermal fluctuations and flows were developed.† Further development of

---

† Interestingly, the early inference by Deubner, Rhodes & Ulrich (1979) of solar internal rotation
based on the high-degree five-minute oscillations was formally similar, although expressed in
terms of rotation rather than local flow fields.

the analysis was reported by Hill, Gough & Toomre (1984), who carried out inversions to infer the radial variations of subsurface velocity and detected possible evidence for temporal variations tentatively identified as the effect of giant convective cells.

Ring analysis, like other local techniques, has only really come into its own in the second half of the 1990s with the availability of high quality, high-degree data from the MDI instrument on the SOHO spacecraft and the high-resolution ground-based Taiwanese Oscillation Network (TON) and GONG. Douglas has continued to contribute to the development of the method as a practical technique through his collaborations with Toomre, Haber, Hill and colleagues. Another local technique which has been very successful at producing maps of subsurface flows and wavespeed anomalies due to thermal fluctuations and magnetic fields is time-distance helioseismology, introduced by Duvall *et al.* (1993). Analogous to travel-time tomography in geoseismology, the subphotospheric structure and flows are deduced from the travel times of waves propagating through the subsurface region. The travel times are inferred from cross-correlating the oscillatory signal at different points or regions of the sun's surface. Some of the results of this approach are discussed in the chapter on 'telechronohelioseismology' by Kosovichev. Although these are probably the most developed local methods, others are being developed and applied. One such is acoustic holography (Lindsey & Braun 1997), which *inter alia* was used by Lindsey & Braun (2000) to obtain the first images of active regions on the far side of the sun before they were carried by rotation onto the visible disk. Over a number of years, Douglas has worked on another method that seeks to exploit the phase distortion experienced by waves propagating through a locally inhomogeneous medium (Gough, Merryfield & Toomre 1991; Julien, Gough & Toomre 1995a,b). The approach has a mathematical elegance and has been demonstrated with success in 1-D problems; but on a 2-D surface the interference between waves propagating in nearly parallel directions introduces additional complication, and the method has not as yet been applied successfully to solar data.

The results of local helioseismology present intriguing glimpses of the complex subsurface structure and dynamics in the Sun, including what Toomre has denoted the 'subsurface weather'. Studies of this nature will be a focus for the upcoming Solar Dynamics Observatory, to be launched by NASA in 2007. In parallel with the instrumental developments more detailed investigations of the diagnostic capabilities of local helioseismology have been initiated (*e.g.* Birch *et al.* 2001). In spite of the successes, there is a need for further development of the forward problem in local helioseismology, in order to understand the effects of inhomogeneities and flows on the propagation of waves in the solar interior and the way that these effects are reflected in the observations. This directly affects the interpretation of the results of local helioseismology. There are ongoing efforts by Gough, Toomre and collaborators to model the effects of shearing flows on ring analysis; in time-distance helioseismology, there has been recent work by Gizon & Birch (2002) (see also references therein).

1.8 Prospects for asteroseismic inference

The early successes of helioseismology, including the results based on low-degree modes from observations of the Sun in disk-averaged light as one would obtain for a distant star, immediately raised the prospect of similar *asteroseismic* investigations of other stars (for an etymological discussion of the term 'asteroseismology', see Gough 1996b). Although the results of such investigations are unavoidably less detailed than in the solar case, this is compensated for by the possibility of studying a broad range of stellar types, with physical effects that are not found in the Sun. An important example are the properties of convective cores, which are found in main-sequence stars of masses only slightly higher than solar. The stochastic excitation of solar oscillations by convection, strongly supported for example by the statistical analysis by Chang & Gough (1998), makes it very likely that similar oscillations should be present in stars with vigorous near-surface convection. Also, asteroseismic investigations are possible in the many other classes of stars that show rich spectra of oscillations, such as several types of white dwarfs (*e.g.* Winget 1988, Kawaler & Bradley 1994).

Early observations of stellar oscillation spectra similar to the low-degree five-minute spectrum of the sun were discussed by Gough (1985). In particular, he considered the almost uniform frequency separation deduced from the observations, characteristic of high-order acoustic modes, and noted that this provided constraints on the stellar radii. Two of these stars, including the solar near-twin α Cen A, were so similar to the sun that solar-like oscillations of low amplitude might be expected; however, it is probably fair to say that the claimed detections, which for α Cen A implied a radius in apparent conflict with other data on the star, may have been spurious.

The third case, reported by Kurtz & Seeman (1983), was an early example of the class of so-called *rapidly oscillating Ap stars* (see Cunha, this volume), which are substantially hotter than the Sun; the properties of the oscillations of these stars are closely linked to the large-scale magnetic field in the stars. Further investigations of the star discussed by Kurtz & Seeman, HR 1217, have shown striking departure from the nearly uniform frequency spacing.

Cunha & Gough (2000) carried out a detailed analysis of the effects of the magnetic field on the oscillation frequencies. On this basis Cunha (2001) reconsidered the expected properties of HR 1217, predicting the presence of an additional mode until then unobserved, which would account for the apparent anomalies of the spectrum. Remarkably, this prediction has now been confirmed by Kurtz *et al.* (2002) from an extended series of observations of the star. This indicates the prospects for extending asteroseismic investigations to include detailed studies of effects of magnetic fields.

Much interest continues to centre on studies of solar-like oscillations. The broad-band nature of the stochastic excitation leads to expected spectra of oscillations where most of the low-degree modes in a range of frequencies are likely present, greatly simplifying the task of identifying the modes.† Furthermore, early experience with solar oscillations has lead to a good understanding of the diagnostic capabilities of their frequencies. Gough (1987) reviewed the use of observed frequencies to constrain the properties of main-sequence stars, particularly their age, emphasizing the need to supplement the asteroseismic data with other, more traditional observations. More detailed analyses of the prospects for age determination were provided by, for example, Gough & Novotny (1990). Gough (2002), emphasizing the great prospects for obtaining information about stellar properties from asteroseismology, nevertheless introduced some remarks of realism in reaction to previous possibly exaggerated claims.

The main difficulty in observing solar-like oscillations in main-sequence stars are their expected very low amplitudes, below $1\,\mathrm{m\,s^{-1}}$ in radial velocity and a few parts per million in photometry. Thus, despite intensive efforts, the claimed detections have until recently been somewhat doubtful. This has changed dramatically in the last few years, with the development of very stable and efficient spectrographic techniques aimed at the detection of extra-solar planets. A clear detection of solar-like oscillations in the star β Hydri was reported by Bedding *et al.* (2001); Gough (2001) noted that this could be regarded as the true beginnings of asteroseismology for solar-like stars. Data of even higher quality were obtained by Bouchy & Carrier (2001) for α Cen A, while Frandsen *et al.* (2002) found clear solar-like oscillations in the red giant ξ Hydrae, with periods of a few hours. Although these investigations have been limited to bright stars, data of similar or better quality for a broad range of stars can be expected when the HARPS instrument

† In contrast, the amplitude limitation mechanism, still inadequately understood, for opacity-driven pulsators leads to only a modest selection of the unstable modes reaching observable amplitudes; this greatly complicates the interpretation of data on such stars. See Dziembowski (this volume).

(Queloz *et al.* 2001) enters operation in 2003 on the 3.6-m telescope of the European Southern Observatory.

The potential of space-based asteroseismology in terms of continuity and absence of atmospheric noise was recognized early (*e.g.* Mangeney & Praderie 1984; Gough 1985), leading to a substantial number of mission proposals (see Roxburgh 2002). Two proposals to the European Space Agency for substantial asteroseismic missions reached Phase A studies: PRISMA (Appourchaux *et al.* 1993) and STARS (Badiali *et al.* 1996) but failed to be selected; the projects were rated highly, but it was recommended to secure the maturity of the field by means of smaller precursor projects. Such projects are now being developed, at the national level: the Canadian MOST mission (Matthews 1998) to be launched in the spring of 2003, as well as the French-led COROT mission (Baglin *et al.* 2002) and the Danish-led Rømer mission (Christensen-Dalsgaard 2002), both with launch planned for 2005. Also, in May 2002 ESA finally selected the far more ambitious Eddington mission (Favata 2002), to carry out asteroseismology on a large number of stars and search for extra-solar planets, with a launch no later than 2008.

Given the advances in ground-based instrumentation and the upcoming space missions, the prospects for asteroseismology are indeed excellent. We have no doubt that Douglas will continue his seminal contributions, making the fullest use of these new opportunities, with emphasis on the application of astrophysical data to the better understanding of astrophysical fluid dynamics and other deep and difficult physical issues.

Acknowledgements. This seems an appropriate place to record our deep gratitude to Douglas for the invaluable inspiration that he has provided us, as well as the general community, as a teacher and colleague, as well as for his friendship. The present review has been supported in part by the Danish National Research Foundation, through the establishment of the Theoretical Astrophysics Center. We are grateful to Werner Däppen for his careful reading of and comments on an earlier version.

References

Ahmad, Q. R., *et al.*, 2001, *PRL*, **87**, 071301
Ahmad, Q. R., *et al.*, 2002, *PRL*, **89**, 011301
Appourchaux, T., *et al.*, 1993, *PRISMA: Probing rotation and interior of stars: microvariability and activity*, ESA Report on Phase-A Study SCI(93)3, ESTEC, Noordwijk
Appourchaux, T., *et al.*, 2000, *ApJ*, **538**, 401
Backus, G. & Gilbert, F., 1968, *Geophys. J. R. astr. Soc.*, **16**, 169

Badiali, M., *et al.*, 1996, *STARS: Seismic telescope for astrophysical research from space, ESA Report on Phase-A Study D/SCI(96)4*, ESTEC, Noordwijk

Baglin, A., Auvergne, M., Barge, P., Buey, J.-T., Catala, C., Michel, E., Weiss, W., and the COROT Team, 2002, in Favata, F., Roxburgh, I. W. & Galadí-Enríquez, D., eds, ESA SP-485, *Proc. 1st Eddington Workshop, 'Stellar Structure and Habitable Planet Finding'*, ESA Publications Division, Noordwijk, The Netherlands, p. 17

Bahcall, J. N., Bahcall, N. A. & Shaviv, G., 1968, *PRL*, **20**, 1209

Baker, N. H. & Gough, D. O., 1979, *ApJ*, **234**, 232

Balmforth, N. J., 1992, *MNRAS*, **255**, 603

Balmforth, N. J. & Gough, D. O., 1990, *Solar Phys.*, **128**, 161

Balogh, A., Bonnet, R. M., Delache, P., Fröhlich, C. & Harvey, C. C. (DISCO Science Team), 1981, *DISCO Re-assessment Study*, ESA Sci(81)6, European Space Agency, Paris.

Bastin, J. A. & Gough, D. O., 1969, *Icarus*, **11**, 289

Bedding, T. R., Butler, R. P., Kjeldsen, H., Baldry, I. K., O'Toole, S. J., Tinney, C. G., Marcey, G. W., Kienzle, F. & Carrier, F., 2001, *ApJ*, **549**, L105

Berthomieu, G., Cooper, A. J., Gough, D. O., Osaki, Y., Provost, J. & Rocca, A., 1980, in Hill, H. A. & Dziembowski, W., eds, *Lecture Notes in Physics*, **125**, Springer-Verlag, Berlin, p. 307

Birch, A. C., Kosovichev, A. G., Price, G. H. & Schlottmann, R. B., 2001, *ApJ*, **561**, L229

Böhm-Vitense, E., 1958, *Z. Astrophys.*, **46**, 108

Bonnet, R. M., Crommelynck, D., Delaboudinière, J. P., Delache, P., Fossat, E., Fröhlich, C., Gough, D., Grec, E., Simon, P. & Thuillier, G. (DISCO Science Team), 1981, *DISCO Assessment Study*, ESA Sci(81)3, European Space Agency, Paris.

Bos, R. J. & Hill, H. A., 1983, *Solar Phys.*, **82**, 89

Bouchy, F. & Carrier, F., 2001, *A&A*, **374**, L5

Boury, A., Gabriel, M., Noels, A., Scuflaire, R. & Ledoux, P., 1975, *A&A*, **41**, 279

Brodskiĭ, M. A. & Levshin, A., 1977, *Dodl. Akad. Nauk. SSSR*, **233**, 312

Brookes, J. R., Isaak, G. R. & van der Raay, H. B., 1976, *Nature*, **259**, 92

Brown, T. M. & Christensen-Dalsgaard, J., 1998, *ApJ*, **500**, L195

Chang, H.-Y. & Gough, D. O., 1998, *Solar Phys.*, **181**, 251

Christensen-Dalsgaard, J., 2002, in Favata, F., Roxburgh, I. W. & Galadí-Enríquez, D., eds, ESA SP-485, *Proc. 1st Eddington Workshop, 'Stellar Structure and Habitable Planet Finding'*, ESA Publications Division, Noordwijk, The Netherlands, p. 25

Christensen-Dalsgaard, J. & Gough, D. O., 1976, *Nature*, **259**, 89

Christensen-Dalsgaard, J., Cooper, A. J. & Gough, D. O., 1983, *MNRAS*, **203**, 165

Christensen-Dalsgaard, J., Dilke, F. W. W. & Gough, D. O., 1974, *MNRAS*, **169**, 429

Christensen-Dalsgaard, J., Duvall, T. L., Gough, D. O., Harvey, J. W. & Rhodes Jr, E. J., 1985, *Nature*, **315**, 378

Cooper, A. J., 1981, *PhD Dissertation*, University of Cambridge.

Cunha, M. S., 2001, *MNRAS*, **325**, 373

Cunha, M. S. & Gough, D., 2000, *MNRAS*, **319**, 1020

Cuypers, J. 1980, *A&A*, **89**, 207

Däppen, W., 1983, *A&A*, **124**, 11

Davis, R., Harmer, D. S. & Hoffman, K. C., 1968, *PRL*, **20**, 1205

Deubner, F.-L., 1975, *A&A*, **44**, 371

Deubner, F.-L., Ulrich, R. K. & Rhodes, E. J., 1979, *A&A*, **72**, 177

Dicke, R. H., 1970, in Slettebak, A., ed., *Proc. IAU Colloq. on Stellar Rotation*, Reidel, Dordrecht, p. 289

Dicke, R. H., 1978, *Nature*, **276**, 676

Dilke, F. W. W. & Gough, D. O., 1972, *Nature*, **240**, 262

Duvall, T. L., 1982, *Nature*, **300**, 242

Duvall, T. L., Jefferies, S. M., Harvey, J. W. & Pomerantz, M. A., 1993, *Nature*, **362**, 430

Dziembowski, W., 1983, *Solar Phys.*, **82**, 259

Ezer, D. & Cameron, A. G. W., 1968, *Astrophys. Lett.*, **1**, 177

Favata, F., 2002, in Favata, F., Roxburgh, I. W. & Galadí-Enríquez, D., eds, ESA SP-485, *Proc. 1st Eddington Workshop, 'Stellar Structure and Habitable Planet Finding'*, ESA Publications Division, Noordwijk, The Netherlands, p. 3

Frandsen, S., Carrier, F., Aerts, C., Stello, D., Maas, T., Burnet, M., Bruntt, H., Teixeira, T. C., de Medeiros, J. R., Bouchy, F., Kjeldsen, H., Pijpers, F. & Christensen-Dalsgaard, J., 2002, *A&A*, **394**, L5.

Frazier, E. N., 1968, *Z. Astrophys.*, **68**, 345

Gavryusev, V., Gavryuseva, E., Delache, Ph. & Laclare, F. 1994, *A&A*, **286**, 305

Gizon, L. & Birch, A. C., 2002, *ApJ*, **571**, 966

Gough, D. O., 1965, *Geophysical Fluid Dynamics II*, Woods Hole Oceanographic Institution, p. 49

Gough, D. O., 1969, *J. Atmos. Sci.*, **26**, 448

Gough, D. O., 1977a, *ApJ*, **214**, 196

Gough, D. O., 1977b, in Spiegel, E. A. & Zahn, J.-P., eds, *Problems of stellar convection, IAU Colloq. No. 38, Lecture Notes in Physics*, **71**, Springer-Verlag, Berlin, p. 15

Gough, D. O., 1977c, in Bonnet, R. M. & Delache, P., eds, *Proc. IAU Colloq. No. 36: The energy balance and hydrodynamics of the solar chromosphere and corona*, G. de Bussac, Clairmont-Ferrand, p. 3

Gough, D. O., 1978a, In *Pleins feux sur la physique solaire. Proc. 2me Assemblée Européenne de Physique Solaire*, CNRS, Paris, p. 81

Gough, D. O., 1978b, in Belvedere, G. & Paterno, L., eds, *Proc. Workshop on solar rotation*, University of Catania Press, p. 255

Gough, D. O., 1981a, in Sofia, S., ed., NASA Conf. Publ. 2191, *Variations in the Solar Constant*, Washington, p. 185

Gough, D. O., 1981b, *MNRAS*, **196**, 731

Gough, D. O., 1982, *Nature*, **298**, 334

Gough, D. O., 1984a, *Mem. Soc. Astron. Ital.*, **55**, 13

Gough, D. O., 1984b, in Ulrich, R. K., Harvey, J., Rhodes, E. J. & Toomre, J., eds, *Solar seismology from space*, NASA, JPL Publ. 84 – 84, p. 49

Gough, D. O., 1984c, *Phil. Trans. R. Soc. London, Ser. A*, **313**, 27

Gough, D., 1985, *Nature*, 314, 14

Gough, D. O., 1986, in Osaki, Y., ed., *Hydrodynamic and magnetohydrodynamic problems in the Sun and stars*, Department of Astronomy, University of Tokyo, p. 117

Gough, D. O., 1987, *Nature*, **326**, 257

Gough, D. O., 1993, in Zahn, J.-P. & Zinn-Justin, J., eds, *Astrophysical fluid dynamics, Les Houches Session XLVII*, Elsevier, Amsterdam, p. 399

Gough, D. O., 1996a, in Jacobsen, B. H., Moosegard, K. & Sibani, P., eds, *Inverse methods, Lecture Notes in Earth Sciences*, **63**, Springer-Verlag, Berlin, p. 1

Gough, D. O., 1996b, *Observatory*, **116**, 313

Gough, D., 2001, *Science*, **291**, 2325

Gough, D. O., 2002, in Favata, F., Roxburgh, I. W. & Galadí-Enríquez, D., eds, ESA SP-485, *Proc. 1st Eddington Workshop, 'Stellar Structure and Habitable Planet Finding'*, ESA Publications Division, Noordwijk, The Netherlands, p. 65

Gough, D. O. & Lynden-Bell, D., 1968, *JFM*, **32**, 437

Gough, D. O. & Novotny, E., 1990, *Solar Phys.*, **128**, 143

Gough, D. & Parkinson, J., 1983, *New Scientist*, **98**, 882.

Gough, D. O. & Spiegel, E. A., 1977, in Spiegel, E. A. & Zahn, J.-P., eds, *Problems of stellar convection, IAU Colloq. No. 38, Lecture Notes in Physics*, **71**, Springer-Verlag, Berlin, p. 57

Gough, D. O. & Tayler, R. J., 1966, *MNRAS*, **133**, 85

Gough, D. O. & Toomre, J., 1983, *Solar Phys.*, **82**, 401

Gough, D. O., Merryfield, W. J. & Toomre, J., 1991, in Gough, D. O. & Toomre, J., eds, *Challenges to theories of the structure of moderate-mass stars, Lecture Notes in Physics*, **388**, Springer, Heidelberg, p. 265

Gough, D. O., Ostriker, J. P. & Stobie, R. S., 1965, *ApJ*, **142**, 1649

Gough, D. O., Pringle, J. E. & Spiegel, E. A., 1976, *Nature*, **264**, 424

Gough, D. O., Spiegel, E. A. & Toomre, J., 1975, *JFM*, **68**, 695

Hansen, C. J., Cox, J. P. & van Horn, H. M., 1977, *ApJ*, **217**, 151

Hill, F., 1988, *ApJ*, **333**, 996

Hill, F., Gough, D. & Toomre, J., 1984, *Mem. Soc. Astron. Ital.*, **55**, 153

Hill, H. A., Stebbins, R. T. & Brown, T. M., 1976, in Sanders, J. H. & Wapstra, A. H., eds, *Atomic Masses and Fundamental Constants*, **5** Plenum Press, p. 622

Houdek, G., 2000, in Breger, M. & Montgomery, M. H., eds, *Delta Scuti and related stars*, ASP Conference Series, **210**, San Francisco, p. 454

Iben, I. & Mahaffy, J., 1976, *ApJ*, **209**, L39

Julien, K. A., Gough, D. O. & Toomre, J., 1995a, in Hoeksema, J. T., Domingo, V., Fleck, B. & Battrick, B., eds, ESA SP-376, *Proc. Fourth SOHO Workshop: Helioseismology*, vol. 2, ESA Publications Division, Noordwijk, The Netherlands, p. 155

Julien, K. A., Gough, D. O. & Toomre, J., 1995b, in Ulrich, R. K., Rhodes Jr, E. J. & Däppen, W., eds, *Proc. GONG'94: Helio- and Astero-seismology from Earth and Space*, ASP Conf. Ser., **76**, San Francisco, p. 196

Kawaler, S. D. & Bradley, P. A., 1994, *ApJ*, **427**, 415

Kurtz, D. W. & Seeman, J., 1983, *MNRAS*, **205**, 11

Kurtz, D. W., *et al.*, 2002, *MNRAS*, **330**, L57

Latour, J., Spiegel, E. A., Toomre, J. & Zahn, J.-P., 1976, *ApJ*, **207**, 233

Ledoux, P., 1951, *ApJ*, **114**, 373

Leibacher, J. & Stein, R. F., 1971, *Astrophys. Lett.*, **7**, 191

Leighton, R. B., Noyes, R. W. & Simon, G. W., 1962, *ApJ*, **135**, 474

Lindsey, C. & Braun, D. C., 1997, *ApJ*, **485**, 895

Lindsey, C. & Braun, D. C., 2000, *Science*, **287**, 1799

Lubow, S. H., Rhodes, E. J. & Ulrich, R. K., 1980, in Hill, H. A. & Dziembowski, W., eds, *Lecture Notes in Physics*, **125**, Springer-Verlag, Berlin, p. 300

Mangeney, A. & Praderie, F., (eds), 1984, *Space Research Prospects in Stellar Activity and Variability*, Paris Observatory Press.

Matthews, J. M., 1998, in Korzennik, S. G. & Wilson, A., eds, ESA SP-418, *Structure and dynamics of the interior of the Sun and Sun-like stars; Proc. SOHO 6/GONG 98 Workshop*, ESA Publications Division, Noordwijk, The Netherlands, p. 395

Merryfield, W. J., Toomre, J. & Gough, D. O., 1990, *ApJ*, **353**, 678

Merryfield, W. J., Toomre, J. & Gough, D. O., 1991, *ApJ*, **367**, 658

Morrow, C. A., 1988, *PhD Dissertation*, University of Colorado and National Center for Atmospheric Research, Boulder.

Noël, F., 1997, *A&A*, **325**, 825

Pap, J. M. & Fröhlich, C., 1999, *J. Atmos. Solar-Terr. Phys.*, **61**, 15

Parkinson, J. H., Morrison, L. V. & Stephenson, F. R., 1980, *Nature*, **288**, 548

Pijpers, F. P., 1998, *MNRAS*, **297**, L76

Queloz, D., *et al.*, 2001, *ESO Messenger*, No. 105, 1

Rhodes, E. J., Ulrich, R. K. & Simon, G. W., 1977, *ApJ*, **218**, 901

Rouse, C. A., 1977, *A&A*, **55**, 477

Roxburgh, I. W., 2001, *A&A*, **377**, 688

Roxburgh, I. W., 2002, in Favata, F., Roxburgh, I. W. & Galadí-Enríquez, D., eds, ESA SP-485, *Proc. 1st Eddington Workshop, 'Stellar Structure and Habitable Planet Finding'*, ESA Publications Division, Noordwijk, The Netherlands, p. 11

Scuflaire, R., Gabriel, M., Noels, A. & Boury, A., 1975, *A&A*, **45**, 15

Severny, A. B., Kotov, V. A. & Tsap, T. T., 1976, *Nature*, **259**, 87

Shibahashi, H., Osaki, Y. & Unno, W., 1975, *PASJ*, **27**, 401

Spiegel, E. A., 1963, *ApJ*, **138**, 216

Toomre, J., Gough, D. O. & Spiegel, E. A., 1977, *JFM*, **79**, 1

Toomre, J., Zahn, J.-P., Latour, J. & Spiegel, E. A., 1976, *ApJ*, **207**, 545

Ulrich, R. K., 1970, *ApJ*, **162**, 993

Ulrich, R. K. & Rhodes, E. J., 1977, *ApJ*, **218**, 521

Willson, R. C. & Hudson, H. S., 1991, *Nature*, **351**, 42

Winget, D. E., 1988, in Christensen-Dalsgaard, J. & Frandsen, S., eds, *Proc. IAU Symposium No 123, Advances in helio- and asteroseismology*, Reidel, Dordrecht, p. 305

Wolff, C. L., 1972, *ApJ*, **176**, 833

I Stellar convection and oscillations

2

On the diversity of stellar pulsations

WOJCIECH A. DZIEMBOWSKI

Warsaw University Observatory and
Copernicus Astronomical Centre, Polish Academy of Sciences, Warsaw, Poland

Pulsation is a common phenomenon in stars. It occurs in a wide range of their masses and in all evolutionary phases, exhibiting large variety of forms. Stochastic driving and just two distinct instability mechanisms are the cause of the widespread phenomenon. The diversity of pulsation properties in stars across the H-R diagram is partially explained in terms of differences in the ranges of unstable modes and in terms nonlinear mechanisms of amplitude limitation. Still a great deal remains to be explained.

2.1 Introduction

Excitation of the fundamental radial mode was the essence of the pulsation hypothesis when it was first proposed by Ritter in 1879, as an explanation of periodic variability in stars. Radial symmetry of the motion was confirmed for a number of objects by means of observational tests. Excitation of the same, presumably fundamental, mode in all δ Cephei type stars got support in the discovery of the period-luminosity relation, which at some point seemed unique. Soon, the hypothesis that only the fundamental radial mode may be excited became a dogma like the earlier one that stars do not vary.

Referring to Schwarzschild's (1942) suggestion that RRc stars might be first overtone pulsators, Rosseland (1949) wrote: *This hypothesis involves the very difficult problem of how to excite a higher mode to pulsation while leaving the fundamental mode unexcited.* Referring to Ledoux's (1951) proposal that nonradial modes are excited in β Canis Majoris stars, Chandrasekhar and Lebovitz (1962), though not questioning the claim, still had this comment: *... one is generally reluctant to accept suggestions to appeal directly to the excitation of non-radial modes (besides the radial modes) on the grounds that such modes should be highly damped relative to radial modes*

*and, further, that their excitation would be "difficult" in view of the possible
source of such excitation being deep in the interior.*

Today we know that in many stars nonradial modes are excited by the
same mechanism as radial ones. In many others only the former are unstable. Firm evidence for overtone pulsation was found among most classical
pulsators such as Cepheids and RR Lyrae stars. In agreement with theoretical predictions, even second pulsators have been identified. The latter
finding is relatively new. It came as a by-product of massive photometric
surveys aimed at the detection of microlensing events.

Astronomers were aware of diversity in the form of stellar pulsation from
the very beginning of astrophysics. Baily introduced his division of RR
Lyrae stars into subtypes a, b, and c already in 1899. Eight years later
Blazkho discovered amplitude variations in one of the RRa stars. Not long
after, Hertzsprung described variations of the shape of light curves with
period in Cepheids. With the progress in observational methods we have
learned about a much larger variety of stellar pulsation. We only partially
understand how it comes about. Remarkably, we still do not have a fully
satisfactory model for the effect discovered by Blazkho.

2.2 Types of stellar pulsation

One natural division of pulsating stars is into *stochastically driven pulsators*
and *unstable-mode pulsators*. There are only few distant stars for which we
have information about stochastically excited modes. The pattern of mode
excitation is the same as in the sun. For the rest of present review I will be
concerned only with the latter type and I will subdivide it into *giant-type*
and *dwarf-type*.

2.2.1 Giant-type pulsators

It is only after data from massive photometric surveys became available
that we have a fair statistics of pulsational behaviour in classical pulsating stars. Table 2.1, which is based on data from Udalski et al. (1999)
and Udalski et al. (2000), gives the percentage of various types of pulsating
Cepheids in, respectively, the Small and Large Magellanic Clouds as determined from the OGLE II project. We see that the fundamental mode is
the most frequent choice but the first-overtone pulsators are common too.
Few pure second-overtone Cepheids are found and only in the SMC. Also
double-mode pulsators are very rare. These facts call for an interpretation.

No firm evidence as yet has been found for nonradial modes in Cepheid

Table 2.1. *Magellanic Cloud Cepheids pulsating in various modes*

modes	SMC (%)	LMC(%)
fundamental	58.5	56.9
fund. + first ov.	1.2	1.4
first overtone	35.9	37.5
second + first ov.	3.6	4.2
second overtone	0.7	0

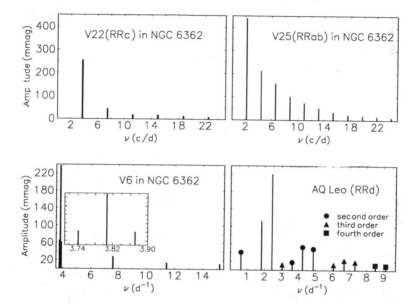

Fig. 2.1. Oscillation spectra for RR Lyrae stars. Data for the NGC 6362 stars are from Olech et al. (2001) and those for AQ Leonis are from Jerzykiewicz & Wentzel (1977). The two stars in the upper row are monomode pulsators. All peaks shown there are at multiples of the pulsation frequencies, which is that of the first overtone for V22 and the fundamental mode for V25. For V6 we see two close side peaks causing Blazkho-type amplitude modulation. In AQ Leonis both the fundamental and first overtone of radial pulsation are excited. The remaining peaks are at various combinations of the two basic frequencies.

pulsation. However, a few cases of long-time amplitude and phase changes were found and remain unexplained. In contrast, long-time modulations are rather common among RR Lyrae stars. The first evidence for nonradial mode excitation in RR Lyrae was found by Olech et al. (1999). The evidence was based on frequency analysis of RR Lyrae light curves, which has revealed the presence of closely spaced peaks. Subsequent analyses (see Kovács, 2002 for a summary) performed on large samples of light curves added many new objects with the same property.

Still the majority of RR Lyrae stars are apparently monoperiodic and pulsating in the fundamental mode (RRab) or first overtone (RRc). The upper panels of Figure 2.1 show examples of oscillation spectra for objects of these two subtypes. The lower panels show the spectra of two types of multiperiodic pulsation. In the left panel we see three closely and equally spaced peaks, which certainly cannot be attributed to different radial modes. The right panel shows the spectrum for AQ Leonis – the first discovered RRd star, which is the adopted designation for objects with the fundamental and first overtone simultaneously present.

Actually, V6 in NGC 6362 is not a strong case for nonradial mode excitation. Although, as we shall see later, its spectrum may be explained in such terms it may also be interpreted in terms of a single radial mode with periodically modulated amplitude. The strong case from the same cluster is the object V37, which has only two close peaks. The observed modulation is then the result of two-frequency beating like in RRd stars but with a longer period. According to surveys summarized by Kovács (2002) the cases of two close peaks are more common.

2.3 Dwarf-type pulsators

Along the main-sequence band there is only one star, BW Vul, that mimic the behaviour of Cepheid and RR Lyrae stars in its pulsation form. It is monoperiodic and of high amplitude, 0.3 mag in the V-band. All remaining pulsators have amplitudes of individual modes below 0.1 mag. Typically, more than one mode is detected if observations are carried out for a long time. A good example is FG Virginis, a δ Scuti type star whose oscillation spectrum is shown in Figure 2.2.

Stars of this type lie in the low-luminosity extension of the Cepheid instability strip. High-amplitude pulsators are found in this type but all lie above the main-sequence band. There is a clear correlation between the pulsation form and the evolutionary status.

Along the main-sequence band, both below and above δ Scuti stars, there are pulsating stars showing striking diversity in the radial orders n of the excited modes. In δ Scuti stars we find p-modes of orders from $n = 1$ up to 7 and some low-order g-modes. Magnetic stars occupying part of the δ Scuti domain choose p-modes of much higher orders ($n > 20$). Immediately below there is a domain of γ Doradus star, which are high-order g-mode pulsators. Above δ Scuti stars, after short break around spectral type A0

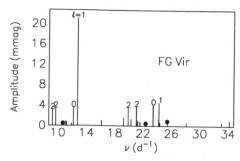

Fig. 2.2. Oscillation spectrum for FG Virginis. Amplitudes and frequencies are from Breger et al. (1998). Numbers on top of the bars indicate the spherical harmonic orders of modes as determined by Viskum et al. (1998). The three bars marked with filled circles are at nonlinear combination frequencies. The remaining 22 bars correspond to eigenmode frequencies.

we have SPB stars, which also choose high-order g-modes. At still higher luminosity is the β Cephei domain, where stars pulsate again in low-order p- and g-modes.

There are three domains of g-mode pulsation along the white dwarf cooling sequence and a domain in the hot extension of the horizontal branch of sdB stars which show a similar mode preference as δ Scuti stars. They all, like the main-sequence stars, are multimode low-amplitude pulsators.

2.4 Inference from linear theory

Linear stability calculations for stellar models predict simultaneous instability of a large number of modes leaving to speculation the problem of the final amplitude outcome of the instability. Nonetheless, such calculations yield an important step towards understanding stellar pulsation. Their results that may be directly compared with observations are the ranges in frequency and spherical-harmonic degree ℓ of unstable modes. The agreement is a support for the model in which the driving effect may then be easily identified. On such grounds, we may claim that we understand the origin of oscillations in nearly all types of objects mentioned in the previous section. It is remarkable that, despite the whole richness in the pulsational behaviours, there are only two driving mechanisms that seem to account for all the cases.

We now have a satisfactory interpretation in terms of the opacity mechanism for the two large instability domains in the H-R diagram: the classical Cepheid instability strip and the newly explained B star instability strip. In the first case the driving effect arises in the hydrogen and helium ionization layers. In the second case it arises in a local maximum of opacity caused by iron lines at a temperature of about 2×10^5 K. Finding this new instability strip, which includes main-sequence and subdwarf B stars, followed improvement in stellar opacities. In fact, finding instability in models of B subdwarf models preceded the discovery of oscillations in these stars (Charpinet et al., 1996).

Even at the linear theory level there are unsolved problems. They concern role of convection, which is far more important for stars of the Cepheid instability strip, especially but not only in determining its red edge. The pioneering efforts by Gough (1977) to include the effects of turbulent convection in stellar pulsation were followed by many. Nonetheless an accurate and credible modeling is missing. In consequence, our understanding of variability in stars cooler than Cepheids is the poorest. In many cases we are not even sure whether the variability is due to pulsation and, if it is, whether it is due to unstable or to stochastically driven modes.

Convection does not always exert a damping effect on oscillations. Brickhill (1983) first noted that in the case of slow modes, modulation of the convective flux during the pulsation cycle promotes driving. He proposed this effect as the cause of g-mode excitation in ZZ Ceti stars – oscillating white dwarfs. There are recent developments of this idea by Goldreich & Wu (1999). This driving mechanism is less common in stars than the classical opacity-mechanism but it is the only alternative mechanism leading to mode instability that may be associated with the observed stellar variability. In addition to ZZ Ceti stars the mechanism may work in oscillating DB (helium) white dwarfs and possibly in γ Doradus stars (Wu, 2002).

Typically, in both giant- and dwarf-type pulsators models, there is a large number of unstable modes. Figures 2.3 and 2.4 show the driving rates as a function of frequency for low-degree modes in representative models of δ Scuti and RR Lyrae stars. The former, which approximately fits data on FG Vir (see Figure 2.2), describes a star in an advanced main-sequence evolutionary phase. The latter is a model of a horizontal-branch star in the advanced core helium burning phase. Let us note the differences. The frequency range of the unstable modes is significantly wider in the main-sequence star. The instability of radial modes extends in this case from radial order $n = 1$ to 7. The wide range of mode frequencies is indeed observed in δ Scuti stars. Our selected model reproduces very well the frequency range of modes detected in FG Vir. In helium burning pulsators, instability goes at most up to $n = 3$ and this is the highest-order mode detected in such stars.

More important differences are seen in the nonradial mode properties.

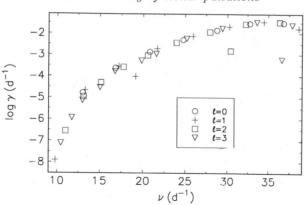

Fig. 2.3. Growth rates and frequencies of low-degree ($\ell \leq 3$) modes in a model of a δ Scuti star. The mass is 1.8 times solar, the initial composition is similar to that of the sun and the evolutionary status is an advanced phase of hydrogen burning in the convective core (the central abundance X_c by mass of hydrogen is reduced to 0.085).

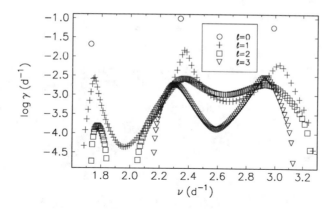

Fig. 2.4. Growth rates and frequencies of low-degree ($\ell \leq 3$) modes in a model of a RR Lyrae star. The mass is 0.67 times solar, the initial composition is typical for population II objects and the evolutionary status is an advanced phase of helium burning in the convective core (the helium abundance is reduced to 17% of the original value). Lack of symbols corresponding to $\ell = 2$ and 3 in certain frequency ranges means lack of unstable modes.

There are many more nonradial modes between consecutive radial modes in the RR Lyrae model than in the δ Scuti model. Spectra of nonradial modes are actually denser than shown in Figures 2.3 and 2.4, as each of the modes is split into $2\ell + 1$ components by rotation. The much greater mode density in the RR Lyrae model is a consequence of the much larger values

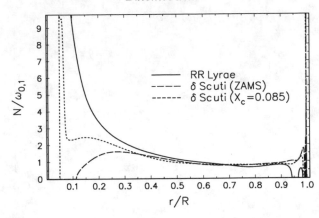

Fig. 2.5. The ratio of the Brunt-Väisälä to the fundamental radial mode frequency in the models used in Figures. 2.3, 2.4 and in a model ZAMS star. In the ZAMS model the central hydrogen abundance $X_c = 0.7$. The maximum value of the ratio in the RR Lyrae model is about 300 and it is reached at $r/R = 0.007$.

of the Brunt-Väisälä frequency N in the interior of this star. In Figure 2.5 we see the behaviour of $N(r)$ in the two models discussed and in a ZAMS star of the same mass as our selected δ Scuti star. The effect of evolution is a growth of N in the interior and a development of a gravity wave (G) cavity there.

The radial order n_g associated with the G cavity is approximately given by

$$n_g \approx \frac{\sqrt{\ell(\ell+1)}}{\pi} \int \sqrt{\left(\frac{N}{\omega}\right)^2 - 1} \frac{\mathrm{d}r}{r} \qquad (2.1)$$

(e.g. Van Hoolst et al., 1998), where $\omega = 2\pi\nu$ is the angular frequency of the mode. The integral should be taken over the G cavity. The frequency distance between consecutive modes of the same degree is thus estimated as

$$\Delta_{\ell,n_g} \approx \frac{\omega}{n_g}. \qquad (2.2)$$

In the RR Lyrae star model, at $\ell = 1$ and a frequency corresponding to the fundamental radial mode, we get $n_g = 180$.

Already in the evolved main-sequence star, some additional nonradial modes result from the growth of the G cavity around the shrinking convective core. Also in this model we see structures in the $\gamma(\nu)$ dependence reflecting mode trapping effect. Local minima correspond to modes partially trapped in the G cavity. The effect is much more dramatic in the RR Lyrae star. For all nonradial modes most of the oscillation energy is confined to the

G cavity. The relative contribution from the acoustic cavity is only about 10% for the $\ell = 1$ modes corresponding to local maxima. At this degree still most of contribution to damping and driving arises in the outer layers so that the difference between in driving rates between the $\ell = 0$ and $\ell = 1$ modes is mainly due to the difference in inertia. At $\ell = 2$ and 3, damping in the G cavity results in mode stability in certain frequency ranges.

Even for the most strongly trapped nonradial modes, the driving rates, γ, are significantly lower than those for the closest radial modes, which is not true in the δ Scuti star. We may be tempted to take this fact as the explanation why RR Lyrae and other evolved stars exhibit preference for radial pulsation. However, more detailed comparison with observations warns us against such inference. The modes detected in FG Virginis at $\nu \approx 10\mathrm{d}^{-1}$ have their driving rates lower by six orders of magnitude than those at $\nu \approx 30\mathrm{d}^{-1}$. The dependence of amplitude on frequency shown in Figure 2.2 bears no resemblance to the $\gamma(\nu)$ dependence shown in Figure 2.3. Clearly, we cannot rely on the driving rates for predicting amplitudes of modes surviving in the nonlinear development.

2.5 Saturation of the linear instability

Christy (1964) was the first to construct fully nonlinear models of Cepheids and RR Lyrae stars. His models converged to a periodic constant amplitude pulsation state, which – according to his interpretation – was reached through a saturation, that is, through modification induced by pulsation in the mean structure, leading to zeroing driving rate. Depending on mean value of L and T_{eff}, the terminal state was either fundamental or first-overtone pulsation. Later on Stobie (1969) found also second-overtone pulsation in his Cepheid models. It took 30 years to find such a form of pulsation in real objects. Stellingwerf (1975) with his novel method was able to determine exclusive domains in the H-R diagram of first overtone and fundamental mode pulsation and an intermediate domain, which he named the EO (either-or), where both modes were possible depending which one was excited first.

The origin of double-mode pulsation was not understood for a long time and even now the problem is not fully clarified. It has been approached in a number of works by means of numerical solution of the full nonlinear problem and with the amplitude equation formalism. Nonlinear saturation shows up at the cubic order in pulsation amplitudes. With our cubic order formalism we (Dziembowski & Kovács, 1984) derived a simple criterion for double-mode pulsation. Here I outline our analysis.

The nonlinear driving rates were written in the form

$$\gamma_{j,N} = \gamma_j \left(1 + \sum_k \alpha_{jk} A_k^2 \right). \tag{2.3}$$

Only the case of two linearly unstable modes was considered. It was assumed that all saturation coefficients α_{jk} are negative, which is necessary for saturation and confirmed by subsequent numerical calculations. The terminal amplitudes are determined by the set of two equations (2.3) with $\gamma_{j,N} = 0$. For monomode solutions, which with our assumptions always exist, we have

$$A_j^2 = -\frac{1}{\alpha_{jj}}. \tag{2.4}$$

It is stable if $s_k \equiv \alpha_{kj}/\alpha_{jj} - 1 > 0$, where $k \neq j$, which means that the mode is more effectively saturating instability of the competing mode than that of its own. If $s_1 > 0$ and $s_2 > 0$ then we are in the EO domain. The double-mode solution

$$A_1^2 = -\frac{1}{\alpha_{11}} \frac{s_1}{s_1 + s_2 - s_1 s_2}, \quad A_2^2 = -\frac{1}{\alpha_{22}} \frac{s_2}{s_1 + s_2 - s_1 s_2}$$

exists if $s_1 s_2 > 0$. However, it is stable only if $s_1 < 0$ and $s_2 < 0$. Thus, monomode and double-mode pulsations are mutually exclusive.

The fact that that double-mode pulsations are so rare may be interpreted in two ways. Either the range of parameters leading to $s_1 < 0$ and $s_2 < 0$ is very narrow or we have $s_1 > 0$ and $s_2 > 0$ but the amplitude of mode 1 cannot reach its saturation value given in equation (2.4) due to an accidental resonance with a damped mode. We preferred the second way and suggested that there is a 2:1 resonance with a higher-order radial mode. The problem with this idea, which was realized later, was lack of required resonance in realistic models of double-mode pulsators.

Successful numerical simulations of double-mode pulsation in Cepheids (Kolláth et al., 1998) and RR Lyrae stars (Feuchtinger, 1998) were obtained only after effects of convection were included. The resonance played no role in this models. In a recent paper Kolláth et al. (2002) interpreted these results with the amplitude equations. They found that the condition $s_1 < 0$ and $s_2 < 0$ was satisfied in their double-mode pulsators. This was never the case in purely radiative models. Unfortunately, they did not identify the specific effect of convection responsible for the enhancement of self-saturation, causing stabilization of the double-mode pulsation.

An unpleasant aspect of this solution is the fact that it rests on a crude description of convection in which there are four adjustable parameters. Furthermore, there is a problem of nonradial modes, whose presence has been ignored in all numerical models so far. We will see later that a resonant coupling with those modes may play a role for the properties of radial pulsation. There is also a question regarding the role of strongly unstable high-degree modes. Many such modes, having driving rates similar to the radial modes, exist in all models of RR Lyrae star and Cepheids (e.g. Van Hoolst et al., 1998). If the saturation is the dominant amplitude-limiting effect, then we should expect that often one of such modes wins the competition. The resulting pulsation would be undetectable by means of photometry. Observations do not indicate that it may be the case. The RR Lyrae strip seems filled up with the pulsators. Thus, we have to admit that we still do not understand why radial monomode pulsation is preferred by stars in the upper Cepheid instability strip.

2.6 Amplitude limitation by resonances

That resonances may play a role in stellar pulsation has been realized well before numerical modeling became possible. This is what Rosseland (1949) wrote about about early attempts to explain the shapes of Cepheid light curves: *The conclusion seems unavoidable that some particular property of the star plays an active role in shaping the curves. It may be something like a resonance between the fundamental and higher mode suggested by Woltjer (1935, 1937).* Modern investigations have confirmed this.

2.6.1 The 2:1 resonance

The Hertzsprung sequence of Cepheid light curves has been successfully explained in terms of the frequency distance from the 2:1 resonance centre between the fundamental mode ($n = 1$) and the second overtone ($n = 3$) (Simon & Schmidt, 1976, Buchler et al., 1990).

Also Woltjer was the first to point out that the 2:1 resonance causes amplitude limitation. Since it is a lower-order effect in terms of amplitude, one might expect that it should be more efficient in amplitude limitation than saturation. However, it does not seem to be the case in Cepheids. The amplitudes at 10-day period, which is the resonance centre, are not markedly lowered. The point is that this resonance may be a sole amplitude-limiting effect only if the damping of modes of higher radial order is fast enough. A stable double-mode solution exists only if $2\gamma_1 + \gamma_3 < 0$.

A critical role for the 2:1 resonance in the amplitude limitation was found in limiting the growth of the ϵ-mechanism-driven instability of high-mass stars by Papaloizou (1973). He showed that the resonance between the unstable fundamental mode and the stable first overtone is very effective preventing a catastrophic mass loss, which was suggested by earlier investigators.

2.6.2 Parametric resonance and dwarf and giant dichotomy

Three-mode coupling caused by the parametric resonance is another lowest-order nonlinear effect leading to amplitude limitation. In this case the effect is due to dissipation of energy by a pair of linearly stable (daughter) modes for whose sum are close to frequency of an unstable (parent) mode. Denoting with subscripts a and b the daughter modes and with c the parent mode we have

$$\omega_c = \omega_b + \omega_a + \Delta\omega ,$$

with $|\Delta\omega| \ll \omega_c$ and

$$\gamma_c > 0 , \quad \gamma_a < 0 , \quad \gamma_b < 0 .$$

An exponential growth of modes a and b occurs if the amplitude of mode c exceeds the critical value, which (e.g. Vandakurov, 1981) is given by

$$A_{c,\mathrm{crit}} = \sqrt{\frac{\gamma_a\gamma_b}{C_{abc}}\left[1 + \left(\frac{\Delta\omega}{\gamma_d}\right)^2\right]} , \qquad (2.5)$$

where $\gamma_d = \gamma_a + \gamma_b$. The coupling coefficient C_{abc} is a volume integral with integrand containing products of eigenfunctions of the three involved modes. The general expression is complicated (Dziembowski, 1982) but it is easy to show that $C_{abc} \neq 0$ only if the azimuthal orders satisfy the condition $m_c = m_a + m_b$ and the difference between the two highest ℓ is not larger than the lowest one. For instance, if the parent mode is radial then the daughter modes must have $m_a = -m_b$ and $\ell_a = \ell_b$.

Freedom in choosing ℓ_a and m_a allows fine frequency tuning. The frequency distance, Δ_{ℓ,n_g}, decreases approximately as ℓ^{-1} (see equations 2.1 and 2.2), hence considering daughter modes with $\ell \to \infty$ we may approach $\Delta\omega = 0$. This favours high-ℓ mode excitation. The opposite effect is that of damping. If the quasi-adiabatic approximation applies then we have approximately (e.g. Van Hoolst et al., 1998)

$$\gamma \approx \frac{\bar{N}^2}{\tau_g}\frac{\ell^2}{\omega^2} , \qquad (2.6)$$

where τ_g is the thermal time scale of the G cavity. In main-sequence stars the instability first appears at certain intermediate though still rather high

ℓ values implying that the daughter modes are most likely undetectable. Important observable consequences occur for the parent mode, whose amplitude may be reduced to the level not much exceeding $A_{c,crit}$ and may be modulated.

The character of the terminal pulsation state resulting from the interaction between the parent and the daughter modes depends on the mismatch, $\Delta\omega$, and the driving (damping) rates γ. Stable stationary solutions with the parent-mode amplitude given by the right-hand side of equation (2.5), but with γ_d replaced by $\gamma_s = \gamma_c + \gamma_d$ exist in wide range of parameters (see Wersinger et al., 1980, Dziembowski, 1982). Outside that range, in particular for a close resonance, only time-dependent amplitude limitation is possible. The solution may take a form of a single- or multi-periodic limit cycle and, going through a series of period doubling, become chaotic. Still equation (2.5) may be used for a crude estimate of the mean amplitude. If $\gamma_s < 0$ then amplitude limitation in any form by the sole effect of the parametric resonance is not possible.

My first application of the theory of parametric resonance was to estimate the amplitude of an $\ell = 1$ g-mode in the sun. Dilke & Gough (1972) showed that the mode may be driven by the ϵ mechanism and speculated that it might reach high amplitude, high enough to mix the solar interior. This was an ingenious idea invented to solve the neutrino deficit problem. The results of my calculations (Dziembowski, 1983) were unfortunately discouraging. The parametric instability was found to set in at very low amplitudes, far lower than needed for mixing. The next application was to explain the low pulsation amplitudes of δ Scuti stars (Dziembowski & Królikowska, 1985). We found that the amplitudes of unstable modes were limited by the three-mode interaction to the level of 1 - 10 mmag, which was in a rough agreement with observations. The values are well below the ones needed to saturate the instability.

It seemed that we were on the road toward explaining the systematic difference between giant and dwarf pulsators. In Cepheids and RR Lyrae stars the parametric resonance does not prevent high pulsation amplitudes for two reasons. Damping rates of daughter modes are much higher than in δ Scuti stars due to the much higher $\bar{N}$ (see Figure 2.5) and τ_g is shorter. The second reason is a weaker coupling (smaller C_{abc}) between the parent radial and the potential daughter modes. The latter are trapped in the deep interior, where the former ones have very low amplitudes. The truth is, however, that not much happened after those works. The difficulty is that most likely much more than just one pair of daughter modes is excited at the onset of the parent-mode instability. It is easy to show that constant-amplitude solutions do not exist if there are more than two pairs. Beyond that, the problem is difficult and to my best knowledge, was never solved.

2.6.3 Higher-order parametric resonance and the Blazkho effect

Resonant coupling between radial and nonradial modes of similar frequencies is a third-order effect in the amplitude expansion. Nonetheless, it may have a greater impact on RR Lyrae pulsation than the lower-order resonant coupling due to the properties of the mode-trapping pattern. We have seen in Figure 2.4 that nonradial modes with frequencies close to those of radial modes are also close to the maxima of the driving rates. This means relatively large amplitudes in the acoustic cavity, hence stronger coupling to radial modes and lower damping rates. Both effects promote parametric instability. Figure 2.4 also shows that the trapping favours excitation of $\ell = 1$ modes.

Van Hoolst et al. (1998) derived the following expression for the amplitude of the parent mode amplitude at the onset of the instability:

$$A_0^2 > \sqrt{\frac{\Delta\omega^2 + \gamma_{\ell,N}^2}{C_{00\ell\ell}}} \, .$$

It is similar to equation (2.5), but specialized to a radial mode and it takes into account saturation of driving by the radial mode, which makes $\gamma_{\ell,N}$ negative. Our survey (Dziembowski & Cassisi, 1999) of realistic models of RR Lyrae stars has revealed that excitation of radial modes is quite likely. The probability ranges from 0.3 to 0.9. A study of the nonlinear development (Nowakowski & Dziembowski, 2001) has shown that in this case there is always a constant-amplitude pulsation state. If a single $\ell = 1, m = 0$ mode is excited then the effect may easily escape detection. The phases of the two modes are locked so that variability remains monoperiodic with only slightly modified period. The effect on the amplitude is more significant but always the radial component strongly dominates in the light and radial velocity variations.

A more interesting situation arises if a pair of $\ell = 1$ modes is excited. The following possibilities exist: the two modes may still be (i) axisymmetric but belong to different triplets; it may be a $m = \pm1$ pair of (ii) the same triplet or; (iii) different triplets. What matters is only that the frequency mismatch

$$\Delta\omega = 2\omega_0 - (\omega_{1,k,m_k} + \omega_{1,n,m_n}) \, ,$$

where subscripts k and n stand for radial orders of $\ell = 1$ modes while m_k and m_n are the corresponding azimuthal orders, is small and that the modes are located possibly close to maxima of the linear driving rate. In this case, the phase lock produces an equidistant triplet with the central peak corresponding to the radial mode, giving rise to a phase and/or amplitude modulated pulsation.

The model predicts equal amplitudes of the side peaks reaching up to 0.35 of the central-peak amplitude. It is a viable model for stars like V6 in NGC 6362 and it is appealing because, if correct, then from the Blazkho period we get valuable information about the deep stellar interior. Unfortunately, the model does not explain all RR stars with variable amplitudes. It is certainly not applicable to cases when only two close peaks are detected. It is also not readily applicable if two side peaks are seen but with too large or very unequal amplitudes.

2.7 Final remarks

Today the main emphasis in studies of pulsating star research is put on asteroseismology, that is on using pulsation data to constrain stellar models, as well as on other applications such as determination of distances to stellar systems. The physics of the pulsation phenomenon is often regarded as sufficiently well understood. In this review I have been advocating an opposite view. I presented a number of problems posed by observations where we are lacking physical understanding.

Even though the domains of occurrence of various types of oscillations in the H-R diagram are well reproduced with the results of linear stability analyses, still certain problems within the linear stability theory remain, awaiting progress in the treatment and understanding of the interaction between convection and pulsation. Among the unsolved problems, the outstanding one is the cause of the universal variability in red giants.

Modern observations do not challenge the hypothesis of pure radial monomode pulsation for the majority of Cepheids and RR Lyrae stars. These objects seem indeed to be amazingly simple natural heat engines. Why this simplest form of motion is chosen we do not quite understand. Many other pulsation modes are unstable. Explaining complexity in nature has became fashionable in science. In my view, priority should be given to explaining simplicity. The question why pulsation is so simple was not answered by our predecessors in the field of stellar pulsation research. Also our generation did not give a fully satisfactory answer. I hope that one of our younger colleagues will tell us why.

Acknowledgement: This work was supported in part by the Polish grant KBN 2P03D 030 20.

References

Breger, M. et al., 1998, *A&A*, **331**, 271.
Brickhill, A. J., 1983, *MNRAS*, **204**, 537.
Buchler, J. R., Moskalik, P. & Kovács, G., 1990, *ApJ*, **351**, 617.
Chandrasekhar, S. & Lebovitz, N. R., 1962, *ApJ*, **136**, 1105.
Charpinet, S., Fontaine, G., Brassard, P. & Dorman, B. 1996, *ApJ*, **471**, L103.
Christy, R. N. 1964, *Rev. Mod. Phys.*,, **36**, 555.
Dilke, F. W. W. & Gough, D. O., 1972, *Nature*, **240**, 262.
Dziembowski, W. A., 1982, *Acta Astron.*, **32**, 147.
Dziembowski, W. A., 1983, *Solar Phys.*, **82**, 259.
Dziembowski, W. A. & Cassisi, S., 1999, *Acta Astron.*, **49**, 371.
Dziembowski, W. A. & Królikowska, M., 1985, *Acta Astron.*, **35**, 5.
Dziembowski, W. A. & Kovács, G., 1984, *MNRAS*, **196**, 731.
Feuchtinger, 1998, *A&A*, **337**, L29.
Goldreich, P. & Wu, Y., 1999, *ApJ*, **511**, 904.
Gough, D. O., 1977, *ApJ*, **214**, 196.
Jerzykiewicz, M. & Wentzel, W., 1977, *Acta Astron.*, **27**, 35.
Kolláth, Z., Beulieu, J. P., Buchler, J. R. & Yecko, P., 1998, *ApJ*, **502**, L55.
Kolláth, Z., Buchler, J. R., Szabó, R. & Csubry, Z., 2002, *A&A*, **385**, 932.
Kovács, G., 2002, in Aerts, C., Bedding, T. R. & Christensen-Dalsgaard, J., eds, *IAU Colloquium 185, Radial and Nonradial Pulsations as Probes of Stellar Physics*, ASP Conf. Ser., **259**, p. 396.
Ledoux, P., 1951, *ApJ*, **114**, 373.
Nowakowski, R. M. & Dziembowski, W. A., 2001, *Acta Astron.*, **51**, 5.
Olech, A., et al., 1999, *AJ*, **162**, 442.
Olech, A., et al., 2001, *MNRAS*, **321**, 421.
Papaloizou, J. C. B., 1973, *MNRAS*, **162**, 143.
Rosseland, S., 1949, *The pulsation theory of variable stars*, The Clarendon Press (Oxford).
Schwarzschild, M., 1942, *ApJ*, **94**, 241.
Simon, N. R. & Schmidt, E. G., 1976, *ApJ*, **205**, 162.
Stellingwerf, R. F., 1975, *ApJ*, **195**, 441.
Stobie, R. S., 1969, *MNRAS*, **144**, 511.
Udalski, A., et al., 1999, *Acta Astron.*, **49**, 1.
Udalski, A., et al., 2000, *Acta Astron.*, **50**, 307.
Van Hoolst, T., Dziembowski, W. A. & Kawaler, S. D., 1998, *MNRAS*, **297**, 536.
Vandakurov, Yu.V, 1981, *Soviet Astron. Letters*, **7**, 128.
Viskum, M., et al., 1998, *A&A*, **335**, 549.
Wu, Y. V., 2002, in Aerts, C., Bedding, T. R. & Christensen-Dalsgaard, J., eds, *IAU Colloquium 185, Radial and Nonradial Pulsations as Probes of Stellar Physics*, ASP Conf. Ser., **259**, p. 506.
Wersinger, J. M., Finn, I. M. & Ott, E. 1980, *Phys. Fluids*, **23**, 1142.

3

Acoustic radiation and mode excitation by turbulent convection

GÜNTER HOUDEK

Institute of Astronomy, University of Cambridge, Cambridge CB3 0HA, UK

For many years the principle source of excitation of oscillations in solar-like stars was under considerable debate. This was related to the fact that mode stability in such stars is governed not only by the perturbations in the radiative fluxes (via the κ-mechanism) but also by the perturbations in the turbulent fluxes (heat and momentum). The study of mode stability therefore demands a theory for convection that includes the interaction of the turbulent velocity field with the pulsation. It is now widely believed that the observed low-amplitude oscillations in the Sun are determined by the balance between the stochastic driving due to the acoustic radiation by turbulent convection and the damping of the intrinsically stable p modes. Acoustic radiation by turbulence also effects the stratification of the equilibrium model by reducing the estimated convective velocities and consequently influencing oscillation properties. In this contribution I review the mechanisms responsible for mode damping in solar-type stars and for stochastic driving by turbulent convection. Amplitude predictions for models of the Sun and β Hydri are compared with observations. Finally I discuss the effect of acoustic radiation by turbulence on the retardation of the convective velocities in solar-type stars with masses $1.0 - 1.9\,M_\odot$ and on mode stability in a Delta Scuti star of $1.65\,M_\odot$.

3.1 Introduction

Attempts to explain the principal source of excitation of solar p modes were carried out first by e.g., Ulrich (1970) and Ando & Osaki (1975) who argued that most of the solar modes are overstable in linear theory. If solar p modes were indeed overstable, some nonlinear mechanism must limit their amplitudes to the low values that are observed; however, no such mechanism has been found to date (Kumar & Goldreich 1989). The problem of identifying a saturation mechanism does not arise if the modes are intrinsically stable but excited stochastically by the acoustic noise generated by the turbulent convection (Goldreich & Keeley 1977). The amplitudes are determined by the

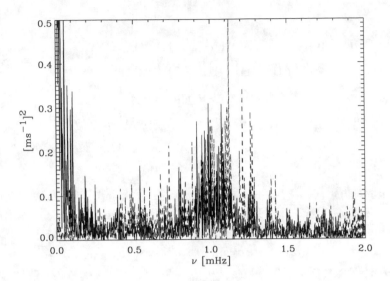

Fig. 3.1. Velocity power spectrum of beta Hydri. The solid spectrum displays the Doppler measurements by Bedding et al. (2001). The dashed spectrum represents a theoretical expectation, obtained by first rescaling solar data (Chaplin et al. 1998) in frequency in the manner of Houdek et al. (1999) and then reducing the amplitudes by a factor of 1.9 (adopted from Gough 2001).

balance between the excitation and damping, and are expected to be rather low. The turbulent-excitation model predicts not only the right order of magnitude for the p-mode amplitudes (Gough 1980), but it also explains the observation that millions of modes are excited simultaneously. Using a time-dependent generalization (Gough 1977) of a local mixing-length model, Gough (1980) found most of the radial p modes in the Sun to be stable. Balmforth (1992a) obtained better agreement between theoretical damping rates and solar linewidth measurements using the nonlocal generalization of the mixing-length model by Gough (1976). Moreover, with the help of these improved damping rate estimates Balmforth (1992b) obtained reasonable agreement between computed and measured solar velocity amplitudes.

The most convincing evidence to date of solar-type oscillations in other stars comes from recent observations of β Hydri (Bedding et al. 2001) and α Cen A (Bouchy & Carrier 2001). In Figure 3.1 the power spectrum of the Doppler velocity measurements of β Hydri (solid spectrum) is compared with a theoretical expectation (dashed spectrum). The theoretical spectrum was obtained by first scaling solar data (Chaplin et al. 1998) in frequency by the ratio of the acoustic cut-off frequencies between a model of β Hydri and the Sun and in amplitude in the manner of Houdek et al. (1999); the amplitudes were then reduced by a factor of 1.9 to render the total power in the frequency interval $0.67\,\mathrm{mHz} < \nu < 1.5\,\mathrm{mHz}$ the same as that of the measured β Hydri spectrum.

In this contribution I shall adopt Balmforth's excitation model to predict amplitudes of radial p modes for the Sun and β Hydri. Stellar models and damping rates are computed in the manner of Houdek et al. (1999).

3.2 Linear damping rates, Γ

Basically, the damping of stellar oscillations arises from two sources: processes influencing the momentum balance, and processes influencing the thermal energy equation. Each of these contributions can be divided further according to their physical origin (see Houdek et al. 1999 and references therein). Here we limit the discussion to the convection dynamics only.

Vibrational stability is influenced crucially by the exchange of energy between the pulsation and the turbulent velocity field. The exchange arises either via the pulsationally perturbed convective heat flux, or directly through dynamical effects of the fluctuating Reynolds stress. In fact, it is the modulation of the turbulent fluxes by the pulsations that seems to be the predominant mechanism responsible for the driving and damping of solar-type acoustic modes. It was first reported by Gough (1980) that the dynamical effects arising from the turbulent momentum flux (also called turbulent pressure p_t) perturbations contribute significantly to the mode damping (see Γ_t in the bottom panel of Figure 3.2). Moreover, he predicted a characteristic plateau in the damping rates centred near $2.9\,\mathrm{mHz}$ (see Figure 3.2). This plateau was later confirmed observationally by Libbrecht (1988) (see also Christensen-Dalsgaard, Gough & Libbrecht 1989). Detailed analyses (Balmforth 1992a) reveal how damping is controlled largely by the phase difference between the momentum perturbation and the density perturbation. Therefore, turbulent pressure fluctuations must not be neglected in stability analyses of solar-type p modes.

Solar p-mode frequencies vary with the solar cycle, the frequencies being largest at sunspot maximum. Theoretical studies have shown that variations in the thermal structure of the superficial layers of the Sun cause positive p-mode frequency variations to be associated with a reduction in the effective temperature, and hence irradiance (Gough & Thompson 1988; Gough 1990; Goldreich et al. 1991; Balmforth, Gough & Merryfield 1996). Additional information of variations of p-mode properties over the solar cycle can be gained by studying theoretical p-mode damping rates. Today fairly accu-

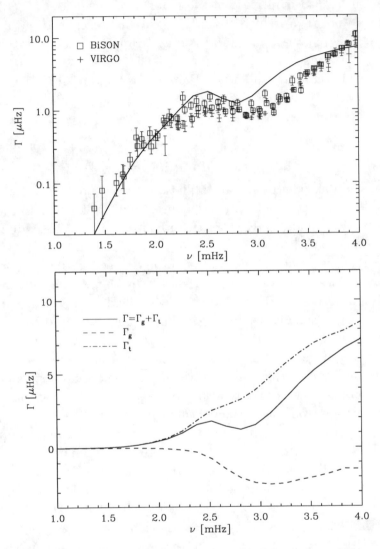

Fig. 3.2. Top: Linear damping rates for a solar model, computed (curve) in the manner of Houdek et al. (1999), are compared with linewidth measurements (symbols) by BiSON (Chaplin et al. 1998) and from the LOI instrument (Appourchaux et al. 1998). **Bottom:** Contributions to the computed damping rates arising from the gas (Γ_g) and turbulent (Γ_t) pressure perturbations.

rate linewidth measurements of the spectral peaks in the acoustical power spectrum are provided by ground-based and space-born instruments, covering a substantial time period. Over the 11-year period of the solar cycle, the radiative interior of the sun will barely be affected by thermal disturbances, because it relaxes diffusively on a Kelvin-Helmholtz time of about

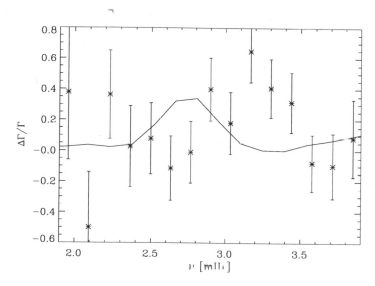

Fig. 3.3. Solar-cycle changes in linewidths from activity minimum to maximum. The plotted data (with error bars) were obtained from BiSON (Chaplin et al. 2002) measurements. The solid curve shows the modelled variations which result from decreasing the eddy shape parameter Φ by 6 per cent, while keeping the entropy in the deep layers of the convection zone constant during the cycle (Houdek et al. 2001).

3×10^7 years. As discussed by Gough (1981) the convective envelope adjusts itself internally to any perturbation on a timescale of a month (which is much less than 11 years), whereas its thermal cooling timescale is about 2×10^5 years (which is much greater than 11 years). Consequently, on a timescale of 11 years, the heat flux remains divergence-free, and the entropy is essentially invariant. In modelling the solar-cycle variation, one has to keep the entropy in the deeper convectively unstable layers constant in models for which the structure is varied to simulate the changes. In Figure 3.3 simulation results of damping rate variations are compared with measurements of linewidth variations over the cycle 23. The solar-cycle changes are modelled by varying the horizontal extend of the convective eddies, represented by the eddy shape parameter Φ (Gough 1977), which is of order unity. The parameter Φ had to be reduced in order to obtain a fair agreement with the BiSON measurements, which suggests that the horizontal granule size is decreasing with activity. This is indeed in agreement with the observations by Muller (1988). Moreover, the decrease of 6 per cent in Φ is in fair agreement with the measured value of $5-10$ per cent (Roudier & Reardon 1998).

3.3 Stochastic excitation

Huge progress on stellar convection has been made recently with the help of hydrodynamical simulations of the outer parts of the convection zone in solar-type stars (e.g., Stein & Nordlund 2001). However, such simulations are very time consuming and consequently simple convection models are still needed in evolutionary computations. Essentially in all such simple models the anelastic (or Boussinesq) approximation (Gough 1969) to the fluid equations is assumed. In this approximation the time derivative of the density fluctuation in the continuity equation is neglected, which is equivalent to filtering out high-frequency phenomena such as sound waves. Consequently a separate model is needed to describe approximately the acoustical noise generated by the turbulent motion of the convective eddies. Such a model was proposed by Lighthill (1952): in this model the density fluctuations are the same between a real fluid with highly nonlinear motion and a fictitious acoustic medium with linear motion upon which an external stress system is acting. Balmforth (1992b) reviewed the theory of acoustical excitation in a pulsating atmosphere, and, following Goldreich & Keeley (1977), he derived the following expression for the rate of energy injected into a mode with frequency ω by quadrupole emission through the fluctuating Reynolds stress:

$$
P_{\mathrm{R}} = \frac{\pi^{1/2}}{8I} \int\limits_0^M \left(\frac{\partial \xi}{\partial r} \right)^2 \rho \ell^3 w^4 \tau \mathcal{S}(m, \omega) \, \mathrm{d}m \, ,
\tag{3.1}
$$

where r is distance to the centre, m is the mass interior to r, and M is the total mass; also ℓ, w, τ are respectively the length, velocity and correlation time scales of the most energetic eddies, determined by the mixing-length model, I is the mode inertia and ρ is the density. The function $\mathcal{S}(m, \omega)$ accounts for the turbulent spectrum, which approximately describes contributions from eddies with different sizes to the noise generation rate P_{R}, and which we implemented as did Balmforth (1992b). The displacement eigenfunction of a p mode is described by ξ.

A similar expression is obtained for the emission of acoustical radiation by low-order multipole sources through the fluctuating entropy. The ratio of the noise generation rate between the fluctuating entropy, P_{s}, and Reynolds stress is (Goldreich, Murray & Kumar 1994):

$$
\left(\frac{P_{\mathrm{s}}}{P_{\mathrm{R}}} \right)^{1/2} \approx \frac{4}{\alpha \Phi} \gamma_1 \, ,
\tag{3.2}
$$

where α is the mixing-length parameter (which is the ratio between the mixing length ℓ and the local pressure scale height) and γ_1 is the first adiabatic

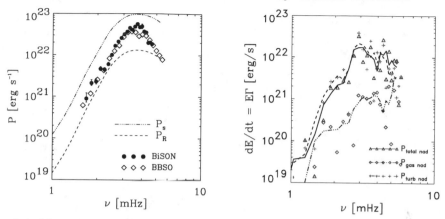

Fig. 3.4. Noise generation rate as a function of frequency for a solar model. **Left:** Results, obtained with equation (3.1), are compared with observations by BiSON (Chaplin et al. 1998) and from the BBSO (Libbrecht 1988). The contribution from the fluctuating entropy P_s is about one order of magnitude larger than the contribution from the fluctuating Reynolds stress P_R. **Right:** Results are obtained from hydrodynamical simulations (adopted from Stein & Nordlund 2001). The contribution from the fluctuating Reynolds stress (P_{turb}) is on average about four times larger than the contribution from the fluctuating entropy (P_{gas}).

exponent. Assuming typical values for α, γ_1 and Φ for the solar case, the value of this ratio is ~ 3. Consequently the noise generation rate due to the fluctuating entropy is about one order of magnitude larger than the contribution from the fluctuating Reynolds stress (see left panel of Figure 3.4). In contrast, the hydrodynamical simulations by Stein & Nordlund (2001), depicted in the right panel of Figure 3.4, found the Reynolds-stress contribution to be the larger. From this comparison it is obvious that there is still controversy as to whether the fluctuating entropy or Reynolds stress is the dominating source of excitation and consequently further studies seem warranted.

With the estimates of Γ and P_R, the mean-square oscillation amplitudes, V_s, are obtained from the expression $V_s^2 = P_R / (\Gamma I)$ (e.g. Houdek et al. 1999). Figure 3.5 shows the mean-square velocity amplitudes for a model of the Sun and β Hydri. For the Sun results are plotted for computations in which both observed (solid curve) and theoretical (dashed curve) damping rates were assumed. Both results are calibrated to the BiSON (Chaplin et al. 1998) observations (symbols) by scaling the maximum values of the computed amplitudes to the measurements. For β Hydri the mean-square velocity amplitudes are scaled by the factor 1.57, obtained from the scaled solar model using the theoretical damping rates: the estimated peak value

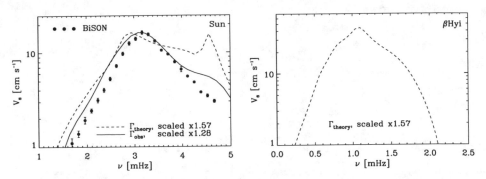

Fig. 3.5. Mean-square velocity amplitudes for models of the Sun and β Hydri, assuming equation (3.1) for the total noise generation rate.

of $65 \, \mathrm{cm \, s^{-1}}$ for the velocity amplitude is in reasonable agreement with the measured value of $60 \, \mathrm{cm \, s^{-1}}$ by Bedding et al. (2001).

3.4 Acoustic radiation in the equilibrium model

In the previous section acoustic radiation was discussed in a pulsating atmosphere. This very mechanism is also working in the static model: through the generation of sound waves, kinetic energy from the turbulent motion will be converted into acoustic radiation and thus reduce the efficacy with which the motion might otherwise have released potential energy originating from the buoyancy forces. This will result in a different stratification of the convectively unstable layers.

The implementation of the acoustic flux in the dynamical equations describing the convective motion of the turbulent eddies is accomplished in a straightforward way by adopting the phenomenological picture of an overturning eddy. In this picture the fluid element maintains balance between buoyancy forces and turbulent drag by continuous exchange of momentum with other elements and its surroundings (e.g., Unno 1967). The equation of motion for a turbulent element can then be written as

$$\frac{2w^2}{\ell} = g\frac{\delta}{T}T' - \Lambda\frac{w^2}{\ell}M_t^{\mu}, \tag{3.3}$$

where T and T' denote the mean temperature of the background fluid and the convective temperature fluctuations, respectively, g is the acceleration due to gravity, $\delta = -(\partial \ln \rho / \partial \ln T)_{p_g}$, and the constants Λ and μ are the emissivity coefficient and Mach-number dependence, respectively. The turbulent Mach number is defined as $M_t = w/c$ (c being the adiabatic sound speed). The last term on the right-hand side of equation (3.3) is derived

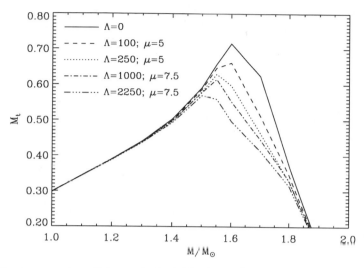

Fig. 3.6. Maximum value of the turbulent Mach number as a function of model mass along the ZAMS. Results are displayed for model computations in which acoustic radiation was either omitted ($\Lambda = 0$) or included assuming different values for the acoustic emissivity Λ and Mach-number dependence μ.

from the Lighthill-Proudman formula (Proudman 1952) and represents a drag due to the acoustic radiative losses, additional to the turbulent drag expressed by the left-hand side, which together are balanced by the potential energy coming from the buoyancy forces (first term on the right-hand side).

We consider two simple models for the emission of acoustic waves by homogeneous, isotropic turbulence: in the first model the acoustic emission is dominated by the energy-bearing eddies and is thus scaled by a Mach-number dependence of $\mu = 5$ (Lighthill 1952). In the second model the acoustic radiation is predominantly emitted by inertial-range eddies, as suggested by Goldreich & Kumar (1990), who derived a Mach-number dependence of $\mu = 15/2$. For the emissivity coefficient we adopt for the model with $\mu = 5$ the value $\Lambda = 100$, as suggested by Stein (1968) for a solar model, and for the model with $\mu = 15/2$, a value of $\Lambda = 1000$, which provides a value for the acoustic flux similar to the model with $\mu = 5$.

The turbulent Mach numbers for models along the ZAMS with masses of $1.0-1.9\,M_\odot$ are depicted in Figure 3.6. The results are displayed for computations in which the acoustic flux was either omitted ($\Lambda = 0$; continuous curve) or included assuming different values for the emissivity coefficient Λ and the Mach-number dependence μ (see eq. 3.3). The effect of acoustic radiation is essentially negligible for models with masses less than

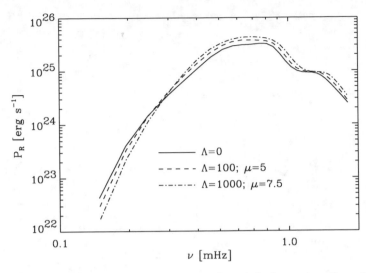

Fig. 3.7. Noise generation rate as a function of p mode frequency for a $1.6\,M_\odot$ star of age $1.49\,\mathrm{Gy}$. Results are displayed for various values of Λ and μ.

about $1.5\ M_\odot$, and becomes largest for models with masses M in the range $1.5\,M_\odot \lesssim M \lesssim 1.8\,M_\odot$. The effect of acoustic radiation in the equilibrium model on the rate at which energy is injected into the individual p modes, P_R, is displayed in Figure 3.7 for a $1.6\,M_\odot$ model of age 1.49 Gy. Interestingly, at frequencies near where P_R is largest, it becomes larger for models in which acoustic radiation was included in the computations; however, the differences are still small. This comes about because P_R depends also on the shape of the modal eigenfunctions (particularly at the top of the convection zone), which are slightly modified in a compensating way between models obtained with and without the inclusion of acoustic radiation in the equilibrium structure.   Computed damping/growth rates for a model of the Delta Scuti star BW Cnc are portrayed in Figure 3.8. These results can be compared with observations from campaigns of the STEPHI network in the Praesepe cluster (Michel et al. 1999). Michel et al. suggest for the observed Delta Scuti star BW Cnc a mass of $1.6 - 1.65\,M_\odot$ and an age of $700 - 800$ My. The top panel of Figure 3.8 displays the results for a model in which acoustic radiation was omitted in the computations ($\Lambda = 0$) and for which the same value $\alpha = 1.89$ was adopted as suggested by a calibrated solar model, assuming the same input physics. Michel et al. (1999) suggest that for the star BW Cnc the observed radial modes have radial order n in the range $1 \lesssim n \lesssim 5$: comparing the observed range with the results of the stability analyses in Figure 3.8, better agreement is obtained for those model

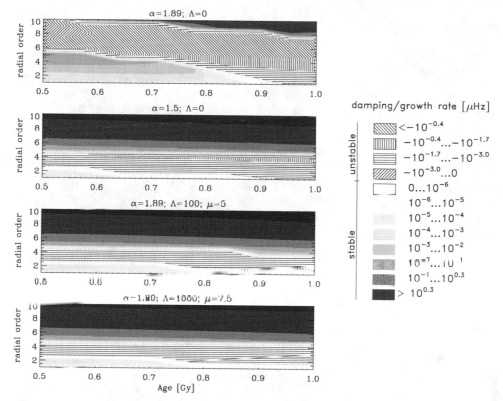

Fig. 3.8. Damping/growth rates, $\Gamma/2$, for an evolving 1.65 M$_\odot$ Delta Scuti star as a function of radial order n and model age. The results are depicted as contour plots for model sequences with different values for the mixing-length parameter α and with the acoustic flux either omitted ($\Lambda = 0$; top panel) or included assuming $\Lambda = 100$ and $\Lambda = 1000$.

sequences in which either the acoustic flux is included in the computations of the mean stratification (Λ=100 and 1000) or in which the mixing-length parameter is reduced ($\alpha = 1.5$).

Acknowledgements. Support by the Particle Physics and Astronomy Research Council of the UK is gratefully acknowledged.

References

Ando, H. & Osaki, Y., 1975, *PASJ*, **27**, 581.

Appourchaux, T. et al. (the VIRGO team), 1998, in Korzennik, S. G. & Wilson, A., eds, *Structure and dynamics of the interior of the Sun and Sun-like stars; Proc. SOHO 6/GONG 98 Workshop*, ESA SP-418, ESA Publications Division, Noordwijk, p. 37.

Balmforth, N. J., 1992a, *MNRAS*, **255**, 603.

Balmforth, N. J., 1992b, *MNRAS*, **255**, 639.

Balmforth, N. J., Gough, D. O. & Merryfield, W. J., 1996, *MNRAS*, **278**, 437.

Bedding, T. R., Butler, R. P., Kjeldsen, H., Baldry, I. K., O'Toole, S. J., Tinney, C. G., Marcy, G. W., Kienzle, F. & Carrier, F., 2001, *ApJ*, **549**, L105.

Bouchy F. & Carrier F., 2001, *A&A*, **374**, L5.

Chaplin, W. J., Elsworth, Y., Isaak, G. R., Lines, R., McLeod, C. P., Miller, B. A. & New, R., 1998, *MNRAS*, **298**, L7.

Chaplin, W. J., Elsworth, Y., Isaak, G. R., Miller, B. A. & New, R., 2002, *MNRAS*, **330**, 731.

Christensen-Dalsgaard, J., Gough, D. O. & Libbrecht, K. G., 1989, *ApJ*, **341**, L103.

Goldreich, P. & Keeley, D. A., 1977, *ApJ*, **212**, 243.

Goldreich, P. & Kumar, P., 1990, *ApJ*, **363**, 694.

Goldreich, P., Murray, N., Willette, G. & Kumar, P., 1991, *ApJ*, **370**, 752.

Goldreich, P., Murray, N. & Kumar, P., 1994, *ApJ*, **424**, 466.

Gough, D. O., 1969, *J. Atmos. Sci.*, **26**, 448.

Gough, D. O., 1976, in Spiegel, E. & Zahn, J.-P., eds, *Problems of stellar convection, IAU Colloq. No. 38, Lecture Notes in Physics*, vol. **71**, Springer-Verlag, Berlin, p. 15.

Gough, D. O., 1977, *ApJ*, **214**, 196.

Gough, D. O., 1980, in Hill, H. A. & Dziembowski, W. A., eds, *Nonradial and Nonlinear Stellar Pulsation, Lecture Notes in Physics*, vol. **125**, Springer-Verlag, Berlin, p. 273.

Gough, D. O., 1981, in Sofia, S., ed., *Variations in the Solar Constant*. NASA Conf. Publ. 2191, Washington, p. 185.

Gough, D. O., 1990, in Osaki, Y. & Shibahashi, H., eds, *Progress of seismology of the sun and stars, Lecture Notes in Physics*, vol. **367**, Springer, Berlin, p. 283.

Gough, D. O., 2001, *Science*, **291**, 2325.

Gough, D. O. & Thompson, M. J., 1988, in Christensen-Dalsgaard, J. & Frandsen, S., eds, *Proc. IAU Symposium No 123, Advances in helio- and asteroseismology*, Reidel, Dordrecht, p. 175.

Houdek, G., Balmforth, N. J., Christensen-Dalsgaard, J. & Gough, D. O., 1999, *A&A*, **351**, 582.

Houdek, G., Chaplin, W., Appourchaux, T., Christensen-Dalsgaard, J., Däppen, W., Elsworth, Y., Gough, D. O., Isaak, G. R., New, R. & Rabello-Soares, M. C., 2001, *MNRAS*, **327**, 483.

Kumar, P. & Goldreich, P., 1989, *ApJ*, **342**, 558.

Libbrecht, K. G., 1988, *ApJ*, **334**, 510.

Lighthill, M. J., 1952, *Proc. Roy. Soc. London*, **A211**, 564.

Michel, E., Hernández, M. M., Houdek, G., Goupil, M. J., Lebreton, Y., Pérez Hernández, F., Baglin, A., Belmonte, J. A. & Soufi, F., 1999, *A&A*, **342**, 153.

Muller, R., 1988, *Adv. Space Res.*, **8**, No 7, (7)159.

Proudman, I., 1952, *Proc. Roy. Soc. London*, **A214**, 119.

Roudier, Th. & Reardon, K., 1998, in Balasubramaniam, K. S., Harvey, J. W. & Rabin, D. M., eds, *Synoptic Solar Physics, ASP Conf. Ser. vol 140*, Astron. Soc. Pac., San Francisco, p. 455.

Stein, R. F., 1968, *ApJ*, **154**, 297.

Stein, R. F. & Nordlund, Å., 2001, *ApJ*, **546**, 585.

Ulrich, R. K., 1970, *ApJ*, **162**, 993.

Unno, W., 1967, *PASJ*, **19**, 140.

4

Understanding roAp stars

MARGARIDA S. CUNHA

Centro de Astrofísica da Universidade do Porto, rua das Estrelas, 4150-762 Porto, Portugal and Instituto Superior da Maia, Maia, Portugal

Rapidly oscillating Ap stars have proved to be extremely interesting objects, for they combine in a unique way different physical properties, like stellar magnetism and abnormal chemical abundances, with important physical phenomena, like acoustic oscillations. In this paper we will discuss how the indirect effect of the magnetic field and the presence of chemical peculiarities may influence different aspects of the pulsations in roAp stars and will try to discuss their implications to our understanding of the latter.

4.1 Introduction

Rapidly oscillating Ap stars (hereafter roAp stars) have now been known for a couple of decades (Kurtz 1982). They are located in the main-sequence part of the classical instability strip, close to the δ Scuti stars, but unlike the latter, roAp stars are small-period pulsators, oscillating with periods that vary typically from 5 to 15 minutes. They are found among the coolest subgroup of classical Ap stars, and, thus, not only are they chemically peculiar, but also they have strong large scale magnetic fields, with typical intensities of a few kG. This combination of properties makes roAp stars extremely interesting targets for asteroseismology. Moreover, the oscillations they exhibit are interpreted as high-order, low-degree modes, opening the possibility of applying asymptotic techniques.

The magnetic fields present in roAp stars influence the oscillations both directly and indirectly. In particular, the magnetic field modifies the frequencies of the oscillations as well as the corresponding eigenfunctions, through the direct effect of the Lorentz forces (e.g. Dziembowski & Goode 1996, Cunha 1999, Bigot et al. 2000, Cunha & Gough 2000, Cunha 2001). Consequently, the process by which one may infer information about roAp stars from the study of their pulsations has to be adapted to account for these effects. However, the effect of the magnetic field is not confined to its direct influence on the oscillations. By interfering with convection, for instance, or with the process of chemical settling, the magnetic field may change the structure of the star, and, thus, indirectly influence the pulsations. In the

51

following sections we will look into some of the possible indirect effects of the magnetic field on the oscillations of roAp stars and their implications.

It is not our intention to write a review on general aspects of roAp stars, for reviews on this subject can be found in recent literature (e.g. Kurtz 1990, Martinez & Kurtz 1995, Martinez 1996, Cunha 1998, Cunha 2002a). The idea is, instead, to try to understand different aspects concerning roAp stars which may be related to the indirect effect of the magnetic field on the oscillations, in the light of a particular theoretical model in which Douglas Gough has been involved for many years (Dolez & Gough 1982, Dolez et al. 1988, Vauclair et al. 1991, Balmforth et al. 2001 - hereafter BCDGV).

In section 4.2 we will discuss, in a simplified approach, the interaction between the magnetic field and the envelope convection, and based on that discussion we will introduce the BCDGV model for mode excitation in roAp stars. In section 4.3 we will discuss some general results of the model described, both concerning mode excitation and the frequency of the excited modes. Finally, in section 4.4 we will predict limits for the instability strip of roAp stars and in section 4.5 we will discuss the implications of the model concerning the differences between roAp and noAp stars (that is, non-oscillating Ap stars). In the last section we will present some conclusions.

4.2   Magnetic field versus convection

In order to try to understand how strongly the magnetic field influences convection, one may compare the force exerted by buoyancy on a fluid element with the magnetic restoring force experienced as that element moves. The buoyancy force per unit mass driving the convection is roughly minus the square of the buoyancy frequency times the displacement of the element: $-N^2\delta r$. The magnetic restoring force per unit mass exerted on the displaced element is estimated by $(|\mathbf{B}|^2 k^2/\mu\rho)\delta r$, where $\mathbf{B}$ is the magnetic field, k is the characteristic wavenumber of the magnetic field-line distortion, μ is the magnetic permeability and ρ is the density. Thus for the magnetic field to inhibit the motion one expects

$$v^2 k^2 \gtrsim -N^2, \tag{4.1}$$

where v is the Alfvén speed. According to equation (4.1), the comparison of forces depends on the length scale, k^{-1}, and thus, on the field geometry.

In particular, for a magnetic field with a dipolar structure one expects that convection may be suppressed in the polar regions, but not necessarily near the equator of these stars. Having this in mind BCDGV constructed a model for roAp stars which incorporated two distinct regions: polar and equatorial. In each region, the magnetic stresses were neglected entirely in the balance of forces, and were acknowledged only through their presumed effect on the convective energy and momentum fluxes. In practice, for each stellar model two envelope models were generated, one with the characteristics of the equatorial region and the other with the characteristics of the polar regions. These two models were matched in the interior to assure that they differed only in their surface layers. This procedure is oversimplified, but without substantial further sophistication it is not easy to proceed.

If the polar regions of stellar models like these occupy the domain $-\vartheta < \theta < \vartheta$ and $\pi - \vartheta < \theta < \pi + \vartheta$ in the spherical polar coordinate system (r, θ, ϕ), with $\theta = 0$ along the magnetic axis, then the frequency ν_{nlm} of a mode of the composite model, that is, the model composed by polar and equatorial regions with the characteristics described above, is (e.g. Dolez et al. 1988, Cunha 1999)

$$\nu_{nlm} \approx \nu_{nl}^{\mathrm{p}} \int_0^{2\pi} \int_{\tilde{\mu}}^1 (Y_l^m)^2 \, \mathrm{d}\mu \mathrm{d}\phi + \nu_{nl}^{\mathrm{eq}} \int_0^{2\pi} \int_0^{\tilde{\mu}} (Y_l^m)^2 \, \mathrm{d}\mu \mathrm{d}\phi, \qquad (4.2)$$

where ν_{nl}^{eq} and ν_{nl}^{p} are the oscillation frequencies in envelope models with the equatorial structure and the polar structure, respectively, $\mu = \cos\theta$ and $\tilde{\mu} = \cos\vartheta$.

Equation (4.2) allows us to calculate the real and imaginary part of the frequencies for a given mode in the composite model, from the frequencies of the same mode in spherically symmetric models with the characteristics of the magnetic poles and equator, respectively.

The abnormal chemical abundances of some elements, and their inhomogeneous distribution over the stellar surface, is another characteristic of roAp stars that should be kept in mind when studying the indirect effects of the magnetic field on the oscillations. It is well known that the magnetic field influences the surface and depth distribution of chemical elements in roAp stars, by interacting with the ions. Also, if we are to believe that the magnetic fields suppress convection around the magnetic poles of roAp stars, then the profiles of chemical abundances will also be influenced by that suppression, and further differences will be introduced between the polar and equatorial regions of the star. Having this in mind, BCDGV have used different chemical profiles when computing the oscillations in the envelope models representing the polar and the equatorial regions of the stars.

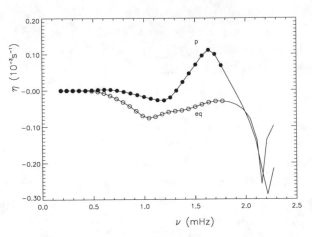

Fig. 4.1. Growth rates in a model with $M = 1.87M_\odot$, $\log T_{eff} = 3.91$ and $\log L/L_\odot = 1.164$. Filled circles show the results in the polar region and open circles show the results in the equatorial region. For details on the chemical profiles see panel (a) of Figure 5 of BCDGV.

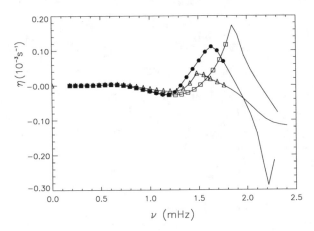

Fig. 4.2. The envelope polar model used in Fig. 4.1 (filled circles) is compared with similar models but with different helium profiles. The open triangles show an envelope polar model with a strong accumulation of helium near the first helium ionization zone, while the open squares show an envelope polar model with homogeneous chemical composition.

4.3   Mode excitation and eigenfrequencies

4.3.1 Excitation

Several ideas concerning the excitation mechanism of roAp stars were proposed over the years (Dolez & Gough 1982, Dolez et al. 1988, Shibahashi 1983, Dziembowski 1984, Dziembowski & Goode 1985, Gautschy et al. 1998,

Matthews 1988). However, with the exception of one of them, these ideas either have not been pursued or have failed to show that models appropriate to roAp stars should be unstable to high frequency oscillations. Gautschy et al. (1998) did find high frequency unstable modes in some roAp star models, but their models assumed the presence of a chromosphere, for which observational evidence has not yet been found (e.g. Shore et al. 1987).

Using models like those described above, BCDGV have found high frequency unstable modes for temperatures and luminosities typical of roAp stars. Because the problem of the excitation mechanism in roAp stars cannot be looked at without considering, simultaneously, the problem of the acoustic cutoff frequency, in an attempt to bracket the physically plausible range of atmospheric conditions, the authors computed pulsational stability with two different outer mechanical boundary conditions. The first boundary condition considered was perfectly reflective, while the second boundary condition was that appropriate to a plane-parallel isothermal atmosphere whose temperature matches continuously with that of the underlying envelope, thus representing a star with no chromosphere. In both cases high frequency unstable modes were found.

In Fig. 4.1 we show the growth rates (imaginary part of the frequency) η_{n0}^{p} and η_{n0}^{eq} of the oscillations of the polar and equatorial models that constitute a composite model with $M = 1.87 M_\odot$, $\log T_{\mathrm{eff}} = 3.91$ and $\log L/L_\odot = 1.164$. All high-order modes are stable in the equatorial model. The polar model, on the other hand, shows unstable subcritical modes of high frequency. A comparison between the growth rates of modes of the same polar model, with the growth rates of similar polar models but with different chemical profiles is shown in Fig. 4.2 . It is clear form these figures that according to the model of BCDGV the possibility of exciting high frequency oscillations in roAp stars depends on the extent of the region in which envelope convection is suppressed and that the frequencies of the modes excited depend on some of the model details, e.g. the chemical composition profile.

4.3.2 Effect on the power spectrum

All of those characteristics of roAp stars that contribute to deviate their structure from spherical symmetry, also contribute to deviate the frequencies of their oscillations away from the values derived from asymptotic theory in spherically symmetric models (Tassoul 1980). The magnetic field is one of the effects that should be kept in mind when such deviations are considered, even if only indirect effects of the latter, like those assumed in the model of BCDGV, are under study.

As illustrated by equation (4.2), in a model like that of BCDGV, in which different angular regions of the star are characterized differently, but within each of them the structure is independent of latitude and longitude, the frequency of a given mode in the composite model is a weighted average of the frequencies that the same mode would have in spherically symmetric models corresponding to each of the regions considered. Because the average is weighted by the spherical harmonic corresponding to the mode considered, different modes will see their frequencies modified differently, when compared to the frequency they would have in the unperturbed spherically symmetric model. Therefore the separation between the modes of oscillation in the power spectrum will also be modified, and should not be forgotten when using the asymptotic theory to infer information about the star.

An example of what has been discussed above concerns the so-called 'second differences' such as $d_n^{l+1,l} \equiv (\nu_{n,l} - \nu_{n-1,l+1}) - (\nu_{n,l+1} - \nu_{n,l})$. In principle, the second differences might be used as a tool to identify the modes of oscillation in roAp stars from a power spectrum containing alternating even- and odd-degree modes (Gabriel et al. 1985; Shibahashi & Saio 1985). However, as argued by BCDGV, if the inhomogeneity introduced by effects like those considered in their models is taken into account, then the second differences may be largely modified (by amounts that can be of the same order of the second differences themselves). Therefore, care should be taken when trying to identify the modes of oscillation in roAp stars by inspecting the second frequency differences.

4.4 Theoretical instability strip

In order to establish a theoretical instability strip for roAp stars, we have carried out linear non-adiabatic calculations for models like those representing the polar regions of the models developed by BCDGV. In Fig. 4.3 we show the models, taken from the evolutionary tracks of Christensen-Dalsgaard (1993), for which pulsational stability calculations have been performed. The large filled circles show models in which no high order unstable modes were found, while small open circles show those in which such unstable modes were present. All models used in these calculations have an envelope which is fully radiative. Thus, in order to know whether the modes found to be unstable in a given model are still unstable in the corresponding composite model, we need to know both the extent of the polar region, that is, the angle ϑ, and the growth rates of the same modes in the corresponding equatorial envelope. Since the former is not available, we take the edges of the instability strip shown in Fig. 4.3 as limits of the true theoretical instability

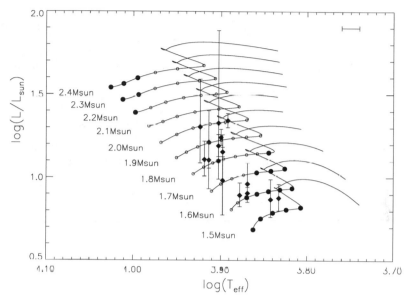

Fig. 4.3. HR diagram in the region where the predicted instability strip for roAp stars is located. Large filled circles show radiative envelope models in which no high order acoustic oscillations were found, while small open circles show the models in which the latter were found. The diamonds show the position of 16 roAp stars (cf. Hubrig et al. 2000). The horizontal bar shown in the top right-hand corner indicates an uncertainty of ± 200 K in the region where most roAp stars are placed and is shown just to guide the eye to what might be a typical uncertainty in the effective temperature of roAp stars.

strip, corresponding to the limit case when the suppression of convection takes place everywhere in the envelope.

Over the theoretical instability strip shown in Fig. 4.3, we have plotted the roAp stars for which Hipparcos parallaxes are available. The effective temperatures and the luminosities of the stars plotted were kindly provided by P. North.

Except for HD 122970 and HD 217522 the observed roAp stars are placed within the predicted instability strip, but they all seem to be concentrated on the lower (fainter) half of the latter. However, most of the candidates to roAp stars observed so far have been chosen through a criterion that introduces a bias towards cooler stars (Don Kurtz, private communication). Therefore, it might be worth trying to improve our knowledge of the observational blue edge for these pulsators, by carrying out a survey having as targets stars of higher effective temperature, before concluding about the predicted theoretical blue edge.

4.5   roAp stars versus noAp stars

Not all candidates for roAp stars observed thus far have been found to pulsate in high frequencies (e.g. Martinez & Kurtz 1994). In fact, no high frequency oscillations have been found in most of the candidates for roAp stars, which means that either most of them do not pulsate in high frequencies, or the amplitude of their oscillations is below the observational threshold. Thus, finding systematic differences that may exist between roAp stars and Ap stars in which high frequency pulsations have been searched for and not found (noAp stars) might be important to the understanding of the mechanism driving the oscillations observed.

That importance has been acknowledge by several authors as is illustrated in several previous studies (e.g. Nelson & Kreidl 1993, Mathys et al. 1996, North et al. 1997, Handler & Paunzen 1999, Gelbmann et al. 2000, Hubrig et al. 2000, Ryabchikova et al. 2000). As result of these studies, some evidence was found for systematic differences in their luminosity and evolutionary status (North et al. 1997, Handler & Paunzen 1999, Hubrig et al. 2000), in the relative abundance of different ions of some chemical elements, like Pr and Nd, (Gelbmann et al. 2000, Ryabchikova et al. 2000), and in the incidence of spectroscopic binaries (Hubrig et al. 2000).

In discussing the implications of models like those of BCDGV to this particular issue we will consider two separated cases: first we will discuss what could prevent high frequency pulsations from being driven in some of the observed stars; secondly we will consider what could prevent high frequency oscillations present in some of these stars from being observed.

4.5.1 noAp stars: are they stable against high frequency pulsations?

What could prevent high frequency pulsations from being driven in some of the observed stars, in the light of models like those described above?

According to the results of BCDGV, in order for high frequency modes to be excited in roAp stars, the magnetic field has to be strong enough to suppress convection, at least in some region of the star, and the positive growth rate of a given mode in the polar region has to be enough to compensate the corresponding negative growth rate in the equatorial region.

In order to inspect whether the ability of the magnetic field to suppress convection in a given star depends on the position occupied by that star in the HR diagram, we compared the force exerted by buoyancy on a fluid element with the magnetic restoring force experienced as that element moves, in several of our theoretical models. In doing that, we have estimated the

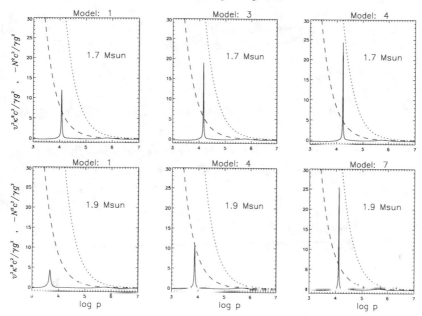

Fig. 4.4. Convective stability characteristics of some of the models shown in Fig. 4.3. In each panel we indicate the mass of the model and the model number, which is equal to 1 when the star is at the ZAMS and increases as the star evolves, corresponding to the circles marked in Fig. 4.3. For each model we show, as function of the logarithm of pressure, the left-hand side of equation (4.1) for two magnetic field intensities [2500 G (dotted curve) and 1000 G (dashed curve)] and the right-hand side of the same equation (continuous curve), both multiplied by $c^2/\gamma g^2$. Here c is the sound speed, γ is the first adiabatic exponent, and g is the acceleration due to gravity.

characteristic wavenumber of the magnetic field distortion, κ, by the inverse of the pressure scale hight. Figure 4.4 illustrates that comparison. Each panel is identified by the mass of the corresponding model and by a model number which is equal to 1 when the model is at the ZAMS and increases as the represented star evolves, corresponding to the open circles shown in Fig. 4.3.

From Fig. 4.4 we see that, for a given magnetic field intensity, it is harder to suppress convection in less massive and/or more evolved stellar models. Thus, one could argue that the statistical evidence found by North et al. (1997) (see also Hubrig et al. 2000) in favour of noAp stars being more evolved than roAp stars, might, at least partially, be explained by our prediction that in more evolved stars the magnetic field intensity needed to suppress convection is greater than in less evolved stars of similar mass. However, the statistical evidence found by the same authors in favour of noAp stars being more massive than roAp stars cannot be explained by the same effect, since the ability of the magnetic field to suppress convection increases with increasing mass, for stellar models at the same distance from the ZAMS.

4.5.2 noAp stars: why would we fail to observe their oscillations?

So far we have discussed what, in the light of the excitation theory suggested by BCDGV, could help explaining intrinsic differences between roAp stars and noAp stars. However, there is a possibility that at least in some cases the observed differences do not originate from intrinsic differences, but rather are a result of observational biases.

Hubrig et al. (2000) mentioned that the difference in the apparent magnitudes of the noAp stars and roAp stars used in their analysis could have influenced their results. Here we suggest another bias that might have affected the statistical analysis carried out by these authors. In Fig. 4.5 we have plotted, for each model computed, the frequency of its most unstable mode. According to these results the characteristic frequency of the oscillations excited depends strongly on the position of the model in the HR diagram. We note that most, but not all, of this dependence appears to be related to the change in $(GM/R^3)^{1/2}$, as we consider stellar models representing more evolved stars, where G is the gravitational constant, and M and R are, respectively, the mass and the radius of the model.

The frequencies of the oscillations predicted with our models might be compared with those observed in individual roAp stars. That work is presently being carried out (Cunha 2002b). Nonetheless, if the results shown in Fig. 4.5 are representative of what happens in roAp stars, we should expect the frequencies of the oscillations generally to increase with decreasing mass and, for stars with similar mass, generally to decrease with decreasing effective temperature. Therefore, if we bear in mind that sky transparency is a worry when observing oscillations which frequencies are just below the smallest frequencies currently observed in roAp stars, it seems possible that the oscillations present in more evolved and/or more luminous stars, having smaller frequencies, are missed more often.

4.6 Conclusions

Even though the model developed by BCDGV is, in many aspects, oversimplified, it illustrates well the effect that a possible suppression of convection in the envelope of roAp stars may have on the excitation of high frequency

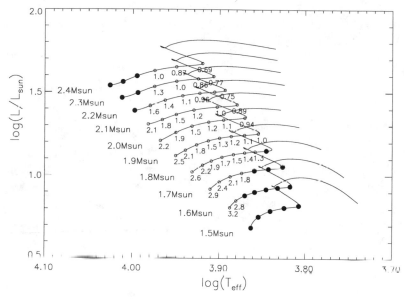

Fig. 4.5. HR diagram in the region where the predicted instability strip for roAp stars is located (symbols are the same as in Fig. 4.3). The numbers printed show, for each model, the frequency, in mHz, of the most unstable mode.

oscillations, and, consequently, the importance of keeping in mind the indirect effects of the magnetic field on pulsations. In fact, as reviewed here, if envelope convection is suppressed in some region of the star, high frequency oscillations may become unstable. Moreover, the 'second differences' might be affected in a way that it may no longer be possible to use them in order to identify the modes of oscillation, like suggested earlier by different authors (Gabriel et al. 1985; Shibahashi & Saio 1985).

Also important is to try to understand whether a model like that suggested by BCDGV is able to explain the observations, both concerning the instability strip for roAp stars and the statistical differences found between roAp and noAp stars. A comparison between the theory and the observations suggests that the two are in relatively good agreement, but that more work is needed in order to fully understand the observations.

Acknowledgements I am grateful to P. North for kindly providing the luminosities and effective temperatures of roAp stars used in the present paper. This work was supported by FCT-Portugal through the grant PD/18893/98 and the project POCTI/1999/FIS/34549 with funds from the European Community programme FEDER.

References

Balmforth, N. J., Cunha, M. S., Dolez, N., Gough, D. O. & Vauclair, S., 2001, *MNRAS*, **323**, 362 [BCDGV]

Bigot, L., Provost, J., Berthomieu, G., Dziembowski, W. A. & Goode, P. R., 2000, *A&A*, **356**, 218

Christensen-Dalsgaard J., 1993, in Baglin A. & Weiss W. W., eds, *Proc. IAU Colloq. 137: Inside the stars*, ASP Conf. Ser., Vol. 40, p. 483

Cunha, M. S., 1998, Contrib. Astron. Obs. Skalnaté Pleso, **27**, 272

Cunha, M. S., 1999, *PhD. thesis*, Cambridge University, U.K.

Cunha, M. S., 2001, *MNRAS*, **325**, 373

Cunha, M. S., 2002a, in Aerts, C., Bedding, T. & Christensen-Dalsgaard, J., eds, *Radial and nonradial pulsations as probes of stellar physics*, ASP Conf. Ser., Vol. 259, p. 272

Cunha, M. S., 2002b, *MNRAS*, **333**, 47

Cunha, M. S. & Gough, D. O., 2000, *MNRAS*, **319**, 1020

Dolez, N. & Gough, D. O., 1982, in Cox, J. P. & Hansen, C. J., eds, *Pulsations in Classical and Cataclysmic Variable stars*, (JILA, Bouder, CO), p. 248

Dolez, N., Gough, D. O. & Vauclair, S., 1988, in Christensen-Dalsgaard, J. & Frandsen, S., eds, *Proc IAU Symp. 123, Advances in Helio- and Asteroseismology*, p. 291

Dziembowski, W. A., 1984, in Gabriel, M. & Noels, A., eds, *Theoretical problems in stellar stability and oscillations*, (Liège), p. 346

Dziembowski, W. A. & Goode, P. R., 1985, *ApJ*, **296**, L27

Dziembowski, W. A. & Goode, P. R., 1996, *ApJ*, **458**, 338

Gabriel, M., Noels, A., Scuflaire, R. & Mathys, G., 1985, *A&A*, **143**, 206

Gautschy, A., Saio, H. & Harzenmoser, H., 1998, *MNRAS*, **301**, 31

Gelbmann M., Ryabchikova T., Weiss W. W., Piskunov N., Kupka F. & Mathys G., 2000, *A&A*, **356**, 200

Handler, G. & Paunzen, E., 1999, *A&AS*, **135**, 57

Hubrig, S., Kharchenko, N., Mathys, G. & North, P., 2000, *A&A*, **355**, 1031

Kurtz, D. W., 1982, *MNRAS*, **200**, 807

Kurtz, D. W., 1990, *ARAA*, **28**, 607

Martinez, P., 1996, *Bull. Astr. Soc. India*, 24, 359

Martinez, P. & Kurtz, D. W., 1994, *MNRAS*, 271, 129

Martinez, P. & Kurtz, D. W., 1995, in Stobie, R. S. & Whitelock, P., eds, *Astrophysical Applications of stellar Pulsation*, ASP Conf. Ser., Vol. 83, p. 58

Mathys G., Kharchenko N. & Hubrig S., 1996, *A&A*, **311**, 901

Matthews, J. M., 1988, *MNRAS*, **235**, 7p

Nelson J. M. J. & Kreidl T. J., 1993, *AJ*, **105**, 1903

North, P., Jaschek, C., Hauck, B., Figueras, F., Torra, J. & Kunzli, M., 1997, in Battrick, B., Perryman, M. A. C., Bernacca, P. L., O'Flaherty, K. S., eds, ESA SP-**402**, *Hipparcos-Venice 97* ESA, Noordwijk, p. 239

Ryabchikova, T. A., Savanov, I. S., Hatzes, A. P., Weiss, W. W. & Handler, G., 2000, *A&A*, **357**, 981

Shibahashi, H., 1983, *ApJ*, **275**, L5

Shibahashi, H. & Saio, H., 1985, *PASJ*, **37**, 245

Shore, S. N., Brown, D. N., Sonneborn, G. & Gibson, D. M., 1987, *A&A*, **182**, 285

Tassoul, M., 1980, *ApJS*, **43**, 469

Vauclair, S., Dolez, N. & Gough, D. O., 1991, *A&A*, **252**, 618

5

Waves in the magnetised solar atmosphere

COLIN S. ROSENTHAL

Søskrænten 58, Stavtrup, 8260 Viby J, Denmark

Oscillations and waves in the quiet and active solar atmosphere constitute a zoo of distinct and overlapping phenomena: internetwork oscillations, K-grains, running penumbral waves, umbral oscillations, umbral flashes etc. The distinctive oscillation spectra associated with the network, the internetwork, and sunspots and pores are a strong indicator that the magnetic field has a significant dynamical effect on wave motions. This immediately raises two questions i) Can waves be used as diagnostic indicators of the magnetic field? and ii) Do the different properties of wave motions in various field geometries have consequences for the efficiency of wave-heating in the atmosphere and corona? I will discuss some new numerical calculations of wave propagation in a variety of model atmospheres, which throw some light on these questions.

5.1 Introduction

The field of helioseismology has shown how waves which propagate through the deep solar interior can be used to determine the internal properties of the Sun – including its stratification, differential rotation, and sub-surface flow fields. Given the wide variety of waves and oscillations observed in the atmosphere of the Sun, in both Quiet and Active Regions, it is natural to ask whether the structures of these regions can also be determined from a wave analysis.

However, a brief consideration of the problem indicates that there are a number of critical differences between the atmospheric-wave problem and the p-mode problem which make the former vastly more difficult to study. The first and foremost is that the solar interior can, to a good approximation, be treated as a spherically symmetric object with small perturbations. This is not true for the solar atmosphere where the characteristic speeds of wave propagation vary strongly in three dimensions. Furthermore, the wave modes are intrinsically more complicated as they include the anisotropic effect of the magnetic field. Further complications arise from non-linearity (e.g. in K-grains and umbral flashes) and radiative effects.

There exists a large body of literature dealing with radiative and/or magnetic effects in one-dimensional atmospheres. In the magnetic case this allows the study of vertical or slanted uniform fields, or height-dependent horizontal fields. Radiative calculations exist at many levels of complexity, culminating in non-LTE radiation hydrodynamical calculations such as those of Carlsson & Stein (1997). Radiative calculations of waves in the solar atmosphere have also been carried out in three dimensions by Skartlien (2000) using a multigroup technique and including the effects of scattering.

However, there is relatively little literature dealing with the magnetohydrodynamics of waves in the solar atmosphere in more than one dimension. One such numerical calculation was carried out as long ago as Shibata (1983), but there have been no further attempts to solve the full wave equations for a two-dimensional field in a stratified atmosphere until the recent work of Cargill et al. (1997).

In this paper I will report on some calculations of the propagation of waves in two-dimensional magnetised atmospheres. We (for a list of collaborators see the author list of Rosenthal et al. 2002) thereby remove the debilitating restriction of considering only one-dimensional atmospheric structures, and we also allow for the possible development of shocks by solving the full (not linearised) equations. At this stage, however, we have not included the effects of radiative transfer in the calculations. I will describe the models in more detail in Section 5.2. In Section 5.3 I report on a calculation of wave propagation in the vicinity of a network element and in Section 5.4 I report on a surprising result on the effect of weak magnetic fields on the efficiency of acoustic sources. Section 5.5 is a summary of the results.

5.2 Description of the models

The models to be described here are strictly two-dimensional in the sense that all the dynamical variables are functions of only two co-ordinates, and the vector fields (velocity and magnetic field) have components only in those two directions. In each case the initial state of the magnetic field is a potential field. This allows us to choose a hydrostatic stratification for the initial pressure and density. We here consider only an ideal-gas equation of state.

The most subtle issues, both physically and numerically, arise in regard to the boundary conditions. At the vertical boundaries we take the straightforward approach of assuming horizontal periodicity. We drive the system by shaking at the lower boundary. For atmospheric models, this is intended to simulate the forcing of the atmosphere by photospheric turbulence. For the models to be presented in this paper, the forcing consists of a vertically-oscillating piston operating at a single frequency, although any combination of time-dependent vertical and horizontal motions could be implemented relatively easily. At the upper boundary, we would like to allow all wave motions to escape without reflection. In practice there exists no perfect way to implement transmitting upper boundary conditions, but we have found that the method of characteristics, in the formalism developed by Korevaar (1989) for a hydrodynamic problem, can be extended to this magnetohydrodynamic case with considerable success.

Our numerical scheme, based on a code by Nordlund and Galsgaard, is described in more detail in Rosenthal et al. (2002). It uses a high order finite-difference scheme for spatial derivatives and a predictor-corrector scheme for time-stepping. A staggered mesh differencing scheme is used to ensure high-accuracy flux-conservation.

5.3 Network and internetwork oscillations

In the quiet Sun, the internetwork chromosphere manifests a filigree pattern of wave like motions. These motions vary greatly in amplitude and in some internetwork regions are entirely absent.

Our model of the network/internetwork consists of an isolated flux tube embedded in an isothermal atmosphere. (Because of the horizontally periodic boundary conditions, this actually corresponds to a "picket fence" structure of flux sheets extending to infinity in each direction.) The properties of the background atmosphere (gravitational acceleration and density scale height) are chosen to resemble those of the Sun. At the atmospheric base, the density and pressure are $2.60 \times 10^{-7}\,\mathrm{g\,cm^{-3}}$ and $1.13 \times 10^5\,\mathrm{g\,cm^{-1}\,s^{-2}}$, respectively. A constant gravitational acceleration of $2.74 \times 10^4\,\mathrm{cm\,s^{-2}}$ is adopted, and the ratio of specific heats is set at $5/3$. The density scale-height is $158\,\mathrm{km}$ and the adiabatic sound-speed is $8.49\,\mathrm{km\,s^{-1}}$.

The vertical magnetic field imposed at the lower boundary consists of a weak flux tube (peak strength 250G) confined to a region approximately $2\,\mathrm{Mm}$ across and separated by $14\,\mathrm{Mm}$ from the next repeat of the unit cell. It is useful to define a plasma-β parameter as the ratio of the square of the sound speed to the square of the Alfvén speed.

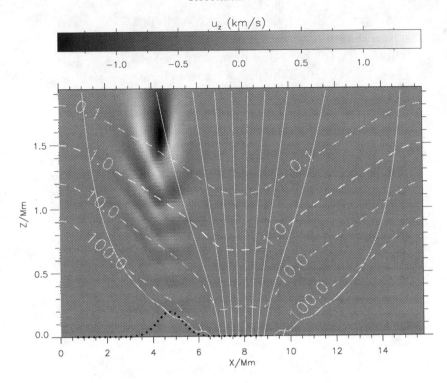

Fig. 5.1. Model of an internetwork oscillation. The solid lines are magnetic field lines and the dashed lines are contours of constant plasma-β. The variable depicted is the vertical velocity (in km s^{-1}) after 220 seconds of simulated time. The heavy dotted black curve at the bottom of the panel marks the distribution of displacement in the piston. Note that the vertical scale has been greatly expanded relative to the horizontal scale.

5.3.1 Internetwork oscillations

To model the internetwork oscillations we drive the system with a simple-harmonic piston located in the region outside the flux sheet. The frequency of the piston is 42 mHz and its amplitude is 3.7% of the sound speed. Figure 5.1 shows a snapshot of the result. The wave generated by the piston at the almost field-free region near the lower boundary is a fast mode which travels upwards initially at the (fixed) speed of sound. As it progresses upwards its amplitude grows due to the decrease in density. The fast mode propagates at the fast speed – the root mean square of the sound speed and the Alfvén speed. Hence, as the wave nears the region where the fast speed and sound speed are comparable (the $\beta = 1$ layer), its speed of propagation and vertical wavelength increase. Above this region, the fast mode is increasingly magnetic in character and its propagation speed increases exponentially with height. This rapid increase in the wave propagation speed causes the waves to be reflected back downwards in much the same way as the increase in sound speed in the solar interior causes non-radial p-modes to be reflected back upwards. In this model, the isosurfaces of constant phase-velocity are not, however, horizontal. The wave packet initially propagates vertically upwards but is therefore reflected back downwards at a slight angle to the vertical – an angle much exaggerated by the false aspect ratio used in plotting Figure 5.1. Interference between the upward and downward propagating components can then give rise to rapid apparent horizontal phase propagation.

The internetwork oscillations are thus trapped in the region below $\beta \approx 1$ and their amplitude is largest close to the reflecting layer. Thus the visibility of the waves will depend strongly on the height of this layer relative to the height of formation of the diagnostic in which the wave are observed This strongly supports the idea that the observed intermittancy in the waves (Carlsson 1999, Judge, Tarbell & Wilhelm 2001) is dependent on the magnetic structure (McIntosh et al. 2001, McIntosh & Judge 2001). Near the reflection layer, the strong interference between the upward and downward propagating wave trains gives rise to very rapid (supersonic) phase speeds, which suggest themselves as a possible explanation for the rapid motions seen in movies of the filigree internetwork oscillation pattern.

5.3.2 *Waves in a network element*

We can model waves in a network element by moving the piston inside the magnetic flux sheet. The results, as shown in Figure 5.2, are quite different from those of the internetwork case. While waves near the edges of the magnetic element continue to show reflection, waves near the centre of the flux sheet propagate upwards to the top of the computational domain without undergoing any significant reflection. The velocity amplitude continues to increase with height but, to a first approximation, the wavelength is unchanged. The simplest description of the waves is therefore that they are acoustic waves everywhere.

This, however, raises a subtle question. At the bottom of the box, the sound speed is higher than the Alfvén speed and sound-like waves are fast modes. But at the top of the box, the Alfvén speed is higher than the sound speed and acoustic waves are slow modes. Has there therefore been some sort of mode-transformation process? The question is not purely a semantic one because it raises the question of the reason for the difference between this

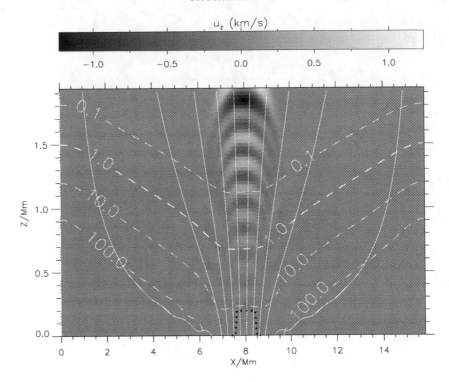

Fig. 5.2. Model of oscillations in a network element. Other than the location and profile of the piston, all parameters are the same as in Figure 5.1.

case and the previous case where the waves remained fast modes and were consequently reflected. The difference apparently arises because in the case where the waves travel parallel to the field there exists a degeneracy between the fast and slow modes at the $\beta = 1$ layer. This allows the waves to cross smoothly from the slow-mode branch of solutions to the fast-mode branch without reflection. In the internetwork case, where the field is significantly inclined to the propagation direction, this degeneracy does not occur and the waves must remain on the fast-mode solution branch and so are reflected.

The principal problem of interpretation with this model is that such oscillations are not seen in the solar network. In fact, the network is a site of considerably decreased acoustic power (relative to the internetwork) at frequencies above the cutoff. The problem may lie with the heavily simplified structure we have assumed for the both the field and the thermodynamic variables in the network. In practice, the gas is certainly hot and the field is probably highly tangled. Several possible explanations for the absence of such waves in network elements present themselves, e.g. that the high-temperature and large scale-height suppress the growth of velocity amplitude with height, or that the tangled field damps or scatters any waves present. A further possibility is that the network fields modify the turbulent spectrum in the underlying photosphere in such a way as to prevent the generation of significant wave power above the cutoff frequency. In the future it will be necessary to carry out simulations of wave propagation in more complex structures in order to resolve this issue of the suppression of wave amplitude in network fields.

5.4 Waves in a weak flux-tube

The next model to be considered was developed as part of an ongoing program to study the properties of waves in sunspots. The particular model to be presented here is not intended as an accurate model of a sunspot, but is included because the waves produced show unusual behaviour which might be relevant to the understanding of waves in atmospheric flux tubes.

The background atmosphere consists of an adiabatically stratified polytropic layer, 1 Mm in depth, surmounted by an isothermal atmosphere of thickness 0.5 Mm. There is no temperature jump between the top of the polytrope and the base of the isothermal layer. The initial magnetic configuration consists of a central strong flux sheet, representing a sunspot umbra, flanked by two symmetrically arranged sheets of oppositely signed flux, resulting in a set of closed flux loops intended to model the sunspot penumbra. The original intent behind the calculation was to model the impingement of external acoustic waves onto the spot, and the source therefore consists of a piston located at the base of the calculation and placed in a region where the flux-density should be negligible. The source itself is, in this case, dipolar and is rather small in horizontal extent (about 0.13 Mm). The frequency is 3 mHz.

The result, shown in Figure 5.3, is immediately surprising. The source is located in a region where the plasma β is of order 100, and one would not therefore expect the magnetic field to exert any significant influence. Yet the solution is clearly entirely dominated by the field and waves from the source are strongly channeled along the closed "penumbral" field lines. The occurence of such waves appears to be robust. Further simulations show that they appear under other atmospheric stratifications and for monopolar as well as dipolar sources.

A tentative explanation for this surprising result comes from consideration of the acoustic dispersion relation. At the base of the numerical domain,

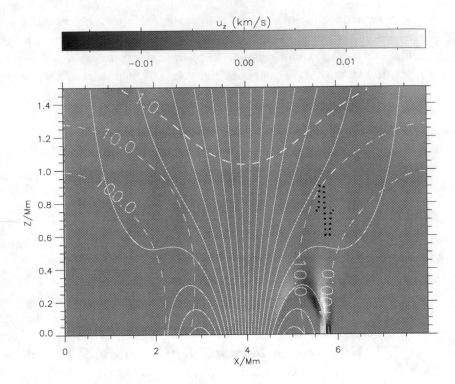

Fig. 5.3. Model of wave propagation in weak flux loop. The vertical component of velocity is shown after 350 seconds of simulated time. For clarity the black dotted curve indicating the piston, a localized dipolar source with zero mean, has been shifted vertically to the middle of the region.

the sound speed is $1.6 \times 10^4 \, \mathrm{m \, s^{-1}}$. At this speed, the characteristic wavelength at $3 \, \mathrm{mHz}$ is about $3 \, \mathrm{Mm}$. In the absence of a magnetic field, the very small dipolar source will therefore overwhelmingly couple to radially evanescent modes – a result confirmed by comparison with a purely hydrodynamic calculation (not shown). Therefore the magnetic field, however weak, provides the only way in which the source can emit propagating waves and so the solution is dominated by magnetic effects, even though the field is very weak in the vicinity of the source. So far this explanation seems reasonable, although what is surprising is the efficiency of the process – the wave amplitude in the penumbral loops is comparable to the imposed amplitude of the source.

What is less easy to understand is the nature of the waves themselves. Analysis of the motions into components parallel and perpendicular to the magnetic field shows that the displacement vector in these waves is polarised predominantly parallel to the field. They are therefore, apparently, compressive sound-like waves channeled along the loop. Typically this is the behaviour one expects of magneto-acoustic slow waves in a strong magnetic field. It is surprising to find it here in what must be considered fast waves in a high-β plasma.

5.5 Conclusions

The most striking result from these calculations, and from other results reported in Rosenthal et al. (2002) and in a forthcoming paper on waves in sunspots, is the very richness of the new phenomenologies opened up by removing the restriction of uni-dimensionality. The propagation of waves through these inhomogeneous and curved fields results in a potentially bewildering variety of refraction, reflection, interference and mode-transformation phenomena. However, with sufficient effort it has so-far proved possible to untangle the various wave modes in most cases we have considered and to present a plausible analysis of the results in terms of familiar magnetoacoustic mode physics.

The network and internetwork calculations shown above represent a good example. The internetwork calculation is particularly interesting because it presents an example of how waves in the magnetised solar atmosphere may be used as a diagnostic of the atmospheric magnetic field. In particular, it appears that the magnetic canopy acts as a surface of reflection of the waves, so that the measured wave amplitude will depend crucially on the relative heights of that canopy and the height of formation of the wave diagnostic – where the canopy is low, the waves will not be seen. Wave amplitude will be largest when the canopy height and the diagnostic-formation height are close to each other. We can hope to test this hypothesis by using extrapolations of photospheric fields to estimate the atmospheric field strength and correlating the height of the $\beta = 1$ layer with the internetwork wave power. Preliminary results by McIntosh et al. (2001) lend support to the idea.

The results reported in section 5.4 are less easy to understand and the interpretation more tentative. It appears that the weak curved "penumbral" field regions in the calculation possess a mode of oscillation unlike the familiar fast and slow modes known in uniform atmospheres. In particular, the fast acoustic-like wave which would propagate in all directions in a uniform high-β plasma is here nearly trapped in a flux sheet. The existence of this mode then allows a source which would normally be a very inefficient radiator of acoustic waves to drive a significant energy flux into the tube.

Although these results are highly preliminary, we should be alive to the possibility that they will amend our thinking about the possible efficiency of wave-heating in the solar atmosphere and corona.

Nevertheless, it must be admitted that at present we simply do not have a wholly satisfactory explanation of this particular phenomenon. As this book is dedicated in honour of the 60th birthday of Douglas Gough, a phenomenon without a clear explanation should not be seen as any cause for alarm or despondency, but should instead be welcomed, in the Goughian spirit, as an occasion for celebration, providing new open questions for future generations of scientists and mathematicians to address and new opportunities to expand our knowledge of the physics of the Sun and stars.

Acknowledgments This is an appropriate place to put on record my thanks to Douglas Gough for the continuing scientific inspiration he has provided throughout my research career as both student and post-doc. His zeal for understanding, his exuberant enthusiasm, and his penetrating insight have made it an honour to have collaborated with him.

I must also offer thanks to my daughter Adelena who, by being born at the appointed time and no earlier, allowed me to attend the meeting at Chateau de Mons where the oral version of this paper was presented.

This research was partially supported by the European Commission through the TMR programme (European Solar Magnetometry Network, contract ERBFMRXCT980190) and by the Norwegian Research Council's grant 121076/420, "Modeling of Astrophysical Plasmas".

References

Cargill, P. J., Spicer, D. S., & Zalesak, S. T. 1997, *ApJ*, **488**, 854

Carlsson, M. & Stein, R. F. 1997, *ApJ*, **481**, 500

Carlsson, M. 1999, in Wilson, A., ed., *Ninth European Meeting on Solar Physics: Magnetic Fields and Solar Processes*, ESA SP-448, ESA Publications Division, Noordwijk, the Netherlands, p. 183

Judge, P. G., Tarbell, T. D., & Wilhelm, K. 2001, *ApJ*, **554**, 424

Korevaar, P. 1989, *A&A*, **226**, 209

McIntosh, S. W., Bogdan, T. J., Cally, P. S., Carlsson, M., Hansteen, V. H., Judge, P. G., Lites, B. W., Peter, H., Rosenthal, C. S., & Tarbell, T. D. 2001, *ApJ*, **548**, L237

Mcintosh, S. W. & Judge, P. G. 2001, *ApJ*, **561**, 420

Rosenthal, C. S., Bogdan, T. J., Carlsson, M., Dorch, S. B. F., Hansteen, V., McIntosh, S., McMurry, A., Nordlund, Å. & Stein, R. F. 2002, *ApJ*, **564**, 508

Shibata, K. 1983, *PASJ*, **350**, 263

Skartlien, R. 2000, *ApJ*, **536**, 465

II Stellar rotation and magnetic fields

6

Stellar rotation: a historical survey

LEON MESTEL

Astronomy Centre, University of Sussex, Falmer, Brighton BN1 9QJ, UK

Prologue

By now I am reconciled to the contemplation of my former undergraduate pupils becoming senior citizens. For Douglas I have always had a particularly soft spot – it was so good for my ego to have daily commerce with a man of the same Napoleonic stature as my own. For decades I have watched his progress, gratified at the successful trailing of my St. John's College gown. I look forward to his reaching three-score-and-ten, confident that like others known to him, he will be treating retirement as a purely notional concept.

6.1 Radiative zones: the Eddington-Vogt-Sweet theory

In a rotating star, with magnetic forces assumed negligible, the equation of hydrostatic support in standard notation is

$$-\nabla p/\rho + \nabla\phi + \Omega^2\varpi = 0, \tag{6.1}$$

where ϕ is the gravitational potential, yielding the gravitational acceleration $g = \nabla\phi$ and satisfying Poisson's equation

$$\nabla^2\phi = -4\pi G\rho; \tag{6.2}$$

p and ρ are the pressure and density; G is Newton's gravitational constant; and in general the angular velocity $\Omega = \Omega(\varpi, z)$ or $\Omega(r, \theta)$ in cylindrical and spherical polars respectively. In a non-degenerate star, p and ρ are related by the equation of state

$$p = p_g + p_r = (\mathcal{R}/\mu)\rho T + aT^4/3, \tag{6.3}$$

where T is temperature, μ is the mean molecular weight, and $\mathcal{R}$ and a are the gas constant and radiation density constant. In all but the most massive stars, with μ lying between 1/2 (pure hydrogen) and 4/3 (pure helium), the radiation pressure p_r is a small correction, and so it will be ignored in most of the subsequent discussion.

The astrophysicist's approach to the problem, going back to von Zeipel, Eddington and Chandrasekhar, and developed later by Sweet and many others, is to treat $\Omega(\varpi, z)$ initially as an arbitrary imposed function, yielding a centrifugal perturbation which distorts the (p, ρ, T) fields from their normal spherical symmetry. It is recognized that the case $\Omega = \Omega(\varpi)$ – which includes uniform rotation – is special, for then there exists a centrifugal potential V, and the curl of (6.1) yields $\nabla p \times \nabla \rho = 0$ ('barotropy'), whence

$$p = p(\Psi), \quad \rho = \rho(\Psi) = \mathrm{d}p/\mathrm{d}\Psi, \tag{6.4}$$

with the joint potential $\Psi = \phi + V$ defining the *level surfaces*. If for the moment the star is assumed to be chemically homogeneous, then $T = T(\Psi)$ also.

To complete the system, one must introduce the energy equation; and as is well known, it is at this point that the problem is qualitatively altered (von Zeipel, in Eddington 1926). If the centrifugal field is conservative, then the radiative flux $\boldsymbol{F}$ through a medium of opacity $\kappa(p, \rho)$ can be written

$$\boldsymbol{F} = -(4acT^3/3\kappa\rho)\nabla T = \left[-(4acT^3/3\kappa\rho)(\mathrm{d}T/\mathrm{d}\Psi)\right]\nabla\Psi \equiv -f(\Psi)\nabla\Psi, \tag{6.5}$$

where c is the speed of light. This variation of $\boldsymbol{F}$ over a level surface, proportional to the effective gravity $\nabla\Psi$, is justly described as 'von Zeipel's theorem'. When Ω is independent of ϖ, $\nabla^2 V = 2\Omega^2$, and the 'von Zeipel paradox' for a uniformly rotating star results if one attempts, as in a non-rotating star, to balance the local radiative efflux of energy by local energy liberation, at a rate ϵ dependent just on the local variables $\rho(\Psi)$ and $T(\Psi)$; i.e. if from (6.5) and (6.2) one writes

$$\rho\epsilon(\rho, T) = \nabla \cdot \boldsymbol{F} = \frac{\mathrm{d}}{\mathrm{d}\Psi}\left(-\frac{4acT^3}{3\kappa\rho}\frac{\mathrm{d}T}{\mathrm{d}\Psi}\right)(\nabla\Psi)^2 \tag{6.6}$$

$$+ \left(-\frac{4acT^3}{3\kappa\rho}\frac{\mathrm{d}T}{\mathrm{d}\Psi}\right)(2\Omega^2 - 4\pi G\rho) .$$

All the quantities in (6.6) are functions of Ψ and so constant on a level surface, except for $(\nabla\Psi)^2$. Thus the coefficient of $(\nabla\Psi)^2$ must vanish separately, so that von Zeipel's theorem (6.5) now holds with the coefficient $f(\Psi)$ forced to be constant over the whole domain. The remaining terms then yield the manifestly spurious result

$$\epsilon \propto \left[1 - (\Omega^2/2\pi G\rho)\right] . \tag{6.7}$$

Von Zeipel's paradox is a mathematical peculiarity. It is unacceptable not only because of the unphysical result (6.7), but even more from the implied non-uniform behaviour as $\Omega \to 0$. In that limit, $\Psi \equiv \phi$, and there is no requirement that the first term in $\nabla \cdot \boldsymbol{F}$ should vanish independently of the second. In fact in a domain in thermal equilibrium but without any active nuclear sources, such as the radiative envelope of a Cowling model star, the two terms on the right of (6.6) are equal and opposite, whereas von Zeipel's argument implies that a small uniform rotation forces the first to vanish, so demanding that the second be balanced by the spurious energy source (6.7). However, as pointed out by Vogt and by Eddington (1929), the correct immediate conclusion from this *reductio ad absurdum* is not that the energy equation *alone* is able to restrict the class of allowed rotation fields, but rather that the net efflux $\nabla \cdot \boldsymbol{F}$ of radiant energy from unit volume is balanced jointly by the local nuclear energy generation, plus the energy transported by a thermally-driven circulation $\boldsymbol{v}$ in meridian planes – in Eddington's words, 'home products plus smuggled goods'.

The energy equation is written

$$c_v \rho \mathrm{d}T/\mathrm{d}t - (p/\rho)\mathrm{d}\rho/\mathrm{d}t = \rho\epsilon - \nabla \cdot \boldsymbol{F} \qquad (6.8)$$

with $\mathrm{d}/\mathrm{d}t$ the derivative following the motion $\boldsymbol{v}$ and c_v the specific heat at constant volume. In a hypothetical steady state, and for the special conservative centrifugal field with Ω constant, (6.8) becomes

$$\rho A(\Psi)(\boldsymbol{v}.\nabla\Psi) = \rho\epsilon + f(\Psi)(2\Omega^2 - 4\pi G\rho) + f'(\Psi)(\nabla\Psi)^2 \qquad (6.9)$$

where $f(\Psi)$ is defined in (6.5) and $A(\Psi) = T\mathrm{d}s/\mathrm{d}\Psi = c_v T\mathrm{d}[\log(T/\rho^{\gamma-1})]/\mathrm{d}\Psi$ with s the specific entropy and γ the usual ratio of specific heats. Division of (6.9) by $|\nabla\Psi|$ and application of the condition of zero net flow of gas across a level surface relates $f'(\Psi)$ and $f(\Psi)$; substitution back into (6.9) then yields for the velocity component normal to the level surface

$$\rho A(\Psi)\boldsymbol{v} \cdot \nabla\Psi = \left[f(\Psi)(4\pi G\rho)(1 - \Omega^2/2\pi G\rho) - \rho\epsilon\right]\left[|\nabla\Psi|^2 \frac{\int \mathrm{d}S/|\nabla\Psi|}{\int |\nabla\Psi|\mathrm{d}S} - 1\right].$$
$$(6.10)$$

Equations (6.9), (6.10) and the continuity equation $\nabla \cdot \rho\boldsymbol{v} = 0$, combined with (6.1), (6.2) and (6.3), suffice to determine the perturbed p, ρ, T, ϕ and $\boldsymbol{v}$ fields in a uniformly rotating radiative zone. In principle, the motions imply a deviation from the strict hydrostatic equation (6.1). From the inertia of the constructed flow, one can compute the deviations of the pressure from the Archimedean value, which for consistency must be small. Note the order of approximation: hydrostatic equilibrium $\to$ thermal imbalance $\to$ velocity field $\to$ dynamical pressure.

In the sub-adiabatic, radiative envelope of a Cowling-type model, the local polytropic index $n > 1/(\gamma - 1)$, yielding $A(\Psi) < 0$. By (6.9), since ϵ is negligible, the flow across a level surface is therefore opposite to effective gravity when the net local heat supply from radiative transfer is positive: in a convectively stable domain, with the specific entropy increasing outwards, energy has to be supplied for gas to flow against gravity. The second bracket in (6.10) is positive at the poles and negative at the equator, and over the bulk of a uniformly rotating envelope, the 'von Zeipel factor' $(1 - \Omega^2/2\pi G\rho) \approx 1$. Thus to first order in the centrifugal perturbation, the theory yields a quadrupolar-type flow, upwards at the poles and downwards at the equator. Note that ρ cancels out, with the predicted vertical velocity staying finite as the density becomes small. However, in a rapid rotator, $\Omega^2/2\pi G\rho$ will reach unity before photospheric densities are reached, so that the theory as developed so far appears to predict a break-up of the circulation into two zones in the low-density surface regions (Gratton 1945; Öpik 1951; Mestel 1999).

Even among the class of conservative centrifugal fields, with $\Omega = \Omega(\varpi)$, uniform rotation is special, for it is the the only one that allows $\nabla^2 V$ to be constant over a level surface, so that this term makes no contribution to the circulation velocity. In general, one writes $\nabla \cdot \boldsymbol{F}$ as the sum of its mean value over a level surface and the variable part $(\nabla \cdot \boldsymbol{F})'$, responsible for driving the circulation (e.g. Mestel, p. 465, in Aller & Mclaughlin 1965). One then finds that there is a term $\propto (\nabla^2 V)'$ and independent of ρ, yielding a *first* order vertical velocity term $\simeq (\bar{\rho}/\rho)$, where $\bar{\rho} = 3M/4\pi R^3$ (Baker & Kippenhahn 1959).

To sum up the essentials of the theory so far: if $\lambda < 1$ is a parameter measuring the ratio of the perturbing (conservative) centrifugal force to gravity, then the velocities generated over the bulk of the radiative zone are $\simeq \lambda(L/Mg)$, yielding a circulation time of the order of the Kelvin-Helmholtz contraction time increased by the factor $1/\lambda$. In the low density surface regions, significantly higher velocities are predicted. Near the domain of overshoot from a contiguous convective zone, the theory yields a vertical velocity becoming large like $1/[n - 1/(\gamma - 1)]$ and a horizontal velocity like its square. The prediction of regions with singular velocities is a sign that other contributions to meridional equilibrium – inertial, viscous, magnetic – though perhaps negligible over the bulk of the zone, are locally important.

The restriction to conservative centrifugal fields can and indeed should be relaxed. The papers by Sweet (1950) and by Sweet and Roy (1953) gave a perturbation procedure for the construction of models of stars and of the associated circulation velocity fields, under a general *prescribed* perturbing force field $\lambda \mathbf{f}$ per unit mass, where again λ is a convenient parameter. The hydrostatic condition (6.1) becomes

$$-\nabla p + \rho \nabla \phi + \rho \lambda \mathbf{f} = 0. \tag{6.11}$$

One writes for ϕ, and analogously for p, ρ, T,

$$\phi = \phi_0(r) + \lambda \phi_1(r, \theta) + \ldots\ldots = \phi_0(r) + \lambda[\phi_{11}(r) + \phi_{12}(r, \theta)] + \ldots, \tag{6.12}$$

where $\phi_{11}(r) = \int_0^\pi \phi_1(r, \theta) \sin\theta d\theta$, and $\phi_{12}(r, \theta)$ is the part that vanishes on averaging over a sphere. There are eight first-order quantities to determine. The r-component of (6.11) imposes two conditions, the θ-component imposes one, and the equation of state (6.3) and Poisson's equation (6.2) impose two each, making seven in all. Imposition of local radiative equilibrium on a star with a *prescribed* perturbing force would over-determine the system by requiring the eight quantities to satisfy nine conditions. It is the vector character of the equation of support which ensures that in a non-spherically symmetric system, local thermal balance in general requires energy transport by a circulation. Equally, as discussed originally by Schwarzschild and later by Roxburgh and by Clement, there exist *circulation-free* models, with the $\Omega(r, \theta)$-field fixed so as to yield (ϕ, p, ρ, T) distributions satisfying radiative equilibrium to first order.

The curl of (6.11) yields the first order equation

$$\frac{\partial}{\partial \theta}\left(\nabla^2 \phi_{12}\right) - \frac{4\pi G \rho_0'}{(-\phi_0')}\frac{\partial \phi_{12}}{\partial \theta} = \frac{4\pi G}{\phi_0'}\left[\frac{\partial}{\partial r}(r\rho_0 f_\theta) - \rho_0 \frac{\partial f_r}{\partial \theta}\right]. \tag{6.13}$$

Once this is solved, e.g. by expansion in Legendre functions, then ρ_{12}, p_{12} and T_{12} can be found; substitution into the steady-state form of (6.8) then yields the first order velocity components

$$\left[c_v \rho_0 T_0 d\{\log(T_0/\rho_0^{\gamma-1})\}/dr\right](v_r)_1 = \lambda(\rho\epsilon - \nabla \cdot \boldsymbol{F})_{12}, \tag{6.14}$$

with $(v_\theta)_1$ fixed by continuity. The method can in principle be continued to higher order in λ. Note that the mean of $(\rho\epsilon - \nabla \cdot \boldsymbol{F})$ over a sphere vanishes only to first order, for e.g. convection of *first*-order perturbations to the thermal field by the *first*-order velocity field contributes *second*-order terms to the energy balance (Sweet & Roy 1953).

The structure of the predicted circulation fields is very sensitive to variations in the assumed perturbing field. Of particular interest is the case of a non-uniform rotation field $\Omega(r)$, which yields $v_r = p(r)P_2(\cos\theta)$, with the deviation from the Eddington-Sweet case determined by the derivatives of Ω. Even if $\varepsilon(r) \equiv \Omega^2 r/|g| \ll 1$ everywhere, the basic assumption of the Eddington-Sweet approach – that the star adjusts its (p, ρ, ϕ, T) field to satisfy hydrostatic equilibrium – will break down locally if the scale of variation d of Ω is too small a fraction of r. Multiply equation (6.1) by ρ and take the curl:

$$\nabla\rho \times \boldsymbol{g} + \nabla\rho \times \Omega^2\boldsymbol{\varpi} + \rho\nabla \times (\Omega^2\boldsymbol{\varpi}) = 0. \tag{6.15}$$

If indeed $\varepsilon \ll 1$ but also $d \ll r$ locally – e.g. in the thin domain between a rapidly rotating stellar core and a slowly rotating envelope – then *prima facie*, the dominant terms reduce (6.15) to

$$\partial \log \rho/\partial\theta \approx -[r^2(\Omega^2)'/g]\sin\theta\cos\theta, \tag{6.16}$$

yielding

$$\partial \log \rho/\partial r \approx \partial(\log\rho(r,0))/\partial r - \sin^2\theta \; \left[r^2(\Omega^2)'/2g\right]'. \tag{6.17}$$

Thus over at least part of the domain of rapidly varying Ω, the last term in (6.17) will be of order $\Omega^2 r^2/gd^2$, while $-\partial \log\rho(r,0)/\partial r$ is of the order of the unperturbed inverse scale-height $1/l = |\rho_0'/\rho_0|$. To avoid an unstable, outwardly increasing density field, it appears that d cannot be less than d_c given by

$$(d_c/l)^2 \simeq \varepsilon(r/l) \tag{6.18}$$

In fact, the largest term dropped in the reduction of (6.15) to (6.16) is $\Omega^2 r\sin\theta(\partial\rho/\partial r)$, and this will not remain small; but as long as $\partial\rho/\partial r < 0$ this term will increase $\partial \log\rho/\partial\theta$ still further; and if $\partial\rho/\partial r > 0$, we in any case have instability. Thus we expect that a local Ω-gradient close to Ω/d_c would in fact be rapidly reduced by spontaneous dynamically-driven motions. However, a rotating star with $\varepsilon < 1$ and $d \simeq |\Omega/\Omega'| \gg d_c$ can *prima facie* be kept in hydrostatic equilibrium by appropriate variations in the (p, ρ, T) fields.

If also $d \ll r$, then the local thermally-driven circulation speeds will be large compared with the standard estimates (Sakurai 1975, Zahn 1992). From the equation of state (6.3) (with radiation pressure neglected), the density perturbation $\rho_{12} \simeq (r^2\rho_0/g_0)(\Omega^2)'$ requires a corresponding temperature perturbation $T_{12} \simeq (T_0/\rho_0)\rho_{12}$. The dominant term in $-\nabla \cdot \boldsymbol{F}$ is then (e.g. Mestel, p. 474, in Aller & Mclaughlin 1965) $\simeq -(L/4\pi r^2 T_0')\partial^2 T_{12}/\partial r^2$, yielding local values $v_r \simeq (L/4\pi\rho g^2)|(\Omega^2)'''| \simeq (L/Mg)(\Omega^2 r/g)(\bar{\rho}/\rho)(r/d)^3$; and by continuity, $v_\theta \propto |(\Omega^2)''''| \propto (r/d)^4$. These comparatively high speeds would be an important part of the input into the equation determining the actual Ω-field (cf. Section 6.4).

6.2 Comparison with geophysical theory

Before considering further the astrophysical problem, it is instructive to contrast the above treatment with that of meteorologists studying the dynamics of the gas in the earth's atmosphere (e.g. Pedlosky 1982, chapters 1 and 2). In a frame rotating with the angular velocity Ω_c of the earth's solid crust, the 'relative fluid velocities' $\boldsymbol{u}$, of order U, satisfy

$$\rho\left(\mathrm{d}\boldsymbol{u}/\mathrm{d}t + 2\Omega_c \times \boldsymbol{u}\right) = -\nabla p + \rho\nabla\Phi + \mathcal{F}, \qquad (6.19)$$

where $\mathcal{F}$ is the viscous force density, $\Phi = \phi + |\Omega_c \times \boldsymbol{r}|^2/2$ is again the sum of the gravitational and centrifugal potentials, and the acceleration is the sum of the relative and Coriolis accelerations. Because of the large scales L of the relative motions, the Rossby number $U/2\Omega_c L$ is small – the accelerations of the winds are small compared with the Coriolis acceleration, so in this sense the earth is a 'rapid rotator'. Note that $\boldsymbol{u}$ includes both the two components in meridian planes – vertical and north-south – and also the azimuthal, east-west component; hence 'U small' implies a small non-uniform rotation as viewed in the inertial frame.

The geophysicist also finds it convenient to take curl of (6.19), which yields the rate of change following the motion $\boldsymbol{u}$ of the relative vorticity $\omega \equiv \nabla \times \boldsymbol{u}$:

$$\mathrm{d}\omega/\mathrm{d}t = (\omega_a \cdot \nabla)\boldsymbol{u} - \omega_a(\nabla \cdot \boldsymbol{u}) + (\nabla\rho \times \nabla p)/\rho^2 + \nabla \times \mathcal{F}/\rho, \qquad (6.20)$$

where

$$\omega_a = \omega + 2\Omega_c \qquad (6.21)$$

is the absolute vorticity, measured in the inertial frame. In the small Rossby number limit, $\omega_a \simeq 2\Omega_c$, and in a friction-free steady state, (6.20) reduces to

$$(2\Omega_c \cdot \nabla)\boldsymbol{u} - 2\Omega_c\nabla \cdot \boldsymbol{u} + (\nabla\rho \times \nabla p)/\rho^2 = 0. \qquad (6.22)$$

In the fluid dynamicist's language, in a steady state, the generation of relative vorticity by baroclinicity is equal and opposite to a similar generation by the relative motions $\boldsymbol{u}$, through stretching and twisting of the already existing vorticity $2\Omega_c$.

As a special case, suppose that the system were symmetric about the axis; then $\nabla\rho \times \nabla p$ is in the azimuthal (ϕ) direction, with (6.22) fixing $\partial u_\phi/\partial z$; and if the baroclinicity is zero, then u_ϕ is a function just of ϖ. In the inertial frame, this is equivalent to a total local angular velocity $\Omega(\varpi) = \Omega_c + u_\phi(\varpi)/\varpi$. Thus we recover (for the case of small shear) the results leading up to the relations (6.4): the existence of a centrifugal potential implies barotropy, and *vice versa*, a special case of the Taylor-Proudman theorem (Pedlosky 1982). In general, the vorticity equation (6.22) fixes the z-variation of the velocity components in terms of the baroclinicity. With the same approximations, the original momentum equation (6.19) yields the 'geostrophic approximation', in which effective hydrostatic equilibrium under the centrifugally modified gravity holds in the vertical direction $\mathbf{n}$, and the meridional and azimuthal components of the horizontal velocity $\boldsymbol{u}_H$ satisfy

$$\boldsymbol{u}_H = (1/f\rho)(\mathbf{n} \times \nabla p), \qquad\qquad (6.23)$$

where f is the local Coriolis parameter $2\Omega_c \cos\theta$. An inertial frame observer would describe one component of (6.23) as balance of the meridional horizontal pressure gradient against the horizontal component of the extra centrifugal acceleration due to the small u_ϕ, and the other as showing how in a steady state, the effect of an azimuthal pressure gradient must be offset by the advection of angular momentum.

The point to emphasize is that in the geophysical approach, the variations in temperature and density are regarded as the *cause* of the whole low-Rossby number motion $\boldsymbol{u}$, which is therefore given the term the 'thermal wind'. This is entirely appropriate in a medium of low optical depth, since the processes directly affecting the temperature – non-uniform solar heating, cloud formation – are only weakly dependent on the motions; rather, it is the motions which respond to the consequent changes in the pressure-density field. If a steady state has been reached, then it is the momentum equation which fixes both the deviation from uniform rotation and also the nearly horizontal meridional velocity. By contrast, in a stellar radiative zone, it is the centrifugal field which is the *cause* of the deviation from spherical symmetry in the (p, ρ) field, and so also of the flow of heat leading to the small pressure variations that drive the meridian circulation. The circulation is present in an axisymmetric star – it does not arise from the the need to balance an azimuthal thermal pressure gradient by advection of angular momentum; but equally, this advection must be included when studying the angular velocity field as a function of position and time (Section 6.4).

6.3 Steady circulation and the mixing problem

Returning to the stellar problem, in the late 40s and 50s interest in the circulation problem was stimulated by studies of stellar evolution. It was often assumed, explicitly or tacitly, that an isolated, rotating, early-type main sequence star would remain chemically homogeneous, in spite of the strong temperature dependence of the rate of energy liberation. The short turnover time of turbulent convective motions was clearly a sufficient argument for effective homogeneity of the core; and from a rather superficial reading of Eddington's 1929 paper, it was supposed that the circulation currents would carry nuclear processed material from the core into the envelope so ensuring that the gradient of mean molecular weight μ remained small throughout the whole star. However, a homogeneous star with steadily increasing μ follows a path in the Hertzsprung-Russell diagram that is up and to the left of the zero-age main sequence. Such models could be appropriate for the comparatively few 'blue stragglers', but not for the far more numerous red giants. It was in fact already clear from pioneering work by Öpik, and later from the studies by Hoyle & Lyttleton, Li Hen & Schwarzschild and Bondi & Bondi, that a non-homogeneous μ-distribution was in principle able to account for the extended envelopes and the associated low surface temperatures of red giant stars. These models consisted essentially of a helium rich, nearly homogeneous inner part, surrounded by a hydrogen-rich envelope.

Such a structure could arise if a highly evolved but homogeneous star – with a convective core, still generating energy by the C-N cycle, and a surrounding radiative envelope – subsequently acquired a hydrogen envelope, e.g. via gravitational accretion of interstellar matter. However, to form a giant with a large radius, the mass accreted during the nuclear lifetime would have to be improbably high – of the same order as that of the accreting star. More plausibly, a giant structure could arise if for some parameter range, rotational mixing was limited to roughly half the envelope. New light was shed on the whole area of stellar evolution from photo-electric studies of the H-R diagrams of evolved globular and galactic clusters, especially by Sandage and collaborators, which suggested strongly that evolution of stars within a given cluster is best described to a zero-order approximation as depending on just one parameter (the stellar mass). Attempts to explain a universal phenomenon like the giant sequence by appealing to an adventitious factor such as mass accretion ceased to carry conviction, just on observational grounds. Following the papers by Sandage & Schwarzschild (1952), Tayler (1954), and especially by Hoyle & Schwarzschild (1955), opinion swung strongly to the view that normal stellar evolution occurs with negligible non-turbulent mixing.

The work of Sweet (1950) and Öpik (1951) brought out clearly that the characteristic time of the Eddington-Vogt currents in a uniformly rotating star is of order $\tau_{\mathrm{KH}}/\varepsilon$, where τ_{KH} is the global Kelvin-Helmholtz time and ε an average centrifugal parameter. It was also early recognized that the theory needed to be completed by consideration of the azimuthal component of the equation of motion. As the first task is to find an upper limit to the efficacy of 'rotational mixing' by the laminar E-V-S circulation, we begin by assuming that a steady circulation is maintained by a torque, due e.g. to a weak magnetic field, able to offset the advection of angular momentum by the circulation itself and so keep Ω nearly uniform.

Prima facie, a rough criterion for rotational mixing to be important is

$$\varepsilon \gg \tau_{\mathrm{KH}}/\tau_{\mathrm{nucl}} \equiv \eta, \qquad (6.24)$$

where τ_{nucl} is the characteristic time for the nuclear evolution of a star, kept homogeneous by mixing currents. Since $\eta \simeq 10^{-3}$, it appeared that whereas slow or moderate rotators would suffer virtually no mixing, a rapid rotator – e.g. an A-type star with a rotation period P of 1 day – would remain homogeneous, and a star with say $P \simeq 2$ days would acquire an inner nearly homogeneous zone of mass well above that of the turbulent convective core. This would imply that the dispersion of rotation periods among the more rapid rotators could yield a noticeable spread in the evolutionary tracks.

However, a crucial new feature is the back reaction of the distribution of μ – brought about by the combined effect of nuclear processing in the core and advection by the circulation – on the star's thermal field and so on the buoyancy forces driving the circulation (Mestel 1953, Mestel & Moss 1986). In a non-rotating star, hydrostatic equilibrium requires that T/μ be spherically symmetric, so that a non-spherical μ-distribution will imply a corresponding θ-dependence in T. As in the theory of Section 6.1, the radiative heat flux $\boldsymbol{F}$ now has both vertical and horizontal components; the divergence $\nabla \cdot \boldsymbol{F}$ in general varies horizontally and does not vanish, so yielding a 'μ-current' velocity $\boldsymbol{v}^\mu$, analogous to the E-V-S velocity $\boldsymbol{v}^\Omega$. The instantaneous μ-current field is a complicated functional of the μ-distribution and its gradients; in particular, its sign at points with the same θ can change from negative to positive as r changes. One simple result emerges at once. If the non-spherical μ-distribution is due just to the distortion by a quadrupolar $P_2(\cos\theta)$ perturbation of an existing $\mu_0(r)$-field with $\mu_0'(r) < 0$, then the sign of v_r^μ is such as to oppose the distortion and so to restore spherical symmetry in μ: a $\mu_0(r)$ field with a negative gradient is in this sense 'secularly stable'.

In a rotating star, one may proceed by superposing the effects of the Ω- and μ-perturbations, yielding a velocity field $\boldsymbol{v}$ depending on both the instantaneous Ω- and μ-fields. With all the nuclear processing taking place in the convective core, then in the radiative envelope, the μ-derivative following the motion vanishes, yielding

$$0 = \mathrm{d}\mu/\mathrm{d}t \equiv \partial\mu/\partial t + (\boldsymbol{v}^\Omega + \boldsymbol{v}^\mu) \cdot \nabla\mu. \qquad (6.25)$$

Equation (6.25) shows clearly the non-linear feed-back of the μ-distribution, set up by the circulation, on the circulation velocity itself. A state of 'steady mixing' has $\partial\mu/\partial t$ uniform over the whole mixing zone, but non-zero spatial μ-gradients, determined in the simplest model just by the travel times from the nuclear-processing convective core to the different points.

The first problem is to determine if any plausible parameter range will allow steady mixing, with μ over the whole star increasing in the time τ_{nucl}, implying from (6.25) a relative spatial variation of μ of order

$$\Delta\mu/\mu \simeq r/v\tau_{\text{nucl}}. \tag{6.26}$$

Near the core, we anticipate that v^μ will oppose the E-V-S velocity v^Ω. We are looking for conditions under which the circulation can continue, with v less than but still of order v^Ω, so that $\Delta\mu/\mu \simeq r/v^\Omega\tau_{\text{nucl}} \simeq \tau_{\text{KH}}/\varepsilon\tau_{\text{nucl}} \equiv \eta/\varepsilon$. In order that $|v^\mu|$ be indeed below v^Ω, we can demand that the μ-perturbation be less than the centrifugal: $\Delta\mu/\mu < \varepsilon$, yielding the tentative, more stringent criterion

$$\varepsilon \gg (\tau_{\text{KH}}/\tau_{\text{nucl}})^{1/2} = \eta^{1/2} \simeq 1/30, \tag{6.27}$$

replacing the earlier estimate (6.24). This turns out in fact to be remarkably close to the result of the detailed treatment (Mestel 1953), which yields $1/30$ for the minimum required value for ε, *computed at the core/envelope interface r_c.*

In itself, the criterion (6.27) is not prohibitive – the local ratio of centrifugal and gravitational accelerations is certainly small enough for perturbation theory to be applicable. But with Ω uniform, $\Omega^2 r^3/GM(r)$ increases outwards, in a Cowling model star by about 30, so that the centrifugal parameter reaches the unacceptably high value of unity at the stellar surface. We have introduced a weak magnetic field to offset the advection of angular momentum, and in fact to keep the rotation uniform, so ensuring that the simple E-V-S contribution v^Ω to the velocity is applicable. The same constraint on the rotation field then shows that one cannot overcome the potential choking by the μ-currents, without simultaneously running into trouble at the surface. Even more stringent conditions hold in an evolved star with a burnt-out core and a shell energy source.

It appears that the steady, E-V-S currents that would be set up in a chemically homogeneous Cowling model star in rapid uniform rotation are prevented from linking the nuclear-processing convective core with the radiative envelope by the choking effect of the μ-currents. The simplest suggestion is that the star evolves with the convective core steadily increasing in μ while the envelope retains its initial composition; the E-V-S circulation is pictured as flowing through the envelope, but is deflected horizontally at the

μ-barrier separating core and envelope. Since the μ-gradient at the barrier will not in fact be infinite, the circulation may still make some penetration (cf. Huppert & Spiegel 1977). This in turn may lead to the circulation's killing itself off through the spread of 'creeping paralysis'. A $\mu_0(r)$-field with a negative gradient is steadily set up, extending from the core through the whole radiative envelope. A slight distortion of the surfaces of constant μ by the P_2 E-V-S flow then yields $\boldsymbol{v}_\mu = -\boldsymbol{v}_\Omega$. It is not clear how long it would take for this state to be reached.

As was noted already in 1953, both thermally-determined processes – the rotational currents and the μ-choking effect – are very delicate. The theory as outlined assumes that the radiative envelope does not suffer any local dynamically-driven mixing, acting to smooth out the distribution of matter set up by the currents (cf. below).

6.4 The angular momentum distribution in a radiative zone

Discussions of the actual rotation law achieved bifurcate sharply, depending on whether or not magnetic effects are supposed significant.

6.4.1 Magnetic radiative zones

Already in Mestel (1953), and in many subsequent papers (e.g. Mestel et al. 1988), it is emphasized that the Alfvèn speed along even a very weak poloidal magnetic field $\boldsymbol{B}_p$ can easily exceed the slow circulation speed: hence there exist steady states in which a very small extra toroidal component $\boldsymbol{B}_t$, maintained by a poloidal current density $\boldsymbol{j}_p = (c/4\pi)(\nabla \times \boldsymbol{B}_t)$, yields a torque density $\boldsymbol{B}_p \cdot \nabla(\varpi B_\phi)/4\pi$ that can easily offset the advection of angular momentum, leaving the rotation with a correspondingly very small deviation from Ferraro's law of isorotation. Note the contrast with Section 6.2, where the ϕ-component of the Coriolis acceleration is balanced by a toroidal thermal pressure gradient, whereas a magnetic field that is axisymmetric can nevertheless exert a toroidal force through the tension along the twisted field lines.

In principle, the postulated magnetic field will contribute also to hydrostatic support. However, even the strongest observed stellar magnetic fields – such as the 34,000 G field of Babcock's star HD215441 – when supposed to increase inwards at a plausible (or even implausible) rate, have a mean energy density far less than that of the rotational kinetic energy. Because the circulation speeds v^Ω are normally so slow, it is very easy to satisfy simultaneously the two inequalities

$$\rho\Omega^2 r^2/2 \gg B^2/8\pi \gg \rho(v^\Omega)^2/2. \tag{6.28}$$

In a simple, axisymmetric system, *over the bulk of the star*, the magnetic torque can effectively dominate the 'toroidal dynamics' in the time available, without there being any significant modification of the 'poloidal dynamics', which remains essentially just hydrostatic pressure balance against gravity and centrifugal force.

Prima facie, the inequality (6.28) could break down near the surface of a radiative zone, where we have seen that the generalized E-V-S theory yields terms $\propto 1/\rho$, either to first or to second order in the centrifugal parameter. However, at such low densities the neglect of the magnetic contribution to the poloidal dynamics will also break down. Approximate models of uniformly rotating magnetic stars by Moss (see Mestel 1999) have $\nabla \times \boldsymbol{B}$ adjusting itself so that (in an obvious notation), $(\nabla F)_\Omega + (\nabla F)_B$ falls off more rapidly than ρ: the Lorentz force kills off the embarrassing $1/\rho$ terms, yielding a net circulation speed that not only stays finite but vanishes as $\rho \to 0$. One would like to see more work on these lines, but it can justly be claimed that this introduction of a weak magnetic field enables one to speak of steady E-V-S circulation models that are both thermally and dynamically self-consistent throughout the radiative zone.

The rather sharp dichotomy between the poloidal and toroidal dynamics is a consequence of the assumption of axial symmetry. A plausible model for the observed strongly magnetic stars is the oblique rotator, with the axis of the large-scale field inclined to the axis of uniform rotation, and this has some properties in common with a top. The Lorentz forces exerted by the field of total flux F cause small but finite density perturbations that are at least approximately symmetric about the magnetic axis. To keep the angular momentum vector invariant in space, there must be superposed on the basic rotation Ω the *Eulerian nutation*, a rotation about the magnetic axis analogous to the geophysicist's Chandler wobble, of order $\omega \simeq (F^2/\pi^2 GM^2)\Omega \ll \Omega$. However, the density-pressure field contains also the much larger perturbations due to the centrifugal forces, which are symmetric about the rotation axis. To maintain hydrostatic equilibrium in a radiative domain, the changes in the (ρ, p) field due to the Eulerian nutation must be offset by nearly divergence-free internal motions (Spitzer, in Lehnert 1958, p. 169; Mestel 1999 and references therein). These dynamically forced, oscillatory 'ξ-motions' have the period of the Eulerian nutation which is much longer than the free oscillation periods of the star, but can be shorter than the Kelvin-Helmholtz or the nuclear time-scales even if the total magnetic flux yields surface field strengths below the detectable limit. In a rapid rotator, the amplitudes of the ξ-motions are large and so could cause significant interchange of material between a convective core and a radiative envelope, so destroying the adverse μ-gradient that is impeding the non-oscillatory E-V-S flow. If so, then this cooperation between thermally-driven rotational currents and rotationally driven dynamical motions could have a significant effect over the leisurely time-scale of main-sequence stellar evolution.

The discussion so far has assumed that the star has already been brought into a state of near uniform rotation, which it can maintain magnetically against advection of angular momentum by the slow circulation. In an aligned magnetic rotator, an initial non-uniform rotation generates torsional oscillations, with exchange of energy between the rotational motion and the toroidal magnetic field generated by the shear. Mestel & Weiss (1987) estimate that even with a generous allowance for an effective macroresistivity due to the hydromagnetic instability of a dominantly toroidal field, a poloidal field of strength as low as 10^{-4}G will still be able to reverse the shear within a stellar lifetime. In the asymptotic steady state, even the limited freedom of isorotation rather than uniform rotation is in fact unlikely to persist. In an axisymmetric state, magnetic field lines which penetrate slightly into a convective zone can interchange angular momentum through coupling with the turbulence; and even a modest departure from axisymmetry will tend to enforce near uniform rotation through the action of toroidal magnetic pressure gradients (Moss et al. 1990).

However, again a large departure from axisymmetry yields qualitatively new results. Whereas in the axisymmetric problem, rotational shearing has only a weak effect on the basic poloidal field, rotational distortion of a highly oblique field can lead to juxtaposition of oppositely directed lines of the basic field, leading to accelerated Ohmic decay (K.-H. Rädler, in Guyenne 1986, p. 569). A weakish field that is initially highly oblique could thus be converted into a more nearly aligned field, simply through the accelerated decay of the perpendicular component; whereas a stronger field will be able to reverse the initially imposed non-uniform rotation before the accelerated Ohmic decay gets under way (Moss et al. 1990)

The postulated presence of some magnetic flux across the whole H-R diagram is by no means unreasonable (e.g. Mestel 2002). In particular, braking of the rotation of late-type stars, either as a consequence of magnetic activity (Schatzman 1962) or through magnetic coupling with a wind (Mestel 1967, 1999; Weber & Davis 1967; Mestel & Spruit 1987) is accepted as a basic phenomenon by all workers. Likewise, the ability of even a weak magnetic field to transport angular momentum efficiently through a stellar radiative zone is recognized (e.g. Mestel & Weiss 1987; Spruit, in Durney & Sofia 1987; Spruit 1999). Certainly, this does not mean that studies outlined in Section 6.4.2 below are redundant, but rather that it should be stressed that neglect of magnetic effects does put severe upper limits on the strength of any field present.

In late-type stars, dynamo action in both the convective zone and in the tachocline, at least in part analogous to that observed and inferred in our Sun, will generate and maintain time-dependent magnetic fields. Such fields will have the topology with poloidal-toroidal flux linkage that appears to be necessary though not sufficient for stability. It is not clear whether the fields of the strongly magnetic early-type stars are slowly decaying 'fossils', or are themselves being maintained by contemporary dynamo action. A fossil field may also be present e.g. in the radiative cores of the Sun and other late-type stars. Magnetic fields are notoriously subject to instabilities, and the considerable literature (e.g. Wright 1973; Acheson 1978; Tayler 1980; Spruit 1999; Mestel 1999 and references therein) has not given an unambiguous answer as to whether non-dynamo maintained fields can persist over stellar lifetimes. Spruit (1999) appeals to observation, arguing that the magnetic white dwarfs are 'fairly strong evidence' that long-term stable fields do exist in main sequence stars.

6.4.2 Non-magnetic radiative zones

Consider now the problems of a rotating radiative zone, with magnetic effects assumed ignorable. The basic dynamical restrictions on the allowed distribution of angular momentum are the Solberg-Høiland criteria (e.g. Tassoul 2000). They are essentially a generalization to a rotating star of the Schwarzschild criterion for stability against convection, restricting the direction of the gradient of the specific angular momentum $h = \Omega\varpi^2$ to a domain bounded by the direction of effective gravity and the outwardly pointing direction of the gradient of specific entropy. They are derived by considering adiabatic perturbations, for which buoyancy forces act to oppose instability. However, as discussed originally by Goldreich & Schubert and by Fricke, and later by Smith & Fricke, on a thermal time-scale a displaced element exchanges heat with its surroundings and so loses its buoyancy. Although, as seen, it is possible in general to find hydrostatic equilibria with arbitrary rotation fields by having appropriate temperature and density variations, it is found that only those with

$$\partial h^2/\partial \varpi > 0, \qquad \partial h/\partial z = 0 \qquad (6.29)$$

are secularly stable (e.g. Tassoul 2000). In an incompressible fluid, the first of these is the Rayleigh stability criterion, while the second is a part of the Taylor-Proudman condition for equilibrium (cf. Section 6.2). In a stably stratified gas, there exist motions that are sufficiently slow for heat exchange to annul the stabilizing buoyancy forces. As summarized by Fricke, one can find which states are *secularly* stable by solving the *dynamical* problem in the corresponding *incompressible* system. (Fricke studies also axisymmetric rotating magnetic stars, satisfying the law of isorotation, finding that only the special case with Ω the same constant on each field line is secularly stable.)

Suppose that Ω is initially uniform and large enough for the E-V-S circulation time-scale to be well below a stellar lifetime. If viscous effects are negligible, the advection of angular momentum changes the Ω-field, according to

$$\partial \Omega/\partial t = -\boldsymbol{v} \cdot \nabla(\Omega \varpi^2)/\varpi^2. \qquad (6.30)$$

It was argued by Busse that the rotation field would evolve towards a circulation-free state (cf. Section 6.1); but these states are themselves secularly unstable according to the criteria (6.29). Again, Osaki and later Kippenhahn & Thomas (in Lamb et al. 1981) argued that in the low-density surface regions of an early-type star, a small deviation from uniform rotation is sufficient to cancel to second order the terms $\propto \bar{\rho}/\rho$, leaving just a velocity of the Sweet order; but again this rotation law is secularly unstable.

The G-S-F thermally-driven instabilities are the seed of a form of weak turbulence, which also acts to redistribute angular momentum, in a time that was early estimated to be at least of the order of the Kelvin-Helmholtz time. James & Kahn (1970, 1971) increased this to the more plausible estimate of the standard E-V-S time, a result supported later by Kippenhahn et al. (1980) and by Kippenhahn & Thomas. Thus it appeared that one should add to (6.30) a quasi-viscous term with a characteristic time of the same order as the circulation time.

Besides the G-S-F instabilities, which will occur under axisymmetric disturbances, there are the possible non-axisymmetric shear instabilities, in which the energy in the differential rotation velocity $v_\phi = \Omega \varpi$ is used to overcome the stabilizing buoyancy forces, and to overturn a stably stratified $\rho(r)$-distribution. There is as yet no consensus as to their likely importance. The standard criterion allowing this Kelvin-Helmholtz-type instability is that the Richardson number

$$J_{\mathrm{R}} = [g(\mathrm{d}\rho/\mathrm{d}r - (\mathrm{d}\rho/\mathrm{d}r)_{\mathrm{ad}})]/\rho(\mathrm{d}v_\phi/\mathrm{d}r)^2 \qquad (6.31)$$

be less than some critical value, usually taken as 1/4 (e.g. Chandrasekhar 1961). This can be written equivalently $\cos\theta(\partial\log\Omega/\partial\log r) > 2N/\Omega$ where N is the Brunt-Väisälä frequency. By allowing for heat diffusion, Zahn (1975) found a much reduced upper limit. Jones and Acheson noted that these results tacitly assume finite amplitude disturbances, whereas for infinitesimal disturbances, a much stronger vertical shear can be tolerated. But assuming shear instability does set in and develop, what rotation law is achieved? Noting that the buoyancy stabilizes only in the vertical direction, Zahn (1975) argued that the turbulence is likely to smooth out the horizontal shear, yielding a rotation law $\Omega(r)$. Following to some extent an earlier treatment by Sakurai, Zahn (1992) then studied the consequent evolution of $\Omega(r,t)$ within a star, the competing processes being magnetic braking by a stellar wind, shear turbulence, and advection of angular momentum by the instantaneous E-V-S circulation, constructed essentially as outlined in Section 6.2, with the μ-current effect included.

However, the assumption $\Omega(r)$ is disputed. Kippenhahn & Thomas cited experiments on rotating liquids as showing that variations of Ω along equipotentials are not spontaneously wiped out, but could retain values that depend on the history of the star. By contrast, Tassoul (2000) bases many studies in this area on an appeal to the analogue of the geophysicist's baroclinic instability, to which the $\Omega(r)$ field is subject for almost all positive values of the Richardson number (6.31). And in a recent paper on the solar tachocline, Gough & McIntyre (1998) appeal to both 'strong theoretical arguments' and a 'wealth of observations' of the terrestrial stratosphere showing that two-dimensional turbulence drives the angular momentum distribution away from rather than towards uniform rotation.

It is both proper and desirable that there be these parallel studies of stellar rotation, with magnetic effects completely ignored; but it should simultaneously be stressed that at least for radiative zones, there is implicit a severe upper limit on the strength of any fields actually present. Likewise, it is incumbent on those who stress the importance of magnetic effects in stellar radiative zones also to keep a watchful eye on potential hydromagnetic instabilities (e.g. Spruit 1999).

A principal aim of these studies is to elucidate any possible effect of rotation on stellar evolution, in particular supplying theoretical background to well-known problems such as the Li abundance in main-sequence stars, and the abundance anomalies in early-type stars (Schatzman & Maeder 1981;

Zahn 1992 and in Durney & Sofia 1987; and many other papers).

6.5 Rotating convective zones

The canonical description of turbulent convection is in terms of the mixing-length l (the macroscopic analogue of the mean-free-path λ in kinetic theory) the associated turbulent velocity v_t, and a corresponding convective 'turnover time' $\tau_c = l/v_t$. The analogy itself suggests the introduction of a kinematic 'eddy-viscosity' $\nu_t \simeq lv_t/3$. Models of the solar convection zone yield as typical values $l \simeq 10^9$ cm, $v_t \simeq 3 \times 10^3$ cm s^{-1}, whence $\nu_t \simeq 10^{12}$cm^{2}s^{-1}, as compared with a microviscosity (kinetic or radiative) $\simeq 10^3$. If the analogy were perfect, the eddy viscosity would smooth out shear over a scale L in the time L^2/ν_t: acting alone, it would enforce uniform rotation of a convective zone in a time much below a stellar lifetime.

Biermann (1951) gave a simple argument to illustrate how this is brought about in a gas with *isotropic* random velocities. In a steady state, equal numbers of particles per second cross unit area of a fixed spherical surface S in each direction, and likewise they must carry the same amount of angular momentum. Clearly, those particles which actually *arrive* at the surface from outside must have in the mean *less* angular momentum than that possessed by the particles in the place of origin. The reason becomes clear if for convenience we use a frame rotating with the assumed uniform angular velocity Ω_c of the star, and for simplicity study the motion of particles near the equatorial plane in between two collisions, separated by the mean-free-path λ. Over the distance $\lambda/2$, an incoming particle with velocity v_m has its component perpendicular to the rotation axis rotated by the Coriolis acceleration at the rate $-2\Omega_c$, acting for the time $\lambda/2v_m$. It follows that a particle arriving at S at right angles must have started moving after its last collision at an angle $\Omega_c\lambda/v_m$ to the radius vector and in the sense opposite to the rotational velocity. The same result holds for all directions to S. This ensures that the particles which actually cross S have on average exactly the same angular momentum as the particles on S, as clearly must be the case in a stationary state.

The crucial point (as noted earlier by Wasiutynski and emphasized by Biermann), is that the argument assumes isotropy of the particle velocities in the rotating frame. If instead, for example, after each collision, the direction parallel to the cylindrical radius vector ϖ were preferred, then the same Coriolis acceleration would give particles crossing S the angular momentum (in the inertial frame) $\Omega_c\varpi^2 + \varpi(\Omega_c\lambda/v_m)v_m \simeq \Omega_c(\varpi + \lambda/2)^2$, just the mean angular momentum in a uniformly rotating star of particles at distance $(\varpi + \lambda/2)$. There would then be an inflow of angular momentum, clearly inconsistent with the assumption of a steady state in uniform rotation.

However, as was first pointed out by by Lebedinski and by Wasiutynski, the turbulence arising from instability under a gravitational field would indeed be anisotropic, with l and v_t having different values in vertical and horizontal directions, and so should lead to a non-uniform rotation field. Generalizing from the treatment of the flow in the equatorial plane, Biermann wrote the turbulent flux of angular momentum as the sum of an isotropic part $-A_1\varpi^2\nabla\Omega$ and a monotropic part $-A_2\hat{\boldsymbol{g}}_\mathrm{e}(\hat{\boldsymbol{g}}_\mathrm{e}\cdot\nabla(\varpi^2\Omega))$, where $\hat{\boldsymbol{g}}_\mathrm{e}$ is the unit vector in the direction of effective gravity. In a steady state (determined by just the action of the eddy viscosity) the two terms must sum to zero, yielding $\nabla\Omega$ parallel to $\hat{\boldsymbol{g}}_\mathrm{e}$; and in a slow rotator like the Sun, $\hat{\boldsymbol{g}}_\mathrm{e}$ is nearly radial, so

$$\Omega \propto r^{-2A_2/(A_1+A_2)} \tag{6.32}$$

The treatment has been generalized by Rüdiger (1989) and Rüdiger & Kitchatinov (in Ribes 1994), following the methods introduced by Osborne Reynolds. The velocity field in a turbulent domain is written as the sum of mean and fluctuating parts:

$$\boldsymbol{v} = (\Omega\varpi\mathbf{t} + \boldsymbol{v}_\mathrm{p}) + \boldsymbol{v}', \tag{6.33}$$

with the mean velocity having in general a meridional (poloidal) circulatory component as well as a rotatory (toroidal) part (in which $\mathbf{t}$ is a unit vector). When substituted into the non-linear inertial term, the mean of the terms in $\boldsymbol{v}'$ yields an effective body force $-\nabla \cdot (\rho\mathcal{Q})$, where $-\rho Q_{ij}$ is the Reynolds stress tensor. The traditional parametrization of Q_{ij} is generalized to

$$Q_{ij} = Q_{ij}^\nu + Q_{ij}^\Lambda. \tag{6.34}$$

Here the diffusive part Q_{ij}^ν depends on the velocity gradients through an eddy viscosity tensor, and the 'Λ-effect' term $Q_{ij}^\Lambda = \Lambda_{ijk}\Omega_k$ is the generalization of the Lebedinski-Biermann terms that depend on Ω but not on its derivatives.

In spherical polar coordinates, the terms in Q_{ij}, that contribute to angular momentum transport are

$$\begin{aligned} Q_{r\phi} &= -\nu_{vv}r\sin\theta\,\partial\Omega/\partial r + Q_{r\phi}^\Lambda, \\ Q_{\theta\phi} &= -\nu_{hh}\sin\theta\,\partial\Omega/\partial\theta + Q_{\theta\phi}^\Lambda. \end{aligned} \tag{6.35}$$

In a slow rotator, $Q_{r\phi}^\Lambda \simeq \Lambda_V\sin\theta\,\Omega$ and $Q_{\theta\phi}^\Lambda \simeq 0$, with Λ_V, ν_{vv} and ν_{hh} all independent of θ. If magnetic torques are ignorable, then in a steady state, angular momentum transport by the Reynolds stresses is balanced by advection by the laminar circulation:

$$\nabla \cdot (\rho \Omega \varpi^2 \boldsymbol{v}_{\mathrm{p}} + \rho \varpi \overline{v'_\phi \boldsymbol{v}'}) = 0. \tag{6.36}$$

If the circulation can be ignored, Reynolds stresses alone fix the Ω-field by

$$r^{-2} \partial(\rho r^3 Q_{r\phi})/\partial r + (\rho/\sin^2\theta) \partial(\sin^2\theta Q_{\theta\phi})/\partial\theta = 0. \tag{6.37}$$

In the slow rotation limit, with $Q_{\theta\phi}^\Lambda \simeq 0$, the zero-stress condition adopted at the surface of a convective envelope and Ω assumed independent of θ at the base, (6.37) yields zero stress throughout the zone, with $\Omega = \Omega(r)$, Biermann's basic result.

Biermann's treatment was extended by Kippenhahn (1963). While crucial for angular momentum transport, the Reynolds stresses of the subsonic turbulence (the 'turbulent pressure') make only a small contribution to hydrostatic support, which is therefore again given by (6.1) in a weakly magnetic domain. And if one assumes the zone to be nearly adiabatic, following the classical Biermann-Cowling analysis for non-rotating systems, then the surfaces of constant ρ and p coincide and the curl of (6.1) again requires the centrifugal field to be conservative. Thus the simplest treatment of the 'poloidal dynamics' enforces Ω constant on cylinders, whereas the above treatment of the 'toroidal dynamics' yielded Ω constant on spheres. Kippenhahn argued that the near barotropy of the (p, ρ)-field and the non-conservative centrifugal field would together enforce a meridian circulation with velocities fixed by making the drag due to the eddy-viscosity contribute to the balance of pressure, gravity and centrifugal force. The constructed $\boldsymbol{v}_{\mathrm{p}}$-field is then fed into (6.36) to yield the deviation of the Ω-field from pure r-dependence. Equatorial acceleration – as observed in the Sun – results when $\boldsymbol{v}_{\mathrm{p}}$ is equatorwards at the surface; this requires Ω to increase outwards, in turn requiring the horizontal mixing-length to be larger than the vertical.

With realistic solar parameters inserted, this iterative scheme did not converge. The form used above for the Λ-effect is valid only if the Coriolis number $\Omega^* = 2\tau_{\mathrm{c}}\Omega_{\mathrm{c}} \ll 1$, whereas the estimated value for the Sun is $\Omega^* > 1$. As Ω^* approaches unity there is now a non-zero contribution to $Q_{\theta\phi}^\Lambda$, so that the Reynolds stresses, *acting alone*, will yield a θ-dependent Ω-law, whereas in the low-Ω^*, Biermann-Kippenhahn approach, it is the circulation that yields departure from the pure r-dependent law (6.32). However, later studies (e.g. Brandenburg et al. 1990, 1992) showed that *as long as adiabaticity is retained*, then even with a more appropriate form for the Λ-effect, the combined effects of the toroidal and poloidal equations still do not allow the isorotation contours to deviate far from Taylor-Proudman cylindrical law. Thus the pioneering Lebedinski-Biermann-Kippenhahn approach has to be modified both for internal self-consistency, and even more in order to tally with the helioseismological data, which yield an Ω-field consistent with neither the Taylor-Proudman nor the Biermann laws as adequate zero-order approximations.

The possible inhibiting effect of rotation on convection had been studied by Cowling (1951). Earlier, Randers and Walén had noted that if the rotation law satisfied the Rayleigh criterion $\mathrm{d}(\Omega\varpi^2)/\mathrm{d}\varpi > 0$, then rotation would act to inhibit the *axially symmetric* modes that would otherwise grow in a superadiabatic region. Cowling derived for a uniformly rotating system the modification to the Schwarzschild instability criterion for local disturbances of the form $\exp[i(l\varpi + m\varpi\phi + nz)]$, stressing that for $m \neq 0$, the constraint of detailed angular momentum conservation is relaxed by the effect of azimuthal pressure gradients. His analysis shows that motions with $m \neq 0$ and $n \simeq 0$ – whirling motions in tubes parallel to the rotation axis – are hardly affected by the rotation. This local linear analysis demonstrates both inhibition and persistence of instability, but cannot predict the expected latitude-dependent reduction in the efficiency of heat transport, with the consequent departure from strict barotropy.

Several workers (Weiss, Durney & Roxburgh, Moss & Vilhu) had emphasized that quite modest departures from adiabaticity could be significant, especially for variation with latitude. More recently, anisotropic turbulent heat conductivity (Brandenburg et al. 1992a) is incorporated into a comprehensive study by G. Rüdiger & L.L. Kitchatinov (Ribes 1994, p. 27). Like the earlier workers, they begin with the toroidal dynamics, ignoring laminar circulation, computing the coefficients in $Q_{r\phi}^\Lambda$, $Q_{\theta\phi}^\Lambda$ from a model of inhomogeneous turbulence, and solving (6.37) for a convective shell subject to stress-free boundary conditions. For their chosen parameters, the resulting Ω-contours are close neither to the Biermann nor the Taylor-Proudman forms, but are approximately radial. For consistency, the circulation $\boldsymbol{v}_{\mathrm{p}}$ driven by the non-irrotational centrifugal field must be slow enough to justify neglect of the laminar advection of angular momentum in (6.36). The curl of the poloidal equation of motion is

$$\mathcal{D}(\boldsymbol{v}_{\mathrm{p}}) = (1/\rho^2)(\nabla\rho \times \nabla p) + \varpi(\partial\Omega^2/\partial z)\mathbf{t}, \qquad (6.38)$$

where the term on the left is the curl of the frictional drag on $\mathbf{v}_p$ and so can be written in terms of Q_{ij}^ν. After non-dimensionalization, the term in Ω^2 has as coefficient the Taylor number $\mathrm{Ta} = (2\Omega R^2/\nu_{\mathrm{t}})^2$. For a moderate rotator like the Sun, with the circulation slow, the two terms on the right of (6.38) must be in approximate balance: the observed equatorial acceleration then requires the poles to be somewhat hotter than the equator, as observed. The actual temperature distribution is determined by the energy equation (with viscous heating ignored):

$$\nabla \cdot (\boldsymbol{F}^{\mathrm{conv}} + \boldsymbol{F}^{\mathrm{rad}}) + c_p \rho \boldsymbol{v}_{\mathrm{p}} \cdot \nabla(\Delta T) = 0, \tag{6.39}$$

where ΔT is the superadiabaticity, $\boldsymbol{F}^{\mathrm{rad}}$ the usual radiative heat flux, and

$$F_i^{\mathrm{conv}} = -c_p \rho \chi_{ij} \partial(\Delta T)/\partial x_j. \tag{6.40}$$

The anisotropic thermal conductivity tensor χ_{ij} has to be constructed from a model of the influence of rotation on the turbulent convection (Kitchatinov et al. 1994). The simultaneous solution of all the equations then does yield a hotter pole, and confirms that for $\mathrm{Ta} < 10^7$, the slow-circulation model – yielding equatorial acceleration – is indeed a good approximation; but for larger Ta, even though the domain is not strictly adiabatic, the rotation approximates to the Taylor-Proudman law $\Omega = \Omega(\varpi)$, and now with $\Omega' < 0$ – equatorial *deceleration*. General agreement with the solar data is shown when $\Omega^* \simeq 2.3$, as compared with the actual estimated solar value of 6.

The most recent studies of fully three-dimensional compressible convection, do not in fact yield behaviour reminiscent of the phenomenological mixing-length picture, yet it appears that the classical mixing-length formulae remain reliable as a convenient *ex post facto* parametrization for estimating velocities and fluxes. A more sophisticated terminology may emerge from further studies of the solar rotation and any associated meridional motions, crucial for solar dynamo theory, and for the appropriate extensions to the younger, more rapidly rotating late-type stars. The predicted changes in $\Omega(\varpi, z)$ at high Ω^* should already make one cautious about too cavalier an extrapolation of results for the solar dynamo to more rapid rotators.

6.6 The solar tachocline

Let me now refer briefly to the transition from the latitude-dependent rotation of the solar convective zone to the nearly uniform rotation inferred for the bulk of the radiative core. The various studies in the literature illustrate the different possible approaches to stellar rotation, as outlined above. When studying the solar core, most workers are prepared to take the rotation at the base of the zone as prescribed. Spiegel & Zahn (1992) follow Spiegel (in Perek 1968, p. 261) in presenting first what I would call a non-magnetic, inviscid, Eddington-Sweet approach to the history of the inner solar rotation. After a transient phase, the laminar meridional velocity of the gas is related to the local angular velocity field by an equation derived from the quasi-steady (6.14) with $\epsilon = 0$, and with the dominant contribution to v coming from the term in Ω'''' (cf. the end of Section 6.1); while the evolution of the Ω-field is given by (6.30). It is found that after a solar lifetime, during which the convective envelope has been subject to the magnetic solar wind torque, the tachocline would have increased in thickness to $2 - 3 \times 10^5$km, far greater than the observationally unresolved thickness. In the second part, the authors introduce a strong turbulent friction into (6.30). An isotropic friction would only increase the spreading rate of the tachocline, but the authors follow Zahn's earlier argument and take the horizontal component to be far greater than the vertical. They predict a satisfactorily thin tachocline and an internal angular velocity intermediate between the polar and equatorial values in the convection zone, in agreement with the interpretation of helioseismic observations. But as seen, the modelling of shear turbulence as acting to produce Ω horizontally uniform is not universally accepted, e.g. by Gough & McIntyre (1998), who when considering the same problem, find themselves entitling their paper: 'Inevitability of a magnetic field in the Sun's radiative interior'.

Helioseismology and now also asteroseismology are providing more and more stringent tests of stellar astrophysics. Let me conclude by noting with pleasure how much Douglas has contributed to the exploitation of this gold-mine.

References

Acheson, D.J. 1978. Phil. Trans. R. Soc., **A289**, 459.

Aller, L.H. & McLaughlin, D.B. (eds) 1965. *Stellar Structure.* University of Chicago Press.

Baker, N. & Kippenhahn, R. 1959. *Z. Astrophys.*, **48**, 140.

Biermann, L. 1951. Zeits. f. Astrophys., **28**, 304.

Brandenburg, A., Moss, D., Rüdiger, G. & Tuominen, I. 1990. Solar Physics, **128**, 243.

Brandenburg, A., Moss, D. & Tuominen, I. 1992a. In Harvey, K. L., ed., *The Solar Cycle* ASP Conf. Proceedings, **27**, p. 536.

Brandenburg, A., Moss, D. & Tuominen, I. 1992b. *A&A*, **265**, 328.

Chandrasekhar, S. 1961. *Hydrodynamic and Hydromagnetic Stability.* Oxford University Press.

Cowling, T.G. 1951. *ApJ*, **114**, 272.

Durney, B.R. & Sofia, S. (eds) 1987. *The Internal Solar Angular Velocity.* Reidel, Dordrecht.

Eddington, A.S. 1926. *Internal Constitution of the Stars.* Cambridge University Press.

Eddington, A.S. 1929. *MNRAS*, **90**, 54.

Gough, D.O. & McIntyre, M.E. 1998. Nature, **394**, 755.

Gratton, L. 1945. Mem. Soc. Astron. Ital., **17**, 5.

Guyenne, T.D. (ed.) 1986. *Plasma Astrophysics.* ESA SP-251.

Hoyle, F. & Schwarzschild, M. 1955. Astrophys. J. Suppl., No 13.

Huppert, H.E. & Spiegel, E.A. 1977. *ApJ*, **213**, 157.

James, R.A. & Kahn, F.D. 1970. *A&A*, **5**, 232.

James, R.A. & Kahn, F.D. 1971. *A&A*, **12**, 332.

Kippenhahn, R. 1963. *ApJ*, **137**, 664.

Kippenhahn, R. & Thomas, H.-C. 1978. *A&A*, **63**, 265.

Kippenhahn, R., Ruschenplatt, G. & Thomas, H.-C. 1980, *A&A*, **91**, 181.

Kitchatinov, L.L., Pipin, V.V. & Rüdiger, G. 1994. Astr. Nachr., **315**, 157.

Lamb, D.Q., Schramm, D.N. & Sugimoto, D. (eds) 1981. *Fundamental Problems in the Theory of Stellar Evolution.* Reidel, Dordrecht.

Lehnert, B. (ed.) 1958. *Electromagnetic Phenomena in Cosmical Physics.* Cambridge University Press.

Mestel, L. 1953. *MNRAS*, **113**, 716.

Mestel, L. 1967. Mém. Soc. R. Liège, **5**, 15, 351.

Mestel, L. 1999. *Stellar Magnetism.* Clarendon Press, Oxford.

Mestel, L. 2002. In Mathys, G., Solanki, S.K., Wickramasinghe, D.T., eds, *Magnetic Fields across the H-R Diagram*, ASP Conf. Proceedings, **249**, p. 3.

Mestel, L. & Moss, D.L. 1986. *MNRAS*, **221**, 25.

Mestel, L. & Weiss, N.O. 1987. *MNRAS*, **226**, 123.

Mestel, L. & Spruit, H.C. 1987. *MNRAS*, **226**, 57.

Mestel, L., Moss, D.L. & Tayler, R.J. 1988. *MNRAS*, **231**, 873.

Moss, D.L., Mestel, L. & Tayler, R.J. 1990. *MNRAS*, **245**, 550.

Öpik, E.J. 1951. *MNRAS*, **111**, 278.

Pedlosky, J. 1982. *Geophysical Fluid Dynamics.* Springer, New York.

Perek, L. (ed.) 1968. *Highlights in Astronomy.* Reidel, Dordrecht.

Rüdiger, G. 1989. *Differential Rotation and Stellar Convection: Sun and Solar-type Stars.* Gordon & Breach, New York.

Ribes, E. (ed.) 1994. *The Solar Engine and its Influence on Terrestrial Atmosphere and Climate.* NATO Workshop, Paris. Springer, Berlin.

Sakurai, T. 1975. *MNRAS*, **171**, 35.

Sandage, A.R. & Schwarzschild, M. 1952. *ApJ*, **116**, 463.

Schatzman, E. 1962. Ann. Astrophys., **25**, 18.

Schatzman, E. & Maeder, A. 1981. *A&A*, **96**, 1.

Spiegel, E.A. & Zahn, J.-P. 1992. *A&A*. **265**, 106.

Spruit, H.C. 1999. *A&A*, **349**, 189.

Sweet, P.A. 1950. *MNRAS*, **110**, 548.

Sweet, P.A. & Roy, A.E. 1953. *MNRAS*, **113**, 701.

Tassoul. J.-L. 2000. *Stellar Rotation.* Cambridge University Press.

Tayler, R.J. 1954. *ApJ*, **120**, 332.

Tayler, R.J. 1980. *MNRAS*, **191**, 151.

Weber, E.J. & Davis Jr., L. 1967. *ApJ*, **148**, 217.

Wright, G.A.E. 1973. *MNRAS*, **162**, 329.

Zahn, J.-P. 1975. Mém. Soc. Roy. Liège, **8**, 31.

Zahn, J.-P. 1992. *A&A*, **265**, 115.

7

The oscillations of rapidly rotating stars

MICHEL RIEUTORD

Observatoire Midi-Pyrénées and Institut Universitaire de France
14 av. E. Belin, F-31400 Toulouse, France

*We review the effects of rotation on the oscillation spectrum of rapidly ro-
tating stars. We particularly stress the novelties introduced by rotation: for
instance, the disappearance of modes in the low frequency band due to the
ill-posed natured of the underlying mathematical problem. This is mainly an
effect of the Coriolis acceleration. The centrifugal effect changes the shape
of the star in the first place. The possible consequences of this deformation
on the oscillation spectrum are briefly analyzed. We also describe other possibly important effects of the centrifugal acceleration which come about on
the time scale of star evolution.*

7.1 A short introduction to rapidly rotating stars

All stars are affected by rotation but some of them, the rapid rotators, are
more affected than the others! Astronomers usually qualify as rapid rotators
all the stars with $v \sin i \geq 50 \, \mathrm{km \, s^{-1}}$, i.e. those with an equatorial velocity
larger than $50 \, \mathrm{km \, s^{-1}}$. Such a value should be compared to the Keplerian
limiting velocity which is

$$V_{\mathrm{kep}} \sim 440 \, \mathrm{km \, s^{-1}} \, (M/M_\odot)^{0.1}$$

for stars on the main sequence (we used the mass-radius relation given by
Hansen and Kawaler 1994). Thus, for these stars the limiting velocity is
weakly mass-dependent and rapid rotators appear as stars whose centrifugal
acceleration exceeds 10% of the surface gravity; since this ratio measures the
impact of rotation on the star structure, rapid rotators are those stars whose
shape is significantly distorted by rotation.

Although important as we shall see later, the centrifugal deformation of
the star is not the first effect to affect the spectrum of oscillations of a star:

the spectrum is indeed first perturbed by the new time-scale imposed by the period of rotation which interfere with other natural periods of the star. In Figure 7.1, we sketch out the frequency bands of modes typical of a stellar situation. Rotation comes into play through the appearance of the Coriolis acceleration, which ensures for any motion that angular momentum is conserved. Note that it not only modifies already existing modes (p- and g-modes) but also brings in new ones called inertial modes[†]. Since the Coriolis acceleration introduces the time-scale $(2\Omega)^{-1}$ (where Ω is the angular frequency of rotation), one should expect that the spectrum is strongly perturbed around and below this frequency. Hence, from the point of view of the oscillations, rapid rotators would be those stars whose pulsation frequencies are close to 2Ω. However, because of the rather high values of the Brunt-Väisälä frequency N found in stellar radiative zones, stars whose pulsation frequencies are strongly perturbed by rotation need to be rapid rotators in the classical sense.

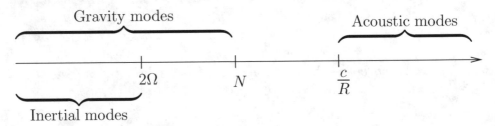

Fig. 7.1. The typical placing of the different spectral bands in the stellar context. Here N and c/R should be understood as the cut-off frequencies of the gravity and acoustic modes respectively, c being a mean value of the sound speed and R the radius of the star.

The foregoing discussion shows that the changes in the spectrum of oscillations of a star brought about by rotation come from two sides: the Coriolis and centrifugal accelerations. As we shall see both are important; the former is certainly better understood and I shall explain the recent progress on this subject, while the latter has mostly indirect consequences, more involved, but likely inescapable for a correct interpretation of rapid rotator pulsations.

Before developing the different aspects of these two non-Galilean effects, I would like to discuss briefly the method to deal with them, perturbative or non-perturbative.

---

† Inertial modes are the vast family of modes whose restoring force is the Coriolis force. They should not be confused with r-modes originally introduced by Papaloizou and Pringle (1978). r-modes are indeed a small subset of inertial modes which have a purely toroidal velocity field; in the field of geophysics they are called Rossby waves, hence the term r-modes which generated much confusion recently in the field of neutron stars (Rieutord 2001).

7.2 Perturbative versus non-perturbative methods

The determination of the eigenspectrum of a star is not an easy task. Although the set of equation is linear, the resolution of an eigenvalue problem is intrinsically a nonlinear operation requiring iterative algorithms to compute some subset of the spectrum. As this is a costly operation it is natural to avoid it or use its results in an optimal way. This is the idea of perturbative methods: using the spectrum and eigenmodes of a non-rotating star, one derives the modifications introduced by rotation to eigenfrequencies and eigenmodes.

In practice, such an approach is useful for slowly rotating stars like the sun. For rapidly rotating stars it is still useful but should be handled with care. Indeed, the main drawback of perturbative methods is their inability to "invent" new modes. They only make an adiabatic transformation of the spectrum of a non-rotating star. Hence they miss modes specific to rotation like inertial modes; they also miss 'singular' modes or modes which do not exist in a perfectly spherical star. For all these modes one should resort to non-perturbative methods which do not assume a separation of variables as done in non-rotating stars.

7.3 The part played by the Coriolis acceleration

The Coriolis acceleration is the term by which the operator governing oscillation modes loses its spherical symmetry. Operator $2\Omega\times$ is indeed not invariant under the rotations, except those around the rotation vector Ω. Hence, when using spherical coordinates, separability of the radial r and polar angle θ variables is lost. This loss shows up in the occurrence of a coupling between the equations governing the lth spherical harmonic of the fields; for instance, if we write the radial velocity $u = \sum_l u^l(r)Y_l^m$, its lth component $u^l(r)$ is coupled to the $(l+1)^{\text{th}}$ and $(l-1)^{\text{th}}$ component of the toroidal velocity components themselves coupled to u^{l+2}, u^l and u^{l-2}; hence an infinite chain of equations appears. These series converge rapidly when the coupling coefficient is small, but at frequencies close to or less than 2Ω this coefficient is of order unity.

However, the loss of separability and the ensuing coupling is only the first difficulty of the problem. The second and main one comes from the ill-posed nature of the mathematical problem: the operator governing the spatial structure of the eigenmodes is of mixed type: in some regions of the star it is hyperbolic and in others elliptic. As shown by Hadamard, hyperbolic operators require initial conditions while here boundary conditions are imposed.

This mathematical property is not specifically associated with the Coriolis acceleration and is already present with the buoyancy force. However, in a non-rotating star, due to the spherical symmetry of the system and the ensuing separability of the variables, the "ill-posedness" does not generate any singularity; but as soon as this symmetry is broken, both terms (Coriolis and buoyancy) introduce hyperbolicity and thus "ill-posedness".

To be more illustrative, let us consider the simplified set up of a rotating spherical shell filled with a stably and radially stratified fluid. To avoid unnecessary complications, we use the Boussinesq approximation. The reader may find in Dintrans and Rieutord (2000) a first investigation of a more realistic case applied to γ Doradus stars. The Boussinesq set up has been used many times in the past (Chandrasekhar 1961, Friedlander and Siegmann 1982, Dintrans et al. 1999, etc.). Neglecting diffusion terms and using non-dimensional variables with the time-scale $(2\Omega)^{-1}$, perturbations of infinitesimal amplitude obey the linear equations

$$\begin{cases} i\omega\boldsymbol{u} + \boldsymbol{e}_z \times \boldsymbol{u} = -\nabla P + \omega^{-1}N(r)^2 iu_r\boldsymbol{e}_r\,, \\ \nabla \cdot \boldsymbol{u} = 0\,, \end{cases} \tag{7.1}$$

where $\boldsymbol{e}_r$ and $\boldsymbol{e}_z$ are unit vectors in the radial direction and along the rotation axis respectively, ω is the frequency of oscillation, $\boldsymbol{u}$ the velocity perturbation (of which u_r is the radial component), and P is the pressure. We recognize the Coriolis acceleration $\boldsymbol{e}_z \times \boldsymbol{u}$ and the buoyancy term $N^2 iu_r/\omega\boldsymbol{e}_r$, $N(r)$ being the Brunt-Väisälä frequency; note that temperature has been eliminated with the energy equation.

To see the mathematical nature of the operator, it is necessary to reduce this system to a single second order equation for the reduced pressure. Keeping only highest order derivatives, one finds

$$(\omega^2 - N^2(r)\cos^2\theta)\frac{\partial^2 P}{\partial s^2} + 2N^2(r)\sin\theta\cos\theta\frac{\partial^2 P}{\partial s\partial z}$$
$$+(\omega^2 - 1 - N^2(r)\sin^2\theta)\frac{\partial^2 P}{\partial z^2} + \cdots = 0$$

as first derived by Friedlander and Siegmann (1982). In this equation s and z are the radial and vertical cylindrical coordinates while θ is the polar angle. This equation shows that the type of the operator changes on the critical surface, whose equation is

$$\omega^4 - (N^2(r) + 1)\omega^2 + N^2(r)\cos^2\theta = 0$$

while the characteristics of the hyperbolic domain are lines (in a meridional plane) given by the equations (assuming $N(r) = r\tilde{N}$):

$$\frac{dz}{ds} = \frac{zs\tilde{N}^2 \pm \xi^{1/2}}{\omega^2 - \tilde{N}^2 z^2}, \qquad \xi = \omega^2 \tilde{N}^2 s^2 + (1 - \omega^2)(\omega^2 - \tilde{N}^2 z^2). \qquad (7.2)$$

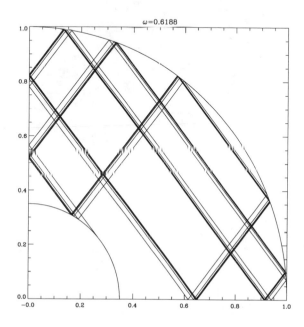

Fig. 7.2. The trajectories of characteristics in the meridional plane of spherical rotating incompressible fluid shell. Note how rapidly the characteristics converge towards the attractor.

These properties have been known for a long time (Friedlander and Siegmann 1982) but it is only recently that their consequences have been fully appreciated (Dintrans et al. 1999, Rieutord et al. 2001). To have a taste of them, it is necessary to understand the dynamics of characteristics which propagate in the hyperbolic regions. These lines physically represent the path of energy of a wave packet travelling inside the domain; they may be compared to the trajectories of a point mass particle in a potential well delimited by the boundaries of the hyperbolic domain. However, unlike a particle which can move in any direction, characteristics at a given point of a meridional plane, have only two possible directions. Thus, their dynamic is much constrained: in fact, the phase space is one-dimensional and no chaos is possible. We illustrate in Figure 7.2 the dynamics of characteristics in the unstratified case.

Fig. 7.3. The kinetic energy of an axisymmetric inertial mode at frequency $\omega =$ 0.7822 associated with an attractor where the shear layers are clearly visible. The fluid is viscous and incompressible. In this numerical solution, $\eta = 0.35$ is the ratio of the shell radii, L = 1000 is the number of spherical harmonics used, Nr = 400 is the number of radial grid points, and the Ekman number measuring viscosity is $E = 1.0 \times 10^{-9}$. Stress-free boundary conditions have been used.

In Rieutord et al. (2000, 2001), we have studied in some details the un-stratified case but we think that the properties thus uncovered are generic and equally apply to the stably stratified case. It turns out that character-istics may follow three types of trajectories: strictly periodic trajectories, quasi-periodic ones or trajectories converging towards an attractor. The first type occurs for some specific frequencies and it can be shown (Rieutord et al. 2001) that they are in finite number; quasi-periodic ones have been demonstrated to exist only in the full unstratified sphere. In fact, converg-ing trajectories is the generic case. The attractor is usually a periodic orbit but may be a point (in the meridional plane when a critical surface meets a boundary, e.g. Dintrans et al. 1999). In all cases, attractors are sin-gularities which make the solution neither integrable nor square-integrable (neither the total momentum nor the total kinetic energy exist). Therefore when attractors are present no eigenmode is possible (the point spectrum of the operator is said to be empty). In fact, attractors control the whole spectrum of gravito-inertial modes in a star: if a star had no diffusion the so-called r-modes would be the only ones to "survive", the spectrum of oscillation would be almost empty. But real stars have diffusion and eigenmodes are still possible: indeed, singularities associated with attractors are regularized into shear layers as shown in Figure 7.3. However, shear layers do appear only when the diffusion is small enough given a length of the attractor. It turns out that long attractors require such a small diffusion that even in stars shear layers would not appear. In such cases, the singularity generated by the inviscid part of the operator is not able to overcome diffusive effects and regular-like modes exist, but it is clear that such modes cannot be obtained with an adiabatic approach.

The foregoing discussion shows that in rotating stars inertial gravity modes should be studied carefully: first the strongest attractors should be localized on the frequency axis as they will correspond to strongly damped modes; surely, no observed period of star pulsation can be within such bands. Secondly, eigenmodes of rotating stars do require diffusion to be properly computed, otherwise results may be resolution dependent.

To conclude this section on the effects of the Coriolis acceleration, we should emphasize that the first step in order to evaluate their importance is to compare the Coriolis frequency 2Ω to that of the lowest order gravity modes ω_g. If $\omega_g \gg 2\Omega$ then perturbative methods are valid for large-scale modes; if $\omega_g \sim 2\Omega$ non-perturbative methods should be used. Presently, γ Doradus stars are the only stars which fall in the second category but others may join the club!

7.4 The part played by centrifugal acceleration

The centrifugal acceleration plays a much more subtle and less understood rôle in the dynamics of rotating stars. Its first effect is obviously to change the shape of the star from a sphere to an oblate axisymmetric spheroid. A second effect appears in radiative zones only: unlike in a non-rotating star, the radiative zone of rotating stars cannot be in hydrostatic equilibrium: an azimuthal flow forced by the baroclinic situation (isobars and isotherms do not match) appears. Such flows, which may be turbulent, are not very strong but are active during a large fraction of the lifetime of the star, their influence is therefore understandable only on a long time scale. It is indeed on such time scales that the element distribution is modified; since this distribution may control the excitation of the modes, via the kappa-mechanism, we see here an indirect influence of the centrifugal acceleration on mode visibility. Another one may be the state of differential rotation which is known to perturb the spectrum of oscillations (Dziembowski and Goode 1992).

Obviously, these effects are still largely unknown and we shall thus restrict our discussion to the first effect which has the advantage of being universal. It indeed affects the whole star and not specifically the radiative or the convective zone.

The centrifugal deformation of a star is an effect controlled by Ω^2 and therefore often thought to be unimportant; for instance, we know that acoustic modes are only slightly perturbed by the Coriolis acceleration and usually first order perturbations in Ω are sufficient. Logically, second order perturbations should be irrelevant.

The relevance of centrifugal corrections comes from the fact that they should be compared to the difference between the frequencies of two neighbouring modes. Indeed, it is clear that mode identification will be vitiated if these corrections are larger than the frequency spacing and are not included.

To make things quantitative somehow, we need to compare the typical wavelength of the mode to the actual centrifugal deformation of a star. We may characterize this deformation by the polar flattening ε defined by

$$\varepsilon = (R_e - R_p)/R_e$$

where R_e and R_p are respectively the equatorial and polar radii. For an $n = 3/2$ polytrope the maximum value of ε is 0.37; we shall take a typical value of $\varepsilon = 0.1$. If we consider the case of a δ-Scuti star with a typical radius of $2R_\odot$, its equatorial distortion amounts to $0.2R_\odot$, while the wavelength of a mode pulsating with a period of 20min is $0.17R_\odot$ (assuming a mean sound speed of $100\,\mathrm{km\,s^{-1}}$). It is therefore clear that the centrifugal deformation is an inescapable feature for a correct identification of acoustic modes in rapid rotators.

In order to better appreciate these effects, Lignières et al. (2001) studied the case of acoustic modes of a perfect gas contained in an axisymmetric ellipsoidal vessel. One interesting property of such a system is that it shows quite clearly the qualitative differences which arise when one goes from the sphere to the ellipsoid. Within a sphere acoustic modes have a maximum pressure on the boundary (recall that their radial structure is described by Bessel functions of semi-integer order). Such modes are called "whispering gallery" modes. Their associated ray path never penetrates a central ball. When one shifts to an ellipsoid, whispering gallery modes still exist but a new family appears: these are the central modes which are associated with ray paths crossing the equatorial plane between the foci of the meridional ellipse. Such modes, which do not exist in a sphere, prove the qualitative differences introduced by a slight change of shape. The two types of modes are shown in Figure 7.4.

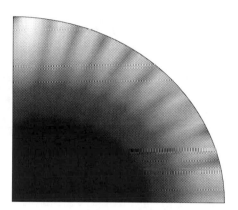 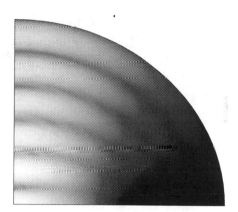

Fig. 7.4. The kinetic energy viewed in a meridional section of an axisymmetric ellipsoid for two types of acoustic modes: on left a whispering gallery mode at $\omega = 25.3$, on right a central mode at $\omega = 21.0$. Both are axisymmetric; note the absence of amplitude at the equator for the central mode. The polar flattening is $\varepsilon = 0.13$ and no dissipation has been included. We used 30 spherical harmonics and 50 radial nodes to compute them.

Now real stars are not true ellipsoids and the sound speed varies radially. Thus acoustic modes live in a cavity which has no special symmetry except the equatorial one. In these conditions it is very likely that the dynamics of acoustic rays is chaotic which would endow the spectrum of acoustic modes with properties of systems experiencing quantum chaos, a possibility which was already pointed out in the masterful work of Gough (1993).

7.5 Conclusions

To conclude this contribution, I would like to stress our progress in the understanding of the effect of rotation in rotating stars but also stress the problems which remain to be solved.

We have seen that the introduction of rotation modifies in many respects the spectrum of oscillations of a star: new modes appear while others disappear. These drastic changes are obviously not captured by perturbative methods.

Rotation introduces through Coriolis acceleration its own modes, namely inertial modes, which are low frequency modes; as such, they strongly interact with gravity modes. This interaction is not visible for slowly rotating stars since only very high order modes are perturbed. However, in rapid rotators the cut-off frequency of inertial modes, 2Ω, may be close to the cut-off frequency of gravity modes; in this case, the low frequency band of the oscillation spectrum of the star is deeply modified: as I have shown, the mathematical nature of the adiabatic problem is such that almost all the low frequency band is 'polluted' by singularities. Some diffusion is therefore required to compute modes in this band.

Besides the Coriolis acceleration, the centrifugal one does not play a less important rôle. Its first effect, which is to change the shape of the "container", may perturb noticeably the acoustic spectrum especially in the high frequency range. However, order-of-magnitude arguments show that even acoustic modes of period similar to those observed in δ-Scuti stars are likely perturbed by the centrifugal effects. Looking back to the gravito-inertial modes, I note that these modes are likely less perturbed by the change of shape since observable ones have a large-scale structure.

Besides these results, it is quite clear that many questions are left open. Especially the effects of the centrifugal acceleration call for more work: the hand-waving arguments need to be refined with more realistic models such as fast rotating polytropes for a first step. Then, a far more challenging issue concerns the evolutionary effects related to the transport and mixing of elements so as to determine the excitation mechanism yielding observable modes. The investigation of such effects clearly calls for two-dimensional stellar models which include transport processes and losses of angular momentum through stellar winds.

Finally, to end with a note of exotism, we should also mention the case of neutron stars which are very rapidly rotating objects. Since the discovery by Andersson (1998) that inertial r-modes are unstable when coupled to gravitational radiation, the oscillation spectrum of neutron stars is actively investigated and there too rotation is an unavoidable ingredient for all types of modes.

Acknowledgements The results presented here have been obtained with the enthusiastic collaboration of B. Dintrans, B. Georgeot, F. Lignières and L. Valdettaro who are all gratefully thanked. Figure 4 owes much to the help of F. Lignières.

References

Andersson N., 1998, *ApJ*, **502**, 708

Chandrasekhar S., 1961, *Hydrodynamic and hydromagnetic stability*. Clarendon Press, Oxford

Dintrans B. and Rieutord M., 2000, *A&A*, **354**, 86

Dintrans B., Rieutord M., and Valdettaro L., 1999, *JFM*, **398**, 271

Dziembowski W.A. and Goode P., 1992, *ApJ*, **394**, 670

Friedlander S. and Siegmann W., 1982, *GAFD*, **19**, 267

Gough D.O., 1993, in Zahn, J.-P. & Zinn-Justin, J., eds, *Astrophysical fluid dynamics, Les Houches Session XLVII*, Elsevier, Amsterdam, p. 399

Hansen C. and Kawaler S., 1994, *Stellar interiors: Physical principles, structure and evolution*, Springer

Lignières F., Rieutord M. & Valdettaro L., 2001, in Combes, F., Barret, D. & Thevenin, F., eds, *Proc. SF2A 2001, Lyon*, EDP Sciences Conf. Ser. in Astronomy & Astrophysics, Les Ulis, p. 127 - astro-ph/0110214

Papaloizou J.C.B. and Pringle J.E., 1978, *MNRAS*, **182**, 423

Rieutord M., 2001, *ApJ*, **550**, 443

Rieutord M., Georgeot B., and Valdettaro L., 2000, *PRL*, **85**, 4277

Rieutord M., Georgeot B., and Valdettaro L., 2001, *JFM*, **435**, 103

8

Solar tachocline dynamics: eddy viscosity, anti-friction, or something in between?

MICHAEL E. McINTYRE

Centre for Atmospheric Science at the
Department of Applied Mathematics and Theoretical Physics,
Centre for Mathematical Sciences, Wilberforce Road, Cambridge CB3 0WA, UK,
http://www.atm.damtp.cam.ac.uk/people/mem/

The tachocline has values of the stratification or buoyancy frequency N two or more orders of magnitude greater than the Coriolis frequency. In this and other respects it is very like the Earth's atmosphere, viewed globally, except that the Earth's solid surface is replaced by an abrupt, magnetically-constrained 'tachopause' (Gough & McIntyre 1998). The tachocline is helium-poor through fast ventilation from above, down to the tachopause, on timescales of only a few million years. The corresponding sound-speed anomaly fits helioseismic data with a tachocline thickness $(0.019 \pm 0.001) R_\odot$, about 0.13×10^5 km (Elliott & Gough 1999), implying large values of the gradient Richardson number such that stratification dominates vertical shear even more strongly than in the Earth's stratosphere, as earlier postulated by Spiegel & Zahn (1992). Therefore the tachocline ventilation circulation cannot be driven by vertically-transmitted frictional torques, any more than the ozone-transporting circulation and differential rotation of the Earth's stratosphere can thus be driven. Rather, the tachocline circulation must be driven mainly by the Reynolds and Maxwell stresses interior to the convection zone, through a gyroscopic pumping action and the downward-burrowing response to it. If layerwise-two-dimensional turbulence is important, then because of its potential-vorticity-transporting properties the effect will be anti-frictional rather than eddy-viscosity-like. In order to correctly predict the differential rotation of the Sun's convection zone, even qualitatively, a convection-zone model must be fully coupled to a tachocline model.

8.1 Introduction

In the quintessential Douglas Gough manner I am going to be provocative straight off and say, in answer to the question in the title, that 'anti-friction' is closer to the mark – flying in the face of classical turbulence theories.

111

How can I make such an outrageous assertion? I can do so because in
significant respects the Sun's interior is very like the Earth's atmosphere,
and we observe the Earth's atmosphere doing it all the time, that is, showing
anti-frictional behaviour. By 'anti-frictional' I mean that if we describe the
fluid system in terms of a differentially-rotating mean state with angular
velocity $\bar{\Omega}(r, \theta, t)$ and azimuthal velocity $\bar{v}_\phi = r \sin \theta \, \bar{\Omega}(r, \theta, t)$, where r, θ, ϕ
are radius, colatitude, and longitude and t is time, plus chaotic fluctuations
$\mathbf{v}'$ about that state, then the averaged effect of the fluctuations is to drive
the system away from solid rotation.

This of course contradicts the classical idea, enshrined in the term 'eddy
viscosity', that chaotic fluctuations by themselves should drive, or rather
relax, the system *toward* solid rotation. The attractiveness of that classical
idea illustrates the perils of conflating 'chaos' with 'turbulence'. The idea
would be correct if another classical idea were correct, namely that turbulence theory should be like gas-kinetic theory, with the turbulent fluctuations
acting like molecular-scale fluctuations about a nearly homogeneous mean
state. Thus the gas-kinetic mean free path is replaced by some 'mixing
length', 'Austausch length', or other lengthscale representative of the irreversible fluctuating displacements of fluid elements. That lengthscale may
or may not be hidden from view within the complexities of a turbulence
theory based on 'closure'. If momentum is transported by the fluctuating
displacements, and if typical displacements are much smaller than the scales
of variation of the mean state – as implied by the stipulation 'nearly homogeneous' – then the effect of the fluctuations on the mean state is like that of
a viscosity, relaxing the system toward solid rotation, essentially because of
the scale separation just mentioned and the implied flux–gradient relations.

The recognition that fluctuations in the Earth's atmosphere often do the
opposite, i.e. drive the system away from solid rotation (though not, of
course, arbitrarily far away), was a major paradigm shift within the terrestrial atmospheric sciences over the past century. That paradigm shift had its
beginnings in the work of Harold Jeffreys in the second and third decades of
the century (e.g. Jeffreys 1933†). It gathered pace in the late 1960s, stimulated by an increasing wealth of observational evidence. It was fundamental

---

† This conference paper, originally from *Procès-Verbaux de l'Assoc. de Météorol.*, UGGI, Lisbon,
Part II (Mémoires), lucidly and cogently summarizes Jeffreys' classic argument, developed
over the preceding decade or more, that observed surface winds imply the existence of what
Victor Starr later called the 'negative viscosity' due to the large-scale eddies, the cyclones and
anticyclones, appearing on weather maps. The reported conference discussion (Jeffreys, op.
cit., pp. 210–11) illustrates that in 1933 no-one, not even Jeffreys, had the faintest idea of
what kind of fluid dynamics might be involved. The 'negative viscosity' phenomenon was still
flagged as a major enigma in the closing pages of the landmark review by E. N. Lorenz (1967).

to solving some of the greatest enigmas with which the atmospheric sciences were confronted in the 1960s.

One of those enigmas was the behaviour of $\bar{\Omega}(r, \theta, t)$ in the tropical stratosphere between 15–30 km, in which the sign of $\partial\bar{\Omega}/\partial r$ reverses quasi-periodically with a mean period around 27 months. This surprising phenomenon was first revealed by radiosonde balloon observations, which had become routine after the second world war in support of operational weather forecasting. The phenomenon is known today as the quasi-biennial oscillation or QBO. Its cause was wholly mysterious in the 1960s. Today, however, the QBO is recognized as one of the clearest illustrations of the point I am emphasizing, the tendency of chaotic fluctuations to drive a stratified fluid system, very often, away from solid rotation; and a further and even clearer illustration can be found in the beautiful laboratory experiment devised and carried out by Plumb & McEwan (1978). A stratified fluid in a large annulus is driven away from solid rotation, $\bar{\Omega} \equiv 0$ in this case, by nothing but the imposition of fluctuations via an oscillating boundary. On a timescale of many boundary oscillations, $\bar{\Omega}$ evolves away from zero, and then develops a pattern of reversals very like that of the QBO.

Together with appropriate conceptual and numerical modelling, the results from the Plumb–McEwan experiment have greatly illuminated our thinking about the QBO, and enriched our repertoire of models of it. The reader interested in the observed phenomena and in today's understanding of them, which is secure, at least qualitatively – and in the history of ideas leading to that understanding – may consult my recent reviews (2000, 2002) together with a major review of research on the QBO by Baldwin et al. (2001), which latter includes an extensive discussion of the observational evidence. Movies of the Plumb–McEwan experiment plus 'technical tips' on how to repeat it are available on the Internet.†

8.2 Long-range and short-range momentum transport

How can the classical turbulence theories be so completely wrong, not just quantitatively but also qualitatively? The answer is not only clear with hindsight but also simple. Because the Earth's atmosphere and the Sun's interior are heavily-stratified, rotating fluid systems, the fluctuations, chaotic though they may be, inevitably feel the wave propagation mechanisms associated with rotation and stratification.

These include the propagation mechanisms of internal gravity waves, Coriolis or 'inertial' (epicyclic) waves, and layerwise-two-dimensional Rossby or

† at http://www.gfd-dennou.org/library/gfd_exp/exp_e/exp/bo/

vorticity waves. By its very nature, any wave propagation mechanism promotes systematic correlations among the fluctuating fields $\mathbf{v}'$, etc. Almost inevitably, the upshot is that momentum and angular momentum are transported over distances far greater than mixing lengths, limited only by the distances over which waves can propagate. Internal gravity waves provide a well known example, in which the most significant correlations are those between the horizontal and vertical components of $\mathbf{v}'$.

Of course there are exceptional cases in which the momentum transports exactly cancel, such as perfect g modes and p modes, in the strict sense of global eigenmodes subject to no excitation or dissipation. The cancellation is not trivial to demonstrate, when $\bar{\Omega} \neq 0$, but it can be demonstrated from the so-called 'nonacceleration theorem' of wave–mean interaction theory (e.g. McIntyre 2000 & refs.), essentially a consequence Kelvin's circulation theorem applied around all latitude circles.

Long-range momentum transports of the kind in question are sometimes called radiation stresses (e.g. Brillouin 1925, on 'tensions de radiation'). They are usually anisotropic, contrary to what might be suggested by the older term 'radiation pressure' still found in the literature. They are related to mean gradients in ways that are anything but local, the global eigenmodes being an extreme case. It is crucial to consider large-scale wavefields and the processes of generation, dissipation, refraction, Doppler-shifting, internal reflection, focusing and defocusing that shape the wavefields.

Classical turbulence theories – all the way from simplistic mixing-length theories to complicated closure theories – take no account of such long-range momentum-transport mechanisms. As already emphasized, the only momentum transport they consider is, by assumption, that arising from local, short-range, Austausch or material-exchange types of process. It is exactly that short-range character, and the implied or hoped-for scale separation, that give rise to 'turbulent stresses' involving flux–gradient relations and eddy viscosities. We may summarize what happens in the Earth's atmosphere by saying that, on a global scale, radiation stresses dominate turbulent stresses. We shall see nevertheless that turbulence can be important in another way, namely through its contribution to shaping the wavefields, as with surf near ocean beaches.

8.3 Potential vorticity

Anti-frictional behaviour is not inevitable when radiation stresses dominate turbulent stresses, but experience has shown it to be commonplace. For instance such behaviour is often produced by broadband internal gravity wave fields – broadband in the sense of having a range of horizontal phase speeds – like those generated by the Sun's convection zone or by the Earth's tropical thunderstorms. In fact the Plumb–McEwan experiment, in which the significant waves are internal gravity waves, shows that even two distinct horizontal phase speeds can be enough.

Anti-frictional behaviour is commonplace, too, in the case of Rossby-wave fields, whether broadband or not, for quite different reasons connected with the properties of the Rossby–Ertel potential vorticity, hereafter 'PV'. Anti-frictional behaviour is especially characteristic of the stresses exerted horizontally by fluctuating layerwise-two-dimensional motion. That is why Gough & McIntyre (1998 & refs., hereafter GM) argued against horizontal eddy viscosity as explaining the thinness of the tachocline.

The PV, denoted here by the symbol Q, is a quantity central to the dynamics of heavily stratified fluid systems, including the dynamics of Rossby waves and other nearly-horizontal, layerwise-two-dimensional motions. Such other motions include layerwise-two-dimensional turbulence, also loosely called 'geostrophic' turbulence despite its possible existence near the equator. The properties of Q will expose the fact that such turbulence is itself intimately bound up with the Rossby-wave mechanism. This will be demonstrated in Sections 8.4 and 8.5. In the dynamical regimes under discussion there is no such thing as turbulence without waves.

In a reference frame rotating with angular velocity $\mathbf{\Omega}_0$ the PV, Q, is defined as

$$Q = \rho^{-1}\left(2\mathbf{\Omega}_0 + \nabla \times \mathbf{v}\right)\cdot\nabla\vartheta , \qquad (8.1)$$

where ρ is mass density and ϑ is potential temperature (materially invariant, $D\vartheta/Dt = 0$, for adiabatic motion; in place of ϑ one may equally well use specific entropy, or any other monotonic function of ϑ alone). For definiteness we identify $\mathbf{\Omega}_0$ with the angular velocity of the Sun's interior just below the tachocline, $|\mathbf{\Omega}_0| \approx 0.27 \times 10^{-5}\,\mathrm{rad\ s^{-1}}$ or $430\,\mathrm{nHz}$, and take the axis of coordinates parallel to $\mathbf{\Omega}_0$. Heavy stratification means that $\nabla\vartheta$ is nearly vertical, $\nabla\vartheta \approx \hat{\mathbf{r}}\,\partial\vartheta/\partial r$, where $\hat{\mathbf{r}}$ is a unit vertical (radial) vector. Heavy stratification also means that the associated buoyancy frequency N greatly exceeds the other reciprocal timescales of interest, including $|\mathbf{\Omega}_0|$ and the vertical shear $r \sin\theta\,\partial\bar{\Omega}/\partial r$. We recall that N is defined by

$$N^2 = g\,\vartheta^{-1}\partial\vartheta/\partial r = g\,\partial(\ln\vartheta)/\partial r , \qquad (8.2)$$

g being the local gravitational acceleration, and that the value of N is of the order of $10^{-3}\,\mathrm{rad\ s^{-1}}$ near the base of the tachocline. The standard measure of stratification against vertical shear, the gradient Richardson number, is defined by

$$Ri = N^2 (r \sin \theta \, \partial \bar{\Omega}/\partial r)^{-2} \, . \tag{8.3}$$

If we use the refined estimate of tachocline depth Δr obtained by Elliott & Gough (1999), $(0.019 \pm 0.001) R_\odot$, about $0.13 \times 10^5 \, \mathrm{km}$, then typical vertical shears $\Delta \bar{v}_\phi / \Delta r \lesssim 10^{-5} \, \mathrm{s}^{-1}$, not much greater than $|\mathbf{\Omega}_0|$. Thus $Ri \gtrsim 10^{-6}/10^{-10} \sim 10^4 \gg 1$ near the base of the tachocline. Even when N is taken to be an order of magnitude smaller, $10^{-4} \, \mathrm{rad \, s}^{-1}$, as near the top of the tachocline, we still have $Ri \gtrsim 10^2 \gg 1$. This says that the tachocline is even more heavily stratified than the most heavily stratified portion of the Earth's stratosphere, where typically $Ri \gtrsim 10$ in a coarse-grain view.

Such Ri values are high enough to enforce layerwise-two-dimensional motion, everywhere including the equator, as pointed out by Spiegel & Zahn (1992, hereafter SZ). A key property of Q during such motion is that not only ϑ but also Q itself is materially invariant, $\mathrm{D}Q/\mathrm{D}t \approx 0$, if the motion can be considered inviscid as well as adiabatic and if MHD (Lorentz) forces can be neglected within the tachocline. Approximately inviscid motion is consistent with large Ri values.†

A second key property of Q, which holds for any motion whatever – even a motion that feels MHD forces – is the integral relation

$$\iint_S Q \, b \, \mathrm{d}A = 0 \, , \tag{8.4}$$

where $\mathrm{d}A$ is the surface area element and where the integral is taken globally over a stratification or isentropic surface S, on which ϑ is constant by definition. The weighting factor $b = \rho/|\nabla \vartheta|$, a positive-definite quantity. It is a stratification-related mass density in the sense that $b \, \mathrm{d}\vartheta$ is the mass per unit area between neighbouring stratification surfaces S; that is, $b \, \mathrm{d}A \, \mathrm{d}\vartheta$ is the mass element. The relation (8.4) is an immediate consequence of Stokes' theorem, the definition of Q, and the fact that each surface S is topologically spherical and has no boundary. For present purposes both the Sun and the Earth are rapidly rotating bodies, with strongly polarized Q fields: except near the equator, $2\mathbf{\Omega}_0$ dominates $\nabla \times \mathbf{v}$ in (8.1). So (8.4) is satisfied through

---

† Note also that tachocline thermal diffusion times estimated as $(\pi^{-1}\Delta r)^2/\kappa$, where the thermal diffusivity $\kappa \sim 10^7 \, \mathrm{cm}^2\mathrm{s}^{-1}$, come out at about 500y. This is well in excess of the likely timescales of months to years for any layerwise-two-dimensional motion that might occur in the tachocline. Viscous and magnetic diffusion times are far longer still. The inviscid, adiabatic material invariance of Q (Ertel's theorem) is easy to verify from $\nabla(\mathrm{D}\vartheta/\mathrm{D}t) = 0$ together with the scalar product of $\nabla \vartheta$ with the inviscid, adiabatic vorticity equation, or alternatively (e.g. McIntyre 2000, Section 9) as a corollary of mass conservation together with Kelvin's circulation theorem applied to small constant-ϑ circuits. The neglect of MHD forces is much more of an open question, but, for what it is worth, the arguments of GM strongly justify such neglect in the downwelling branches of the tachocline ventilation circulation.

cancellation of strong positive and negative contributions from the northern and southern hemispheres respectively.

Owing to the positive-definiteness of the weighting factor b, the relation (8.4) imposes a severe constraint on the possible evolution of the global-scale Q distribution on each surface S. We shall see that (8.4) is almost enough, by itself, to guarantee that layerwise-two-dimensional fluctuations about a mean state of solid rotation will behave anti-frictionally. Consistently with (8.4), one may picture Q as the amount per unit mass of a fictitious 'PV substance' composed of charged particles to which the stratification surfaces S are completely impermeable. They are impermeable even if the motion is not adiabatic. Even if mass leaks across a surface S, through thermal diffusion, the notional particles of 'PV substance' remain trapped on that surface (Haynes & McIntyre 1990). The 'PV charge' is conserved in the same way as electric charge. That is, pair production and annihilation are allowed, but no net charge creation or destruction. Just as $b\,\mathrm{d}A\,\mathrm{d}\vartheta$ is the mass element, $Q\,b\,\mathrm{d}A\,\mathrm{d}\vartheta$ is the charge element. The picture is consistent with (8.4) because the value, zero, of $\iint Q\,b\,\mathrm{d}A$ cannot be changed by pair production and annihilation. Nor can it be changed by the advective rearrangement of the notional particles on each surface S by any layerwise-two-dimensional motion.

A third key property of Q is its 'invertibility'. This says that the isentropic distributions of Q, i.e. the distributions of PV values on the surfaces S, contain nearly all the kinematical information about the layerwise-two-dimensional motion – whether or not MHD forces are significant. At each instant, to good approximation, one can 'invert' the PV field to get the velocity, pressure and density fields. The dynamical system is then completely specified by the PV inversion operator together with a single prognostic equation for the rate of change of Q, supplemented, if necessary, by a prognostic (induction) equation for any magnetic fields that may be present. The equation for Q can be $\mathrm{D}Q/\mathrm{D}t = 0$ or an appropriate generalization, symbolically

$$\mathrm{D}Q/\mathrm{D}t = \text{viscous, diabatic and MHD terms} . \tag{8.5}$$

The single time derivative reminds us that Rossby waves and other layerwise-two-dimensional motions, viewed in the rotating frame, are chiral: they notice the direction and sense of $\mathbf{\Omega}_0$. The mirror-image motion is impossible.

All this is simplest to see in the limiting case of anelastic motion and infinitely heavy stratification, in which $N^2 \to \infty$ and $Ri \to \infty$. The surfaces S become rigid and horizontal – horizontal in the billiard-table sense, with the sum of the gravitational and centrifugal potentials constant – and the flow on each S becomes strictly horizontal and strictly incompressible. Then $\mathbf{v} = \hat{\mathbf{r}} \times \nabla_\mathrm{S}\psi$ for some streamfunction ψ, and

$$Q = b^{-1}(f + \nabla_\mathrm{S}^2\psi) \tag{8.6}$$

with b strictly constant, where f is the vertical component of $2\mathbf{\Omega}_0$ and ∇_S and ∇_S^2 are the two-dimensional gradient and Laplacian on the surface S. We may regard (8.6) as a Poisson equation to be solved for ψ when Q is given. Solving it is a well defined, and well behaved, operation provided that the given Q field satisfies the integral relation (8.4) on each S. Symbolically,

$$\mathbf{v} = \hat{\mathbf{r}} \times \nabla_\mathrm{S}\psi \quad \text{with} \quad \psi = \nabla_\mathrm{S}^{-2}(bQ - f) . \tag{8.7}$$

This expresses PV invertibility in the limiting case. Notice that the limiting case is degenerate in that the radial coordinate r enters the problem only as a parameter. There is no derivative $\partial/\partial r$ anywhere in the problem, either in the Laplacian or in the material derivative $\mathrm{D}/\mathrm{D}t = \partial/\partial t + \mathbf{v}\cdot\nabla$, $\mathbf{v}$ now being strictly horizontal. Not only is the motion layerwise-two-dimensional, but the layers are completely decoupled. There is, therefore, an implicit restriction on magnitudes of $\partial/\partial r$, i.e. an implicit restriction on the smallness of vertical scales, as the limit is taken.

More realistically, when N^2 and Ri are large but finite, $\partial/\partial r$ reappears in the problem and brings back vertical coupling. The motion remains layerwise-two-dimensional in the sense that the notional 'PV particles' move along each stratification surface S, but the surfaces themselves are no longer quite horizontal, nor quite rigid. All the vertical coupling comes from the PV inversion operator. The two-dimensional inverse Laplacian in (8.7) is replaced by an inverse elliptic operator that resembles a three-dimensional inverse Laplacian when a stretched vertical coordinate Nr/f is used.

Here one has to make tradeoffs between accuracy and simplicity. The simplest though least accurate inversion operator is that arising in the standard 'quasi-geostrophic theory', an asymptotic theory for large Ri and $f \neq 0$, valid away from the equator. MHD forces are still absent from the inversion and, if significant at all, enter the problem only through the prognostic equation (8.5). The operator ∇_S^2 becomes $\nabla_\mathrm{S}^2 + \rho^{-1}\partial_r\left(\rho f^2 N^{-2}\partial_r\right)$. We may expect this to be a self-consistent approximation if Alfvén speeds are of the order of $|\mathbf{v}|$ or less. Notice that for tachocline eddies of horizontal scale $10^5\,\mathrm{km}$, say, the vertical coupling extends over a vertical scale $\sim (f/N) \times 10^5\,\mathrm{km} \sim 0.004 \times 10^5\,\mathrm{km}$ at latitude $45°$, fairly small in comparison with a tachocline thickness of $0.13 \times 10^5\,\mathrm{km}$.

Some idea of what is involved in constructing more accurate inversion operators can be gained from the recent work of Ford et al. (2000) and

Mohebalhojeh & Dritschel (2001 & refs.) and summarized in a recent review of mine (2001); see also the earlier discussion by Hoskins et al. (1985). Subtle generalizations of the notions of 'geostrophic balance' and 'magnetostrophic balance' are involved, and there are ultimate limitations on the accuracies attainable and on good mathematical behaviour, owing to phenomena such as Lighthill radiation, equatorial inertial instabilities, symmetric-baroclinic or Høiland instabilities, and magneto-rotational or Chandrasekhar–Fricke–Balbus Hawley instabilities.

Before going further with the theory, let us take note of what layerwise-two-dimensional motion looks like in the real-world example that has been the most thoroughly studied, the Earth's stratosphere. There, thanks to today's observing systems, we can see many of the associated phenomena in remarkable detail, including conspicuous examples of Rossby-wave propagation.

8.4 A glimpse of the Earth's stratosphere

Figure 8.1 presents two snapshots of the stratosphere, showing at a spatial resolution of a few degrees latitude the effects of layerwise-two-dimensional motion on two stratification surfaces S. These surfaces lie at altitudes of about 31 and 37 km. An animated version can be seen on my website.† The figure is reproduced by courtesy of Dirk Offermann, Martin Riese, and the other scientists involved in the CRISTA space-based remote-sensing project; see Riese et al. (2002). The quantity shown is the mixing ratio X_{N_2O} of a biogenic chemical tracer, nitrous oxide, that is destroyed photochemically on a timescale of years but resupplied, across the stratification surfaces S from the troposphere below, on the same timescale of years, by a global-scale circulation called the Brewer–Dobson circulation. This is a stratospheric counterpart of the tachocline ventilation circulation. In the stratosphere the upwelling branch of the circulation is in the tropics; therefore X_{N_2O} values are highest there. White areas are data gaps.

The layerwise-two-dimensional motion has far greater horizontal velocities than the Brewer–Dobson circulation, and far shorter timescales of days to weeks. On such timescales X_{N_2O} is a near-perfect passive tracer, indeed material invariant, $DX_{N_2O}/Dt = 0$ to good approximation. Thus, apart from the overall pole-to-equator gradient, the patterns seen in Figure 8.1 are shaped almost exclusively by the layerwise-two-dimensional motion.‡

† In colour, at www.atm.damtp.cam.ac.uk/people/mem/papers/LIM/index.html#crista-movie
‡ The observational resolution is enough for our purposes, though there must in reality be invisible fine-grain detail, such as the filamentary, cream-on-coffee patterns found in recent high-resolution observational and modelling studies of stratospheric flows at lower altitudes (e.g. Norton 1994, Waugh & Plumb 1994, Waugh et al. 1994, Appenzeller et al. 1996).

Each snapshot shows similar features, notably the well-mixed region (medium gray) on the right, with nearly uniform tracer values, sandwiched between relatively isolated polar and tropical airmasses having very different tracer values, with steep gradients in transition zones between. It is clear from the animated version and from numerical model simulations, which produce generically similar tracer distributions (e.g. Norton 1994), that the layerwise-two-dimensional motion is causing strong mixing on each stratification surface S in an extensive midlatitude region sandwiched between the polar and tropical airmasses. A long tongue of tropical air is being drawn eastward past the tip of South America (light gray, inner band on the left) and marks the early stages of a typical mixing event, in which air is visibly recirculating within the midlatitude region at the instant shown. This horizontal recirculation is conspicuous in the animation. Because of the strong mixing, it is reasonable to regard the motion as fully turbulent, in the layerwise-two-dimensional sense, in middle latitudes.

However, the motion as a whole has not only its turbulent aspect but also the wavelike aspect anticipated theoretically. This too is conspicuous in the animated version of Figure 8.1, which shows the long axis of the central, elongated dark region rotating clockwise through an angle of about $70°$ longitude in 5 days, 10–15 August 1997, relative to the Earth. The central region marks the core of the 'polar vortex', characterized by large negative values of Q. Because of the approximate material invariance of Q, it behaves like an advected tracer on the short timescales of the layerwise-two-dimensional motion, and has a distribution somewhat like that of X_{N_2O} apart from an additive constant.

The rate at which the long axis rotates is determined by a competition between the mean winds – which broadly speaking blow clockwise, at speeds of the order of $80\,\mathrm{m\,s^{-1}}$, about nine times faster than $70°$ in 5 days – and a wave propagation mechanism that powerfully rotates the long axis anticlockwise relative to the air. This is the Rossby-wave or vorticity-wave mechanism.† The phase progression is necessarily one-way (here anticlockwise, or retrograde, relative to the air), as a consequence of the chirality associated with the single time derivative in equation (8.5).

As is well known, the Rossby-wave mechanism operates whenever Q has a mean gradient $\partial \bar{Q}/\partial \theta$ on stratification surfaces S, such as the gradient associated with the global-scale polarization – the positive-to-negative, pole-to-pole variation in Q values due to the rotation of the whole system, Ω_0 say,

---

† As usual, terminology contradicts historical precedent. Carl-Gustaf Rossby was one of the greatest pioneers in atmospheric science, and his memory deserves special honour, but the wave mechanism was noted decades earlier by Kelvin and Kirchhoff, in special cases at least.

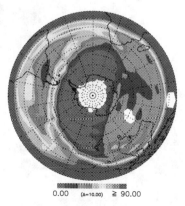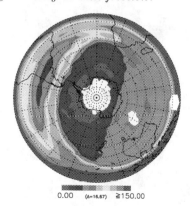

0.00 $(\Delta=10.00)$ ≥ 90.00 0.00 $(\Delta=16.67)$ ≥ 150.00

Fig. 8.1. Nitrous oxide (N_2O) mixing ratios X_{N_2O} observed at two stratospheric altitudes on 11 August 1997 by the CRISTA infrared spectrometer, from Riese et al. (2002). White areas are data gaps. On Rossby-wave timescales of days and weeks N_2O is an accurate passive tracer, though destroyed photochemically on Brewer–Dobson timescales of years. In the right half of each picture X_{N_2O} values increase equatorward nearly monotonically or stepwise monotonically (being nearly constant over the large medium gray regions on the right). Polar-vortex values (dark central regions) are close to zero, and tropical values are high, imported from the troposphere by the Brewer–Dobson upwelling. **At left and right respectively:** pressure-altitudes are 4.64 hPa and 10 hPa, roughly 37 km and 31 km; ranges of mixing ratios in parts per billion by volume are 0–90+ and 0–150+ with contour intervals 10 and 16.67, where '+' signifies that maximum values may slightly overshoot the plotted range; the light band in the subtropics highlights the ranges 60–70 and 100–116.67 ppbv. CRISTA (CRyogenic Infrared Spectrometers and Telescopes for the Atmosphere) detects a number of chemical species through their infrared spectral signatures and is a large (1350 kg) helium-cooled instrument flown from the Space Shuttle.

which must here be taken to be somewhat faster than the rotation of the solid Earth. North–south material displacements across that gradient give rise to a pattern of fluctuating Q anomalies on the surfaces $\mathcal{S}$ that alternate in sign downstream, every 90° of longitude in the case of Figure 8.1. Inversion of that pattern of Q anomalies to obtain the fluctuating velocity field produces north–south velocities that lag north–south displacements by a quarter wavelength in longitude, implying one-way phase propagation with highest $\bar{Q}$ values on the right, i.e. retrograde phase propagation.

To check this qualitative picture in the simplest possible way using the standard Rossby–Haurwitz wave theory, take the limiting case (8.7), linearize the prognostic equation $DQ/Dt = 0$ for small disturbances Q', ψ', $\mathbf{v}'$ about a mean state of solid rotation Ω_0, regard each stratification surface $\mathcal{S}$ as precisely spherical and look for disturbances with complex amplitude $\hat{Q}$ and spherical-harmonic structure $Q' = \mathrm{Re}\{\hat{Q}P_n^m(\cos\theta)\exp(im\phi - i\omega t)\}$.

The linearized prognostic equation is

$$\frac{\partial Q'}{\partial t} + \frac{v_\theta'}{r}\frac{\partial \bar{Q}}{\partial \theta} = 0 \,, \tag{8.8}$$

with $\partial\bar{Q}/\partial\theta = -2\Omega_0 b^{-1}\sin\theta$, $b = $ constant, and $\Omega_0 = |\mathbf{\Omega_0}|$. PV inversion reduces to $\psi' = \nabla_{\mathrm{s}}^{-2}(bQ')$, hence $\hat\psi = -r^2 b\hat{Q}/\{n(n+1)\}$, with $\hat\psi$ the complex amplitude of ψ', i.e., $\psi' = \mathrm{Re}\{\hat\psi P_n^m(\cos\theta)\exp(im\phi - i\omega t)\}$. Noting that $v_\theta' = -(r\sin\theta)^{-1}\partial\psi'/\partial\phi$ and that $\partial/\partial\phi = im$, we have

$$\omega = -\frac{2\Omega_0 m}{n(n+1)} \,. \tag{8.9}$$

This illustrates the qualitative picture sketched above, including the one-way propagation associated with chirality – the single power of ω coming from the single time derivative. Because the angular phase velocity $\omega/m < 0$, the phase propagation is retrograde and the meridional disturbance velocity v_θ', with complex amplitude $\propto -im\hat\psi$, lags the displacement, with complex amplitude $\propto -im\hat\psi/(-i\omega) = m\omega^{-1}\hat\psi$, by a quarter wavelength in longitude.

More realistic models of stratospheric Rossby waves must take account of the turbulent mixing in middle latitudes. The mixing has an obvious qualitative effect: it weakens the PV gradient $\partial\bar{Q}/\partial\theta$ in middle latitudes and strengthens it at the subtropical edge of the midlatitude mixing region (outermost light band, clearest on the right of Figure 8.1) and at the polar edge bounding the vortex core. This characteristic reshaping of the $\bar{Q}(\theta)$ profile is suggested schematically by the cartoon on the left of Figure 8.2, in which y denotes northward distance in arbitrary units, $y \propto -\theta$, in a midlatitude slab model. The dashed and heavy lines represent the $\bar{Q}(y)$ profiles before and after mixing. The middle graph presents the corresponding $\bar{Q}(y)$ profiles in an actual numerical experiment, to be referred to shortly. Thus, in more realistic models, the quasi-elastic resilience associated with the Rossby-wave mechanism tends to be concentrated in transition zones of steep Q gradients, also marked by steep $X_{\mathrm{N_2O}}$ gradients, lying between the tropical, midlatitude and polar airmasses. The same quasi-elastic resilience is part of why the three airmasses are chemically so distinct, with little mixing between them, a phenomenon seen again and again by stratospheric researchers and much studied because of its significance for ozone-layer chemistry. 'Shear sheltering' is also involved (Juckes & McIntyre 1987; Hunt & Durbin 1999).

For our purposes, however, the most important point of all is that the layerwise-two-dimensional mixing in middle latitudes owes its existence to the Rossby waves. In this respect the situation illustrated in Figure 8.1 is fundamentally similar to the ocean-beach situation, in which the turbulence in the ocean-beach surf zone owes its existence to surface gravity waves. That is part of what I meant by the assertion that in the dynamical regimes under discussion 'there is no such thing as turbulence without waves'.

The midlatitude mixing occurs for well-understood reasons associated with flow unsteadiness, hyperbolic points, and so on – a chaotic-advection kinematics very much tied, in this case, to the wave propagation, as analysed in detail by, for instance, Polvani & Plumb (1992). We may say that the turbulent mixing is intimately, and inseparably, part of the wavemotion. It is therefore reasonable to consider these stratospheric Rossby waves to be *breaking* waves. For this reason, the midlatitude mixing region is often called the 'stratospheric surf zone'.

Numerical experiments in which the initial condition is axisymmetric, and in which Rossby waves are then excited somehow, commonly produce surf zones like that seen in Figure 8.1 (e.g. Norton 1994). The formation of surf zones is a very robust feature of such experiments, almost regardless of how chaotic or regular the waves, as such, happen to be. In the Earth's stratosphere the Rossby-wave fields can on occasion be fairly regular, as in the case of Figure 8.1, or, more typically in the northern-hemispheric winter, rather more chaotic.

A fundamentally similar phenomenon of surf-zone formation was demonstrated long ago in the idealized numerical experiments of Rhines, in a classic paper entitled 'Waves and turbulence on a beta-plane' (Rhines 1975). The designations 'Rossby-wave breaking' and 'stratospheric surf zone' can be justified in a very general way, from wave–mean interaction theory (e.g. McIntyre & Palmer 1985), having regard to Kelvin's circulation theorem. This has application to most if not all non-acoustic wave types.

In the Rossby-wave case the whole conceptual picture is illustrated by a specific model of wave breaking in a certain parameter limit, known as the Stewartson–Warn–Warn model, in which the surf zone is narrow and the interplay between the wavelike and turbulent dynamics can be precisely and comprehensively described using matched asymptotic expansions (Haynes 1989 & refs.). This is based on the midlatitude slab model in the limiting case (8.7), and has provided a set of detailed examples including that from which Figure 8.2b is derived. The interplay works both ways, at leading order: not only do the waves create the turbulence – again justifying the idea of 'wave breaking' – but the turbulence, in turn, strongly influences the wavefield, and in particular the systematic correlations between v_θ' and v_ϕ' that are significant for horizontal momentum transport. The wavefield, through the PV inversion operator, senses the horizontal rearrangement of PV substance by the turbulence within the surf zone.

8.5 Turbulence requires waves

There is an alternative, independent justification for the assertion that in the dynamical regimes under discussion 'there is no such thing as turbulence without waves'. The justification follows simply and directly from PV invertibility, involving no restriction to special parameter regimes, and no reliance on particular mathematical techniques such as that of matched asymptotic expansions.

We assume the existence of turbulence without waves, and show that this leads to a contradiction. More precisely, consider a layerwise-two-dimensional PV mixing event like those depicted in Figure 8.2a,b, in which the PV profile $\bar{Q}(\theta)$ or $\bar{Q}(y)$ is changed by a finite increment $\delta\bar{Q}$ within some finite mixing region $y_1 < y < y_2$ or $\theta_1 < \theta < \theta_2$, in such a way as to respect the integral relation (8.4). The dashed lines show the initial $\bar{Q}(y)$ profile. In the case of the cartoon in Figure 8.2a, the profile of $\delta\bar{Q}(y)$ is a simple N-shape, having negative slope within the mixing region.

Imagine that the mixing somehow takes place without any wave mechanism being involved. The PV invertibility principle says that when the $\bar{Q}$ profile changes then the mean velocity profile must change too, by $\delta\bar{v}_\phi$ say. In the limiting case (8.7) the relevant inversion is trivial, b being constant; for instance in the slab model it is simply

$$\delta\bar{v}_\phi(y) \;=\; \int_y^\infty \delta\bar{Q}(\tilde{y})\, b\, \mathrm{d}\tilde{y} \;. \tag{8.10}$$

For the N-shaped $\delta\bar{Q}(y)$ profile, the shape of $\delta\bar{v}_\phi(y)$ is a simple parabola. For the $\delta\bar{Q}(y)$ profile implied by Figure 8.2b, the shape of $\delta\bar{v}_\phi(y)$ is qualitatively the same, the parabola-like shape given by the right-hand plot, Figure 8.2c.

These mean flow changes show a net momentum deficit. Notice that $\int_{y_1}^{y_2} \delta\bar{v}_\phi(y)\,\mathrm{d}y \;=\; \int_{y_1}^{y_2} y\,\delta\bar{Q}(y)\,b\,\mathrm{d}y$ (integrating by parts): the total momentum change, ignoring a constant factor ρ, is equal to the first moment of $\delta\bar{Q}(y)$. This is negative for the N-shaped $\delta\bar{Q}(y)$ profile. The first moment, and the momentum change itself, both have unambiguous meanings in virtue of the integral relation (8.4), which implies that $\int_{y_1}^{y_2} \delta\bar{Q}(y)\,b\,\mathrm{d}y \;=\; 0$ in the present limiting case (8.7), with b constant. So the momentum deficit is indeed a deficit whenever $\delta\bar{Q}(y)$ is such that the mixing event was indeed a mixing event, in the sense of weakening the gradient of $\bar{Q}$ within the mixing region $y_1 < y < y_2$.

The argument just presented can easily be generalized from the slab

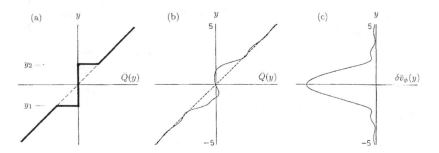

Fig. 8.2. Demonstration that rearrangement of PV substance by layerwise-two-dimensional mixing on a stratification surface S, within some latitude band $y_1 < y < y_2$, must entail momentum transport outside the band hence wavelike as well as turbulent fluctuations. (This follows from PV invertibility, and does not require accurate material invariance of Q.) The quantitative examples in plots (b) and (c) are by courtesy of P. H. Haynes (personal communication); for full mathematical details see Killworth & McIntyre (1985) and Haynes (1989). Plot (a) shows idealized Q distributions before and after mixing; (b) shows the same in an accurate slab-model simulation, using $DQ/Dt = 0$ together with the inversion (8.7); (c) shows the resulting mean momentum change, given by equation (8.10), whose profile would take a simple parabolic shape in the idealized case corresponding to (a).

geometry to the spherical geometry, replacing equation (8.10) by

$$\delta\bar{v}_\phi(\theta) = r\,(\sin\theta)^{-1}\int_0^\theta \delta\bar{Q}(\tilde{\theta})\,b\sin\tilde{\theta}\,\mathrm{d}\tilde{\theta}\;. \qquad (8.11)$$

We can also remove the restriction to the limiting case (8.7), reverting to finite N^2 and Ri. In the most accurate versions it is necessary to redefine the mean $\bar{Q}$ around latitude circles as a weighted 'isentropic' mean at constant ϑ, i.e. following a stratification surface S, with weighting function b, so as to respect the integral relation (8.4). It is then convenient to switch to ϑ as the vertical coordinate, as discussed under the heading 'isentropic coordinates' in the atmospheric-science literature (e.g. Andrews et al. 1987). The main conclusion, that layerwise-two-dimensional PV mixing produces a momentum or angular momentum deficit, still holds good.†

It follows that – whatever the purely turbulent (Austausch) stresses that might be involved – such turbulent stresses cannot satisfy the momentum budget on their own. This point was made long ago by Stewart & Thomson (1977) who, however, used it to claim that turbulent mixing scenarios like

---

† In fact the conclusion holds exactly on each surface S in any thought experiment in which the initial and final states are axisymmetric. The vertical coupling represented by the ∂_r term in quasi-geostrophic theory is incapable, by itself, of transporting absolute angular momentum, and so never enters the calculation.

those of Figures 8.2a,b cannot be realized. This overlooked the possibility that 'the problem of turbulence' might have a wavelike aspect, allowing momentum to be exchanged between the mixing region and its surroundings.

To summarize, then, the implication in reality is that wave-induced momentum transport, not confined to mixing regions such as $y_1 < y < y_2$ in Figure 8.2, is an essential part of the picture – essential to making sense of the fluid dynamics as a whole. The turbulent mixing scenarios can in fact be realized, but only in the presence of waves, which, in the stratospheric case at least, are chiefly Rossby waves.

The Stewartson–Warn–Warn model played an important role in developing the conceptual framework just sketched, by illustrating, with great precision, how everything works and fits together in a particular set of idealized thought experiments. We may note too that the same thought experiments provide especially clear examples of anti-frictional behaviour.

In each case a shear flow $\bar{v}_\phi \propto y$ is disturbed by monochromatic Rossby waves generated by an undulating boundary located at positive y, outside the domain of Figure 8.2. The graphs plotted in Figures 8.2b,c come from one such thought experiment but are qualitatively similar to those from all the others. In each case the surf zone or mixing zone surrounds the location $y = 0$, a so-called 'critical line' where, by definition, $\bar{v}_\phi$ coincides with the longitudinal phase speed of the Rossby waves. The phase speed is retrograde relative to the mean flow throughout $y > 0$, so that Rossby-wave propagation is possible there. The fluid behaviour within the surf zone is complicated and chaotic in most cases; for detailed examples and for a definitive and thorough analysis see Haynes (1989). The behaviour is anti-frictional because the momentum transport that accounts for the momentum deficit in the surf zone, equation (8.10) and Figure 8.2c, is everywhere against the mean momentum gradient $\partial \bar{v}_\phi / \partial y > 0$, through a positive correlation between v_y' and v_ϕ'.

More generally, equations (8.10) and (8.11) show that in any thought experiment starting with solid rotation on a given stratification surface S, such as the solid rotation observed below the Sun's tachocline, the formation of mixing regions like that in Figure 8.2 will drive the system away from solid rotation. This general point is reinforced by the integral relation (8.4). Because b is positive definite, (8.4) tells us at once that the only way to mix Q to homogeneity on a stratification surface S is to make Q zero everywhere – a fantastically improbable state on a planet rotating like the Earth, or in a star rotating like the Sun. Since real Rossby waves do break, and do mix Q, they must be expected to do so imperfectly, mixing more strongly in some places than in others and producing the characteristic spatial inhomogeneity that always seems to be observed, as illustrated by Figure 8.1. Again, the effect is to drive the system away from solid rotation.

To be sure, one can imagine a thought experiment in which the air on and near the stratification surface S begins by rotating solidly, and then has its angular velocity uniformly reduced by breaking Rossby waves. The PV mixing would have to be distributed in just such a way as to give a uniformly reduced pole-to-pole latitudinal profile of Q, keeping it precisely proportional to $\cos\theta$. But the tailoring of a Rossby-wave field to do this would be a more delicate affair than standing a pencil on its tip, and the natural occurrence of such a wave field would be another fantastically improbable thing. A sufficient reason for its improbability is the positive feedback associated with PV mixing. As soon as some region begins to be mixed, PV gradients and Rossby quasi-elasticity are weakened, facilitating further mixing. Conversely, PV gradients are tightened at the edges of the mixing regions, tending to inhibit mixing there, as evidenced for instance by the steep X_{N_2O} gradients at the edges of the surf zone in Figure 8.1.

Nonmagnetic laminar spindown would also produce differential rotation. This was part of why GM argued that an internal magnetic field in the Sun is not merely possible but actually inevitable – the only way to account for the observed near-solid rotation in the radiative, heavily stratified interior.

8.6 Concluding remarks

The general arguments of Section 8.5 are enough to show that in heavily stratified systems there is no such thing as turbulence without waves, and hence that stratification-constrained horizontal eddy viscosities are implausible, if MHD effects are negligible. The general arguments leave open the question of which waves. The reason for focusing here on Rossby waves rather than gravity waves, for instance, is that the timescale of the tachocline ventilation circulation, $\sim 10^6$y (see SZ and GM), while long in comparison with that of the sunspot cycle, ~ 10y, is short in comparison with the Sun's lifetime and spindown time, $\sim 10^{10}$y. A number of published and unpublished estimates of gravity-wave amplitudes in the Sun's interior, based on the reasonable hypothesis that gravity waves are generated mainly by the overlying convection zone, point to momentum transports that could be significant on spindown timescales but fall far short of being significant on tachocline ventilation timescales.

Acoustic waves are still weaker, for this purpose, leaving only the question of Alfvén and other MHD waves. GM's arguments justify the neglect of MHD effects in the downwelling branches of the tachocline ventilation circulation, outside the midlatitude band of upwelling. The downwelling firmly confines the large-scale interior magnetic field to the tachopause and below, as GM showed with the help of an appropriate magnetic boundary-layer theory. Alternating solar-cycle fields could be carried downward from the top of the tachocline but should diffusively self-annihilate far faster than the 10^6y timescale of the downwelling.

In the midlatitude upwelling branch, the magnetic boundary-layer theory fails. Superficial layers of the interior field must be fed into into the tachocline then into the convection zone. Exactly how that happens is uncertain, and no detailed model is available as yet, though it must be the main means whereby the interior field leaks out, contributing to the decay from its primordial state – a scenario that fits well with the fact that interior magnetic diffusion times for the largest possible scales, i.e. for a simple internal dipole, are comparable to the Sun's lifetime. In the tachocline's upwelling branch the presence of large-scale magnetic field may give rise to significant angular momentum transport, either directly via Alfvénic elasticity as field lines are sheared out, or via MHD waves. Or the upwelling may itself be locked into solid rotation. All we know so far is that mass conservation dictates that the upwelling branch must exist, and that the physics of thermal diffusion and the pattern of vertical shear in the tachocline inferred from helioseismic inversions (e.g. Thompson et al. 1996, Kosovichev 1997, Schou et al. 1998) dictates, through a very robust 'thermal wind balance', i.e. through large-scale hydrostatic and cyclostrophic balance, that the upwelling must indeed be taking place in middle latitudes (SZ, GM). The upwelling branch remains the biggest missing piece of the jigsaw puzzle put together in GM.

The 'gyroscopic pumping' of the tachocline ventilation circulation, and of the stratospheric Brewer–Dobson circulation, is a well understood process in stratified, rotating fluid dynamics and has been discussed extensively elsewhere (e.g. McIntyre 2002 & refs.). Persistent westward or eastward forces pump fluid persistently poleward or equatorward, respectively, through Coriolis-induced turning. Ekman pumping is the special case in which the east–west forces are frictional forces near a boundary; in the stratosphere the forces are wave-induced. For the Sun the important points to note are (1) that the only process able to provide east–west forces of sufficient strength is the three-dimensional turbulence in the convection zone, through its Reynolds and Maxwell stresses, (2) that the resulting ventilation circulation tends to burrow downward, (3) that the burrowing can be stopped only by the interior magnetic field, and (4) that a complete model of convection-zone differential rotation must take account of the thermal structure of the tachocline induced by the ventilation circulation, hotter in downwelling and cooler in upwelling regions, with its thermal-wind link to differential rotation. The implied differential rotation must be continuous with that in the convection zone, again by thermal-wind balance. It follows that the tachocline determines the differential rotation at the base of the convection zone – not *vice versa* – and that the convection zone reacts back on the tachocline by reshaping the Reynolds and Maxwell stresses and the consequent gyroscopic pumping. To my knowledge there has been no effort, so far, to construct a model that captures this two-way coupling.

With the inevitability of a poloidal magnetic field in the interior below the tachopause, we may expect magneto-rotational instabilities to be potentially significant in the interior as well as in the tachocline's upwelling branch. As pointed out in Balbus & Hawley (1991 & refs., especially Fricke 1969), magneto-rotational instabilities should operate in stellar interiors with poloidal magnetic fields, in such a way as to prevent Ω from decreasing outwards from the rotation axis. This plus Ferraro's law of isorotation could clamp the upwelling and most of the interior into solid rotation – except in the 'polar pits' at the hairy-sphere defects in the horizontal magnetic field at the tachopause, which as GM pointed out are the only locations where the gyroscopically pumped tachocline circulation can burrow down far enough to burn lithium and beryllium. Notice, incidentally, how the arguments of Section 8.5 are vitiated by the poloidal magnetic field: PV advection is nullified by MHD effects on the right-hand side of equation (8.5). There is no longer any tendency to form PV mixing regions!

If the interior is clamped into solid rotation almost everywhere, it hardly needs saying that there are strong implications both for helioseismic inversion and for understanding primordial spindown. Furthermore, older speculations such as mine of 1994, about QBO-like torsional oscillations in stellar interiors, would now appear to be ruled out.

Acknowledgements I thank Martin Riese for Figure 8.1, Peter Haynes for Figures 8.2b,c, Pascale Garaud for sharing unpublished results, Jørgen Christensen-Dalsgaard, Rosanne Gough, Mike Thompson, and Sylvie Vauclair for organizing a superb conference and for inviting me to take part, Ed Spiegel and Nigel Weiss for stimulating conversations on astrophysics over the years, and last, but certainly not least, Douglas Gough for further such stimulus and encouragement and for his indomitable spirit, sense of fun, and passion for good science. My research owes much to atmospheric-science and fluid-dynamics colleagues too numerous to mention and was supported by the Natural Environment Research Council, the Isaac Newton Institute for Mathematical Sciences, and a SERC/EPSRC Senior Research Fellowship.

References

Andrews, D. G., et al., 1987, *Middle Atmosphere Dynamics*. Academic, 489 pp.

Appenzeller, C., et al., 1996, *J. Geophys. Res.*, **101**, 1435.

Balbus, S. A. & Hawley, J. F., 1991, *ApJ*, **376**, 214.

Baldwin, M. P., et al., 2001, *Revs. Geophys.*, **39**, 179.

Brillouin, L., 1925, *Annales de Physique*, **4**, 528.

Elliott, J. R. & Gough, D. O., 1999, *ApJ*, **516**, 475.

Ford, R., et al., 2000, *J. Atmos. Sci.*, **57**, 1236.

Fricke, K., 1969, *A & A*, **1**, 388.

Gough, D. O. & McIntyre, M. E., 1998, *Nature*, **394**, 755.   [GM]

Haynes, P. H., 1989, *J. Fluid Mech.*, **207**, 231.

Haynes, P. H. & McIntyre, M. E., 1990, *J. Atmos. Sci.*, **47**, 2021.

Hoskins, B. J., et al., 1985, *Q. J. Roy. Meteorol. Soc.*, **111**, 877.

Hunt, J. C. R. & Durbin, P. A., 1999, *Fluid Dyn. Res.*, **24**, 375.

Jeffreys, H., 1933, in Saltzman, B., ed., *Selected Papers on the Theory of Thermal Convection*, New York, Dover (1962), 200.

Juckes, M. N. & McIntyre, M. E., 1987, *Nature*, **328**, 590.

Killworth, P. D. & McIntyre, M. E., 1985, *J. Fluid Mech.*, **161**, 449.

Kosovichev, A. G., 1997, *Sol. Phys.*, **170**, 43.

Lorenz, E. N., 1967, *The Nature and Theory of the General Circulation of the Atmosphere*. Geneva, World Meteorol. Org., 161 pp.

McIntyre, M. E., 1994, in Nesme-Ribes, E., ed., *The Solar Engine and its Influence on the Terrestrial Atmosphere and Climate* (Vol. **25** of NATO ASI Subseries I, Global Environmental Change), Heidelberg, Springer, 293.

McIntyre, M. E., 2000, in Batchelor, G. K., Moffatt, H. K., Worster, M. G., eds, *Perspectives in Fluid Dynamics: A Collective Introduction to Current Research*, Cambridge, University Press, 557.

McIntyre, M. E., 2001, in Hodnett, P. F., ed., *Proc. IUTAM Limerick Symposium on Advances in Mathematical Modelling of Atmosphere and Ocean Dynamics*, Dordrecht, Kluwer Academic Publishers, 45.

McIntyre, M. E., 2002, in Pearce, R. P., ed., *Meteorology at the Millennium*, London, Academic Press and Royal Meteorol. Soc., 283.

McIntyre, M. E. & Palmer, T. N., 1985, *Pure Appl. Geophys.*, **123**, 964.

Mohebalhojeh, A. R. & Dritschel, D. G., 2001, *J. Atmos. Sci.*, **58**, 2411.

Norton, W. A., 1994, *J. Atmos. Sci.*, **51**, 654.

Plumb, R. A. & McEwan, A. D., 1978, *J. Atmos. Sci.*, **35**, 1827.

Polvani, L. M. & Plumb, R. A., 1992, *J. Atmos. Sci.*, **49**, 462.

Rhines, P. B., 1975, *J. Fluid Mech.*, **69**, 417.

Riese, M., et al., 2002, *J. Geophys. Res.*, **107**, no. 8179, p. CRI 7–1.

Schou, J., et al., 1998, *ApJ*, **505**, 390.

Spiegel, E. A. & Zahn, J.-P., 1992, *A&A*, **265**, 106.   [SZ]

Stewart, R. W. & Thomson, R. E., 1977, *Proc. Roy. Soc. Lond.*, **A354**, 1.

Thompson, M. J., et al., 1996, *Science*, **272**, 1300.

Waugh, D. W. & Plumb, R. A., 1994, *J. Atmos. Sci.*, **51**, 530.

Waugh, D. W., et al., 1994, *J. Geophys. Res.*, **99**, 1071.

9

Dynamics of the solar tachocline

PASCALE GARAUD

Department of Applied Mathematics and Theoretical Physics
University of Cambridge, Cambridge CB3 9EW, UK

Douglas Gough & Michael McIntyre proposed, in 1998, the first global and self-consistent model of the solar tachocline. Their model is however far more complex than analytical methods can deal with. In order to validate their work and show how well it can indeed represent the tachocline dynamics, I report on progress in the construction of a fully nonlinear numerical model of the tachocline based on their idea. Two separate and complementary approaches of this study are presented: the study of shear propagation into a rotating stratified radiative zone, and the study of the nonlinear interaction between shear and large-scale magnetic fields in an incompressible, rotating sphere. The combination of these two approaches provides good insight into the dynamics of the tachocline.

9.1 Introduction

The tachocline was discovered in 1989 by Brown et al.; it is a thin shear layer located at the interface of the uniformly rotating radiative zone and differentially rotating convective zone of the sun. Several issues about these observations remain unclear. Why is the radiative zone rotating uniformly despite the latitudinal shear imposed by the convection zone, and why is the tachocline so thin? How can the tachocline operate the dynamical transition between the magnetically spun-down convection zone and the interior? The first model of the tachocline was presented by Spiegel & Zahn (1992). They studied the propagation of the convection zone shear into the radiative zone under various hypotheses; in particular, they showed that in the case where angular momentum in the tachocline was transported only by isotropic viscosity the convection zone shear would propagate deep into the radiative zone within a local Eddington-Sweet timescale (rather than a viscous timescale) contrary to what is suggested by observations (Schou et al.,

131

1998). Very roughly, the mechanism for shear propagation into a stratified region is the following: the existence of shear leads to a slight imbalance in the hydrostatic equilibrium and thereby drives meridional flows; these can burrow into the radiative zone, transporting and redistributing angular momentum deeper and deeper. Spiegel & Zahn then studied ways of confining the shear to a thin tachocline through angular-momentum transport by *anisotropic* Reynolds stresses; however, in a first part of this paper I would like to look a little more in detail at the isotropic case, as it can both be used in further investigations of the Gough & McIntyre model, as well as in more general studies of stellar rotation and rotational mixing.

9.2 One half of the problem: shear propagation into a rotating stratified fluid

In this first part, I will consider solar-type stars only and assume that their radiative zone is a stable, isotropic fluid with uniform viscosity μ_v, and that it has little influence on the dynamics of overlying convection zone. As a result, I will simply assume that the convection zone is imposing a given shear to the underlying stably stratified region. Also, I will assume that the dynamical timescale of this system is short compared to the stellar evolution timescale and the stellar spin-down timescale, so that I can limit my study to the steady-state case. This assumption will be dropped in future works on this subject. The equations describing this steady system are

$$\rho_h \boldsymbol{u} \cdot \nabla \boldsymbol{u} = -\nabla \tilde{p} - \rho_h \nabla \tilde{\Phi} - \tilde{\rho} \nabla \Phi_h + \mu_v \nabla^2 \boldsymbol{u} + \frac{1}{3} \mu_v \nabla (\nabla \cdot \boldsymbol{u}) \ ,$$

$$\rho_h T_h \boldsymbol{u} \cdot \nabla s_h = \nabla \cdot (k \nabla \tilde{T}) \ ,$$

$$\frac{\tilde{p}}{p_h} = \frac{\tilde{\rho}}{\rho_h} + \frac{\tilde{T}}{T_h} \ ,$$

$$\nabla^2 \tilde{\Phi} = 4\pi G \tilde{\rho} \ ,$$

$$\nabla \cdot (\rho_h \boldsymbol{u}) = 0 \ , \tag{9.1}$$

where ρ, p and T are respectively the total density, pressure and temperature, $\boldsymbol{u} = (u_r, u_\theta, u_\phi \equiv r \sin\theta \, \tilde{\Omega})$ is the velocity field with respect to spherical polar coordinates (r, θ, ϕ), Φ is the gravitational potential, and k is the thermal conductivity. These equations are the first-order perturbation around the non-rotating hydrostatic background equilibrium (denoted by suffix h); this is a good approximation, as we will see, provided the centrifugal force is much smaller than the gravitational force. The background quantities ρ_h, p_h, Φ_h and T_h are extracted from the standard solar model calculated by

Christensen-Dalsgaard et al. (1991). The perturbed quantities are denoted by tildes, $\tilde{p}, \tilde{\Phi}, \tilde{\rho}$ and $\tilde{T}$. The full nonlinearity of the momentum advection process is kept.

The boundary conditions used on the system are the following: the convection zone shear (as it is observed in the sun) is imposed at the top boundary and continuity of the stresses across the radiative-convective interface imposes another two conditions (on the continuity of the radial derivatives of the azimuthal and latitudinal velocities). A small impermeable core is removed from the region of computation near the centre to avoid singularities. This core is assumed to be rotating solidly, with a rotation rate Ω_{in} determined through the steady-state condition that the total flux of angular momentum through the boundary is null. The regions outside the domain of simulation are assumed to be highly conductive so that they satisfy $\nabla^2 \tilde{T} = 0$, which provides the thermal boundary conditions to apply to the system.

Using the assumption of axisymmetry, I reduce the momentum equation in (9.1) to:

$$\boldsymbol{u} \cdot \nabla_\xi \left(\xi \sin\theta \, u_\phi \right) = \frac{E_\mu}{\rho_h} \mathrm{D}^2 \left(\xi \sin\theta \, u_\phi \right) , \tag{9.2}$$

$$-\frac{1}{\rho_h} \frac{\partial}{\partial z} \left(\rho_h u_\phi^2 \right) = -\frac{\sin\theta}{\rho_h} \left(\frac{\partial \rho_h}{\partial \xi} \frac{\partial \tilde{\Phi}}{\partial \theta} - \frac{1}{\epsilon} \frac{\partial \tilde{\rho}}{\partial \theta} \right) + \frac{E_\mu}{\rho_h} \mathrm{D}^2 \left(\xi \sin\theta \, \omega_\phi \right) ,$$

where $\xi = r/r_c$ is the new radial coordinate normalized by the radius r_c of the star, z is the normalized cylindrical coordinate that runs along the rotation axis, θ is the co-latitude, $E_\mu = \mu_v/r_c^2 \Omega_c$ is the Ekman number, $\epsilon = r_c^2 \Omega_c^2 (l\Phi_h/l\xi)^{-1}$ is the ratio of the centrifugal to gravitational forces, and $\omega = \nabla \times \boldsymbol{u}$ is the vorticity. In this expression the following normalizations have been applied: $[r] = r_c$, $[u] = r_c \Omega_c$, $[\tilde{\Phi}] = r_c^2 \Omega_c^2$, $[T] = 1\,\mathrm{K}$, $[\rho] = 1\,\mathrm{g\,cm^{-3}}$, where r_c is the radius of the radiative zone and Ω_c is the typical rotation rate of the star. The operator D^2 is defined as

$$\mathrm{D}^2 = \frac{\partial^2}{\partial \xi^2} + \frac{\sin\theta}{\xi^2} \frac{\partial}{\partial \theta} \left(\frac{1}{\sin\theta} \frac{\partial}{\partial \theta} \right) \tag{9.3}$$

The energy equation becomes, to first order in the thermodynamical perturbations

$$\epsilon T_h \frac{\sigma N_h^2}{\Omega_c^2} \frac{\rho_h}{E_\mu} u_r = \nabla_\xi^2 \tilde{T} \tag{9.4}$$

where σ is the Prandtl number, N_h is the background buoyancy frequency.

Finally, the equation of state can be combined with the radial and latitudinal components of the momentum equation to provide an expression for

$\tilde{\rho}$:

$$\frac{1}{\rho_\mathrm{h}}\frac{\partial \tilde{\rho}}{\partial \theta} = \frac{\rho_\mathrm{h}}{p_\mathrm{h}} r_\mathrm{c}^2 \Omega_\mathrm{c}^2 \left[\frac{\cos\theta}{\sin\theta} u_\phi^2 - \frac{\partial \tilde{\Phi}}{\partial \theta}\right] - \frac{1}{T_\mathrm{h}}\frac{\partial \tilde{T}}{\partial \theta} . \tag{9.5}$$

Two standard approximations are often performed. The first one is the Boussinesq approximation, commonly used in studies of the tachocline, which is only justified when the thickness of the layer studied is much smaller than the background density scale-height. The second approximation consists in neglecting the effects of the mean centrifugal force on the system by supposing that its main contribution is a very small (negligible) oblateness of the hydrostatic background.

At the time of the Mons conference I presented numerical and analytical solutions of this system of equations and boundary conditions under both approximations. It has since appeared that both approximations were highly unjustified in this problem (as the bulk of the radiative zone spans many scale-heights, and as the mean centrifugal force creates a global baroclinicity of the system that must be taken into account) and lead to erroneous results. I now present instead the solution to the complete problem, solving the equations presented in (9.1). These equations are solved numerically, and the results suggest a scaling of the unknowns $\tilde{T}, u_r$ and u_θ which depends essentially on the parameter

$$\lambda = \sigma N_\mathrm{h}^2/\Omega_\mathrm{c}^2 . \tag{9.6}$$

9.2.1 Slow rotating case ($\lambda \gg 1$)

In the case of slow rotation, I find by studying the numerical results that $\tilde{T}$ and the poloidal components of the velocity $u_{r,\theta}$ scale the following way:

$$\tilde{T} = \epsilon T_\mathrm{h}\overline{T} ,$$
$$u_{r,\theta} = E_\mu/(\lambda\rho_\mathrm{h})\overline{u}_{r,\theta} , \tag{9.7}$$

where the quantities with bars are the scaled quantities, of order of unity. It is also found that $\tilde{\Phi}$ is always of order of unity, which is expected. Note that the scaling for the meridional motions is a local Eddington-Sweet scaling (see Spiegel & Zahn, 1992). Applying this ansatz to the system of equations given in (9.1), an expansion in powers of $1/\lambda$ reveals that the angular-momentum balance is dominated to zeroth order by viscous transport; thus

$$\mathrm{D}^2(\xi \sin\theta u_\phi) = 0 , \tag{9.8}$$

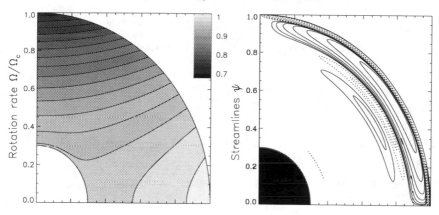

Fig. 9.1. Numerical solution of the system (9.1) for a solar-type star rotating 100 times slower than the sun ($\lambda \simeq 10^4$). The quadrants show the radiative zone only and the imposed shear at the top of the radiative zone is solar-like (i.e. $\tilde{\Omega}_{cz} = (1 - 0.15 \cos^2\theta - 0.15 \cos^4\theta)$. The left panel shows the angular velocity, which is viscously dominated. The interior rotation rate is 0.957 times that imposed at the surface at the equator. The right panel shows the streamlines (dotted lines represent a clockwise flow, and solid lines represent an anti-clockwise flow). The contours are logarithmically spaced. The structure is reminiscent of Holton layers.

which determines the angular velocity profile uniquely. Using this result in the first order equations provides a relation between the temperature and gravitational potential perturbations:

$$\frac{p_h}{\rho_h} \frac{\partial}{\partial z} \left(\frac{\rho_h}{p_h} u_\phi^2 \right) = \sin\theta \left(\frac{\partial \overline{T}}{\partial \theta} - \frac{d \ln T_h}{d\xi} \frac{\partial \tilde{\Phi}}{\partial \theta} \right) ,$$

$$\frac{\partial}{\partial \theta} \nabla_\xi^2 \tilde{\Phi} = -\frac{4\pi G \rho_h r_c}{g_h} \left[\frac{d \ln p_h}{d\xi} \left(\frac{\cos\theta}{\sin\theta} u_\phi^2 - \frac{\partial \tilde{\Phi}}{\partial \theta} \right) + \frac{\partial \overline{T}}{\partial \theta} \right] , \quad (9.9)$$

which can be solved independently for $\overline{T}$ and $\tilde{\Phi}$. Finally, the temperature fluctuations lead to meridional motions through

$$\overline{u_r} \simeq \nabla_\xi^2 \overline{T} . \quad (9.10)$$

Figure 9.1 shows the results of the numerical solutions for the angular velocity profile and the meridional motions corresponding to a slowly rotating solar-type star (for which $\lambda \simeq 10^4$).

9.2.2 Fast rotating case $(\lambda \ll 1)$

In the case of fast rotation it is found that the correct scaling is

$$\tilde{T} = \lambda \epsilon T_{\mathrm{h}} \overline{T} \ ,$$

$$u_{r,\theta} = E_\mu / \rho_{\mathrm{h}} \overline{u}_{r,\theta} \ . \tag{9.11}$$

This time, I perform an asymptotic expansion in the small parameter λ. In this limit the temperature fluctuations are strongly damped by the rapid heat diffusion (as $\lambda \ll 1$ is equivalent to the small Prandtl number limit) and the system reaches an equilibrium which is determined by the zeroth order equations:

$$\frac{p_{\mathrm{h}}}{\rho_{\mathrm{h}}} \frac{\partial}{\partial z} \left(\frac{\rho_{\mathrm{h}}}{p_{\mathrm{h}}} u_\phi^2 \right) = - \sin\theta \frac{\mathrm{d} \ln T_{\mathrm{h}}}{\mathrm{d}\xi} \frac{\partial \tilde{\Phi}}{\partial \theta} \ ,$$

$$\frac{\partial}{\partial \theta} \nabla_\xi^2 \Phi = 4\pi G \frac{\rho_{\mathrm{h}}^2}{p_{\mathrm{h}}} \left[\frac{\cos\theta}{\sin\theta} u_\phi^2 - \frac{\partial \tilde{\Phi}}{\partial \theta} \right] \ . \tag{9.12}$$

These equations can in principle be solved for u_ϕ^2 and $\tilde{\Phi}$ and provide, to the next order in λ, the meridional flow through the advection diffusion balance:

$$\overline{\boldsymbol{u}} \cdot \nabla_\xi (\xi \sin\theta u_\phi) = \mathrm{D}^2 (\xi \sin\theta u_\phi) \ , \tag{9.13}$$

and, finally, the temperature fluctuations through

$$\overline{u} = \nabla_\xi^2 \overline{T} \ . \tag{9.14}$$

The results of the numerical simulations for small lambda $(\lambda \simeq 10^{-2})$ are shown in Fig. 9.2.

9.2.3 Solar rotation rate

In the solar case, the parameter λ varies between 0.1 and 1 in the region between the two boundaries. Although the solution is closer to the fast rotating case, the asymptotic analysis does not apply and the dynamics of the system result from a complex interaction of the momentum balance, the thermal energy advection-diffusion balance and the Poisson equation.

9.2.4 Discussion

I have studied the nonlinear dynamics of the radiative zone of a rotating solar-type star when a latitudinal shear is imposed by an overlying convection zone. This study is valid provided that the star is far from break-up (i.e. that the centrifugal force is small compared to the gravitational potential).

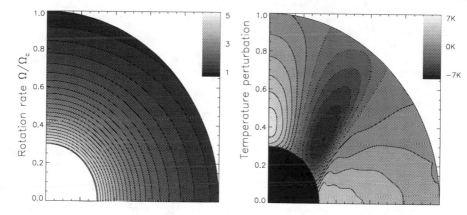

Fig. 9.2. Numerical solution of the system (9.1) for a solar-type star rotating 10 times faster than the sun ($\lambda \simeq 10^{-2}$). The left panel shows the angular velocity, which increases with depth through angular-momentum conservation. Note how the latitudinal variation of the angular velocity is small compared to its radial variation. The interior rotation rate is 5.26 times that imposed at the surface at the equator. The right panel shows the temperature fluctuations. Note that even when the stellar oblateness is of order of 10^{-3}, the temperature fluctuations remain of order of 10^{-6} through efficient heat diffusion.

I found that few approximations can be safely used in this study: the nonlinear advection terms and the effects of the centrifugal force must be carefully included in momentum equation. However, in the limit where the star is far from the break-up point, the perturbations to the hydrostatic background are found to be small indeed, which justifies the linearization of the equation of state.

Two asymptotic limits were found, which depend on the value of the parameter $\lambda = \sigma N_{\mathrm{h}}^2 / \Omega_{\mathrm{c}}^2$. In the case of a slowly rotating star (with $\lambda \gg 1$) the hydrostatic background acquires a small ellipticity and the angular velocity profile is viscously dominated. The meridional flow velocities are of order of the local Eddington-Sweet velocity (e.g. Spiegel & Zahn, 1992) and take the shape of alternating dipolar cells reminiscent of the Holton layer structure. In the case of a fast rotating star (yet far from breakup), the temperature fluctuations are determined by an advection-diffusion balance which limits their amplitude to roughly $\lambda \epsilon T_{\mathrm{h}}$; this value is independent of the rotation rate. The angular velocity profile and the fluctuations in the gravitational potential can be determined independently through the momentum equation and the Poisson equation. It is found that the angular velocity increases sharply with depth, as expected from equation (9.2), and varies little with latitude. Similarly, the perturbation to the gravitational potential vary little with latitude, which suggests the possibility of approximating this limit analytically as a one-dimensional problem. This is under current investigation.

Finally, the effects of the boundary conditions on the problem (and in particular the presence of a lower rigid boundary) remain to be carefully analysed.

To summarize the first part of this paper, I have shown that it is possible to study semi-analytically (in some cases) and numerically the problem of shear propagation into the solar radiative zone in a self-consistent way, when taking into account isotropic viscosity only†. The main result is the following: as Spiegel & Zahn predicted, in this isotropic case the shear imposed by the convection zone penetrates all the way into the solar core. The failure to reproduce observations therefore suggests that other dynamical phenomena must be present in the solar radiative zone.

9.3 The other half of the problem: nonlinear interaction between a large-scale field and flows in a rotating sphere

Having studied the difficulty of hydrodynamical models to explain the structure of the solar tachocline, Gough & McIntyre (1998) suggested an alternative theory, namely that the observations could be reproduced through the existence of a large-scale fossil field in the solar radiative zone. As McGregor & Charbonneau (1999) showed, such a field can indeed impose a uniform rotation throughout most of the radiative zone and confine the shear to a thin tachocline provided none of the field lines are anchored into the convection zone‡: the field must be entirely confined to the radiative zone. Studies in the non-magnetic case following the lines described in the first part of this paper seem to suggest that shear-driven baroclinic imbalance leads to downwelling flows near the poles and the equator, with a localized upwelling in mid-latitudes (in regions of little shear). This phenomenon is illustrated in Fig. 9.3. Gough & McIntyre combined these two results and suggested that baroclinically driven flows could indeed lead to the confinement of the field through nonlinear advection, and proposed a new model of the tachocline based on this idea. However, only a fully nonlinear numerical study can verify whether this dynamical balance could indeed lead to the observed rotation profile.

As a first step towards a complete numerical simulation of the tachocline according to the Gough & McIntyre model, I have looked at the nonlinear

† Incidentally, it is clear that this type of analysis is not limited to the solar case, but can be applied to other stars with a wide range of rotation rates, masses, and ages. It will be interesting to compare the corresponding results to the asteroseismic observations of COROT.

‡ The shear would otherwise propagate along field lines according to Ferraro's isorotation law.

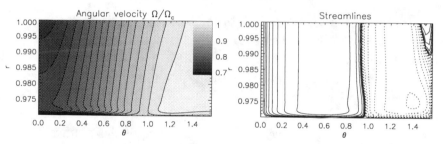

Fig. 9.3. Numerical solution of the system (9.1) for a solar-type star rotating with the observed solar angular velocity. The rigid bottom boundary was artificially placed at $r = 0.97 r_c$ to mimic the presence of a confined large-scale magnetic field. The left panel shows the angular velocity, when the convection zone shear is imposed at the top. The right panel shows the streamlines, with dotted lines representing clockwise flow and solid lines representing anti-clockwise flow. Note the two-cell structure with upwelling in mid-latitudes; note also the presence of an equatorial boundary layer.

interaction of a dipolar magnetic field and shear-driven motions only, when all thermal/compressibility effects are neglected. In these simulations, the fluid is incompressible with constant density ρ. This allows me to determine, through a simplified model, whether the idea of field confinement through meridional motions of the type described by Gough & McIntyre is indeed possible. In order to do this, I have created a numerical model in which meridional flows are created by the shear, not through baroclinic driving but through Ekman pumping on the boundary. The interest of this approach is that the geometry of the flow in this simplified problem is qualitatively similar to that shown in Fig. 9.3: it possesses a downwelling near the poles and the equator, and upwells in mid-latitude.

The numerical procedure is the following. I solve the following system of equations

$$2(\hat{\boldsymbol{e}}_z \times \boldsymbol{u})_\phi = \Lambda(\boldsymbol{j} \times \boldsymbol{B})_\phi + \frac{E_\mu}{\rho}(\nabla^2 \boldsymbol{u})_\phi \,,$$

$$[\nabla \times (\boldsymbol{u} \times \boldsymbol{B})]_\phi + E_\eta (\nabla^2 \boldsymbol{B})_\phi = 0 \,,$$

$$\nabla \cdot \boldsymbol{u} = 0 \,,$$

$$\nabla \cdot \boldsymbol{B} = 0 \,,$$

where $\boldsymbol{B}$ is the magnetic field, E_η is the magnetic Ekman number and Λ is the global Elsasser number defined as $\Lambda = B_0^2/\rho r_c^2 \Omega_c^2$. This system is solved in a spherical shell, where, as in the first part of this paper, the outer boundary corresponds to the bottom edge of the convection zone and the inner core is removed to avoid singularities. The outer boundary is now assumed to be impermeable in order to create artificially an Ekman layer at the interface with the convection zone which will drive the required meridional flows. A point dipole is placed at $r = 0$ such that the radial field at the pole at $r = r_{in}$ is B_0. The regions outside the region of simulation are supposed to be conductive so that the magnetic field satisfies $\nabla^2 \boldsymbol{B} = 0$ in the steady state case. As in the first set of simulations, the inner core is rotating solidly with angular velocity Ω_{in} where Ω_{in} is determined through the steady-state requirement that the angular-momentum flux through the inner boundary is null.

The results are now discussed for fixed diffusive parameters, when only the Elsasser number is varied. For low Elsasser number ($\Lambda \ll 1$), the system is dominated by the Coriolis forces and the magnetic field is mostly passive. The shear imposed by the convection zone propagates deep into the radiative zone along the rotation axis, thereby satisfying Proudman's rotation law. Two meridional circulation cells are created by Ekman pumping on the outer boundary, with downwelling at the poles and the equator and upwelling in mid-latitudes, as required; they burrow deep into the radiative zone. In the limit of strong magnetic field (for $\Lambda \gg 1$) Lorentz forces rule the dynamics of the system; the poloidal field is barely affected by the rotation or the meridional motions and retain a dipolar structure throughout the interior. The shear propagates into the radiative zone along the field lines which have a footpoint in the convection zone, through Ferraro's isorotation theorem. Again this limit fails to reproduce the observations.

Only in the intermediate case ($\Lambda \simeq 1$) does the system begin to show the existence of a tachocline. Indeed, in this limit the field is still strong enough deep in the interior to dominate the dynamics of the system, but the meridional motions driven at the outer boundary manage to advect the field downwards near the equator thereby confining the field *into* the radiative zone in that region. At higher latitudes, however, some field lines retain their footpoints in the convection zone. As the field is mostly confined into the radiative zone, it imposes a state of near-uniform rotation save perhaps in a thin diffusive boundary layer near the top of the radiative zone and near the poles. This structure is reminiscent of the tachocline. Moreover, the meridional motions are confined to the shallower regions of the radiative zone by the underlying field; the consequences of the existence of this shallow mixing layer on the light element abundances is directly observable. The results of the simulations in the intermediate case are shown in Fig. 9.4.

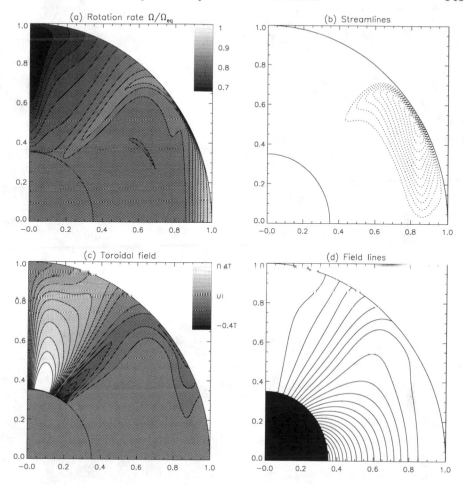

Fig. 9.4. Simulation results for $\Lambda = 1$, $E_\mu = 6.25 \times 10^{-5}$, and $E_\eta = 6.25 \times 10^{-5}$. Panel (a) shows the angular velocity, which is nearly uniform in the radiative zone whereas the shear is confined to the equatorial regions near the upper boundary, and around the poles. Panel (b) shows the meridional motions, which are confined to the upper layers of the radiative zone by the underlying magnetic field. The flow is downwelling at the equator and upwelling in mid-latitudes. Panel (c) shows the toroidal field, which is virtually null in regions of uniform rotation. Finally, panel (d) shows the poloidal field lines, which are confined in the radiative zone by the meridional motions.

9.4 Conclusion

It is now time to combine the results that we have learned from these two separate studies of the rotation profile in the radiative zone and relate them to the tachocline dynamics.

The model of the tachocline proposed by Gough & McIntyre (1998) attempts to solve the following problem: how can we explain that the solar radiative zone is rotating uniformly when a study of standard stellar rotation would normally suggest the existence of strong shear. Gough & McIntyre suggested that the interaction of a large-scale field and baroclinically driven meridional motions in the tachocline could lead to the observed angular velocity profile. This model hangs on two key points: the magnetic field must be entirely confined to the radiative zone to impose uniform rotation, and the meridional motions must be confined to the tachocline to explain both the required two-cell structure (which can then in turn confine the field) and the observed light element abundances (Elliott & Gough, 1999).

By studying the dynamics of a thin layer of stratified fluid representing the tachocline I have shown that baroclinic effects do indeed lead to a two-cell circulation with upwelling in mid-latitudes and downwelling near the poles and the equator. I have then taken a complementary approach and looked at the dynamical interaction between such a two-cell circulation and a dipolar large-scale field. This allowed me to show that the meridional motions can indeed confine this field to the radiative zone for some range of values of the magnetic field strength. The confined field imposes a uniform angular velocity to most of the radiative interior, save in a thin tachocline where all the dynamical interactions described above take place.

To conclude, I believe that these preliminary analyses show that the model proposed by Gough & McIntyre possesses the right physical elements for the description of the dynamics of the tachocline. There remains now only the task of completing this work through the numerical resolution of their whole model.

Acknowledgments I thank Douglas Gough for his help and his encouragements throughout my PhD thesis and in the year that has followed. This work has benefited immensely from his sharp and inspired vision of the subject.

References

Brown, T. M., Christensen-Dalsgaard, J., Dziembowski, W. A., Goode, P., Gough, D. O. & Morrow, C. A., 1989, *ApJ*, **343**, 526

Christensen-Dalsgaard, J., Gough, D. O. & Thompson, M. J., 1991, *ApJ*, **378**, 413

Elliot, J. R. & Gough, D. O., 1999, *ApJ*, **516**, 475

Gough, D. O. & McIntyre, M. E., 1998, *Nature*, **394**, 755

MacGregor, K.B. & Charbonneau, P., 1999, *ApJ*, **519**, 911

Schou, J. et al., 1998, *ApJ*, **505**, 390

Spiegel, E. A. & Zahn, J.-P., 1992, *A&A*, **265**, 106

10

Dynamo processes: the interaction of turbulence and magnetic fields

MICHAEL PROCTOR

Department of Applied Mathematics and Theoretical Physics
University of Cambridge, Silver Street, Cambridge CB3 9EW, UK

This chapter reviews recent research on the interaction of magnetic fields with MHD turbulence, with particular application to the question of the influence of Lorentz forces on the efficiency of large-scale field generation.

10.1 Scales for solar magnetic fields

The solar magnetic field outside the radiative core exists on a great range of length and time scales; these embrace all sizes from that of the disc itself to that of the diffusion length scales of a few km, well below present observational resolution. While it is the largest scales that force themselves on our attention, due to the visibility of sunspots and associated coronal structures, and the coherence of the solar cycle, it is not clear whether these large-scale fields control, or are controlled by, the small-scale fields that have much greater total energy. While the cycle is clearly global in nature, the "magnetic carpet" of small-scale field structures that appear in quiet regions would seem to be a local manifestation of dynamo action due to turbulent stretching.

Linear dynamo theory, in particular the "mean-field" or "α-effect" models, has proved amazingly successful in predicting aspects of the solar cycle such as the butterfly diagram. In fact some of this 'success' has nothing to do with the physics employed, but derives from the symmetry of the underlying geometry. As Knobloch (1994) convincingly argues, all manifestations of oscillatory wavelike processes in a finite domain with reflexional symmetry will take the form at onset of travelling wave structures with either dipole or quadrupole symmetry. In this severe view all that remains is to determine the direction of travel of the wave crests. Furthermore, the physics of the mechanism is by no means fully understood, even in the kinematic limit where Lorentz forces are ignored. While the mechanism for production of

143

toroidal field by differential rotation is fairly secure, there is difficulty in obtaining a convincing and detailed explanation of the return of coherent field to great depths, as is now demanded by 'interface' models of the dynamo (Parker 1993). The favoured thinking at the present time is that the action of the three-dimensional turbulence acts to pump field down to a level where it can be acted on by the zonal flow. Even within this paradigm, however, there is disagreement as to whether the source of the helical flows with strong vertical component are just turbulent, as in the original Parker (1955) picture, or whether the process involves magnetic buoyancy and is essentially nonlinear. (see e.g. Thelen 1997). Large-scale photospheric fields, such as sunspots and pores, have important dynamics of their own, which is outside the scope of this paper. But the cyclical nature of the appearance of bipolar regions and the regularity of their latitudinal distribution indicate that they are part of of the same large-scale process, and are not the products of autonomous dynamo action.

Finally, we come to the magnetic carpet, the small-scale fields clearly shown in recent TRACE observations (Title 2000). These fields are if anything more prominent during times of reduced sunspot activity; they show little or no cyclical behaviour and are seen well away from active regions. It seems likely that these fields are caused by local dynamo action. When trying to understand such dynamo action at very high magnetic Reynolds numbers ($R_m \equiv UL/\eta$, where U, L are velocity and length scales and η is the magnetic diffusivity), it is important to escape from the preconceptions induced by kinematic dynamo studies and the MHD of steady flows. In a turbulent flow the field lines are stretched at almost all locations, as nearby trajectories of the flow particles separate exponentially. Thus there is a strong mechanism for enhancing magnetic energy locally. In the fully developed state, although there will be some intermittency due to cancellation caused by folding of the trajectories, the fluid will be permeated with field to a much greater extent than would appear at the surface. This is because, at a boundary, areas of surface particles are not conserved. We therefore expect to find tangled fields with a fractal dimension between 1 and 2, even in the limit $\eta \to 0$. These fields will exert a significant dynamical influence on the flow, and so can be expected to be at equipartition levels. Such a "magnetic fondue" can be glimpsed in the magnetic carpet, but the basic arguments apply to all scales where the magnetic Reynolds numbers are large, and where the turnover time (which, rather than the diffusion time, is the appropriate time for growth of the fields) are not too large. The mechanisms at larger scales are likely to be affected by Coriolis forces; however the magnetic carpet, and numerical simulation, (Cattaneo 1999, Cattaneo

& Hughes 2001) show that large-scale rotation is not necessary for dynamo action. The problem, therefore, is to understand how the large-scale observed fields can be generated (the "α" part of an α-ω dynamo), in these magnetically dominated flows. The answers are still highly controversial. This paper reviews recent theoretical ideas and associated numerical work in an attempt to throw light on the difficulties.

10.2 Field structure in kinematic dynamos at large R_m

The α-effect, or mean field dynamo, has long been a mainstay of theories of the solar cycle, and it is still widely used today. The text by Moffatt (1978) gives an excellent overview of early applications, while more recent references can be found in Weiss (1994). There are two basic assumptions; that there is scale separation between 'mean' and 'fluctuating' fields; and that the averaged e.m.f. induced by the small-scale fields is a local function of the mean magnetic field and its derivatives. The first assumption would seem reasonable, but the second is harder to justify in the interesting case where the magnetic Reynolds number is very large, even on the smallest scales. (When the small-scale $R_m \ll 1$ then a rigorous theory can be constructed; see e.g. Moffatt's book). There are several important consequences of large R_m (Galloway & Proctor 1992, Cattaneo et al 1995):

- Field structures are highly intermittent, with length scales $\sim R_m^{-\frac{1}{2}}$.
- These structures do not depend much on the value of R_m, but on the topology of the flow pattern. Only the thickness of the structures depends on R_m.
- These small-scale fields can be *self sustaining*; that is, there is a small-scale dynamo.

The smallest scales of the field appear very rapidly in kinematic computations at high R_m – in fact after a few turnover times L/U. However the growth rate of dynamo disturbances does depend on R_m, but typically appears to reach a limiting value independent of R_m as $R_m \to \infty$, though precise computation becomes very difficult owing to to the small length scales involved. This limiting growth rate is typically of order L/U; these are known as *fast dynamos* (see, e.g. Childress & Gilbert 1995). In spite of the difficulty in resolving the smallest structures, we find that scaling laws for the eigenfunctions are established accurately at much lower R_m and can be accurately calculated. Such laws give a power-law distribution for integrated quantities such as $R_1 \equiv |\langle \mathbf{B} \rangle|^2 / \langle |\mathbf{B}|^2 \rangle$, which $\sim R_m^\gamma$, where γ is a constant of order unity depending on the flow, as shown in Figure 10.1.

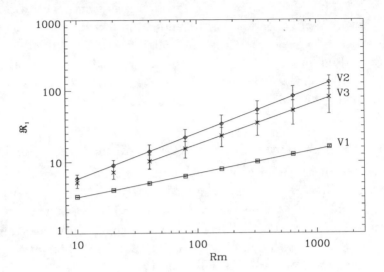

Fig. 10.1. Behaviour of the quantity R_1 defined in the text as a function of R_m for three different dynamo flows (from Cattaneo et al. © 1995 by the American Physical Society). The greater the slope, the greater the energy in the small-scale fields.

These power laws demonstrate that the field distribution is fractal in nature; and indeed if γ is not too small then at large R_m the smallest scales of field are dominant, as is perhaps to be expected. One final aspect of these kinematic fast dynamos deserves attention. When the dynamo field exists on essentially the same scale as the velocity field, helicity of the flow is not necessary for efficient dynamo action. This shows that the fact that the magnetic carpet fields are on too short a timescale to notice the Sun's rotation does not rule out dynamo action as their cause. Such fields will not work as mean-field dynamos (see below), because for them helicity is essential. If buoyancy is the principal driving mechanism then helicity cannot be introduced directly; it follows that for an efficient mean field dynamo we require either rotation or inhomogeneity (giving gradients of large-scale helicity).

10.3 Dynamical equilibration of small-scale dynamos

How large can a small-scale dynamo field get before the growth of the field is halted by the dynamical effects of the Lorentz force? In the solar context, where the viscosity is small, we expect such effects to occur when the magnetic energy density $|\mathbf{B}|^2/2\mu_0$ is comparable with the kinetic energy density

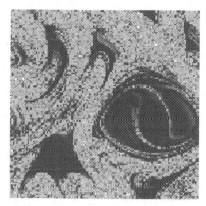

Fig. 10.2. Finite time Liapunov exponents for a simple quasi-two dimensional dynamo (after Cattaneo et al. 1996). Lighter shades indicate greater stretching. (a) Kinematic case, (b) dynamic case when Lorentz forces have reduced the stretching properties of the flow.

$\rho|\mathbf{u}|^2$ (equipartition). This expectation is confirmed by the results of several calculations of model dynamos, and by the full MHD simulation of a convective dynamo by Cattaneo (1999). At high values of R_m, as the field amplitude grows, we must pass from a growth rate comparable with the turnover time to one which is zero! How is this accomplished? One mechanism, which would hold for spatially constrained flows, would be for the kinetic energy to be reduced, thus reducing the magnetic Reynolds number towards the critical value. This is most unlikely to happen when the kinematic R_m is far above critical, since that would demand a huge reduction in the kinetic energy. Instead, these systems equilibrate in a much more subtle way, which is almost invisible in the Eulerian statistics. An example is given for a simplified model by Cattaneo, Hughes & Kim (1996), and examples of finite-time Liapunov exponents for the kinematic and dynamic cases are shown in Figure 10.2. They reduce their efficiency as a dynamo by altering their stretching properties, so that the Liapunov exponents go down, leading to less efficient energy growth, leaving cancellation effects to mop up such growth as remains. (It is possible that in some cases the cancellation is enhanced, rather than the stretching reduced. But the detailed results produced to date do not show this. Such enhancement is more likely to be a consequence of two-dimensionalization of the flow induced by a large-scale field.) How long does it take for equilibration to be achieved? The growth rates at high R_m are fastest for the smallest scales of motion, so one could expect that each scale might become dynamically active after a time proportional to its turnover time. Magnetic energy reaches equipartition successively at longer and longer scales. Finally we have "MHD turbulence" with Lorentz forces important at all scales. The crucial question for the coherent dynamo involved in the solar cycle is: can fields which have a scale much greater than that of the turbulent flows grow at a substantial rate? Thus we need to address the dynamical effect of the Lorentz force on mean field growth.

10.4 Growth and equilibration of mean fields

In this section we discuss the way in which large-scale ("mean") fields can arise as a result of small-scale fluctuating motion. We first note that the distinction between large and small scales is only clear when the small-scale turbulence is homogeneous. Any systematic large-scale inhomogeneity will ineluctably lead to components of the Fourier spectrum of the field on the same scale. These are of a different nature, however, from the independently generated fields that form the cycle. The effects of the small-scale on the large-scale fields may be seen by writing the magnetic field $\mathbf{B} = \overline{\mathbf{B}} + \mathbf{b}$ and the velocity field $\mathbf{U} = \overline{\mathbf{U}} + \mathbf{u}$; then the induction equation for time derivative of the mean field $\overline{\mathbf{B}}$ becomes

$$\frac{\partial \overline{\mathbf{B}}}{\partial t} = \nabla \times (\overline{\mathbf{U}} \times \overline{\mathbf{B}}) + \nabla \times \mathcal{E} - \nabla \times \eta \nabla \times \overline{\mathbf{B}} \text{ , where } \mathcal{E} = \overline{\mathbf{u} \times \mathbf{b}} \text{ .} \quad (10.1)$$

In order to get significant mean fields on a relevant (i.e. non-diffusive) timescale we need the "α-effect" $\boldsymbol{\alpha}$, defined by the *ansatz* $\mathcal{E} = \boldsymbol{\alpha} \cdot \overline{\mathbf{B}}$, to be of order $|\mathbf{u}|$, i.e. independent of η, the magnetic diffusivity. While $\boldsymbol{\alpha}$ is straightforward to calculate when the small-scale R_m is small (see, e.g. Moffatt 1978), it is much harder to see how to proceed when the small-scale R_m is large. In the Parker (1955) picture field lines are twisted and rotated by a helical "cyclonic event". For events shorter than the turnover time we can say that $\mathcal{E}$ is proportional to $-\mathcal{H}$, where $\mathcal{H}$ is the helicity. But if such an event persists longer than a turnover time the constant of proportionality may change sign due to multiple rotations. Thus even the sign of the effect is not certain, and there are other problems associated with the possible nonlocal dependence of $\mathcal{E}$ on $\overline{\mathbf{B}}$. Nonetheless, one can imagine an experiment in which a *uniform* magnetic field $\mathbf{B}_0$ permeates a region of homogeneous MHD turbulence. There is no large scale dynamo but $\mathcal{E}$ can be calculated as a function of $\mathbf{B}_0$. It is crucial to understand how $\boldsymbol{\alpha}$ depends on $\mathbf{B}_0$. We expect it to reduce with increasing field ("α-quenching"), but when do Lorentz forces become important and initiate this quenching? There is considerable controversy over this question. To fix ideas, define B_E, the equipartition field strength, as $(\mu_0\rho\overline{|\mathbf{u}^2|})^{\frac{1}{2}}$. Then we can all agree that because of the symmetry under sign change of $\overline{\mathbf{B}}$, we expect some functional dependence for large R_m of the form

$$\boldsymbol{\alpha}(\overline{\mathbf{B}}) = \mathcal{F}(R_m^a|\overline{\mathbf{B}}|^2/B_E^2). \tag{10.2}$$

The controversy resides in the value of the exponent a. If $a \ll 1$ then the large-scale fields can reach equipartition values with relative ease, while if a is not small the mean field mechanism shuts down when $|\overline{\mathbf{B}}|$ is still well below B_E, making the timescales for the production of large-scale fields inordinately long.

Before looking at recent simulations which cast light on the value of a, we first deal with the formula for $\boldsymbol{\alpha}$ in MHD turbulence originally put forward by Pouquet, Frisch & Léorat (1976) and revisited by Blackman & Field (2000). We begin with fluctuating magnetic and velocity fields $\mathbf{b}$, $\mathbf{u}$. Then a uniform field $\overline{\mathbf{B}}$ is added, and this has the effect of changing the fluctuating fields to $\mathbf{b} + \mathbf{b}'$, $\mathbf{u} + \mathbf{u}'$, where $\mathbf{b}'$, $\mathbf{u}'$ obey the equations

$$\left. \begin{aligned} \frac{\partial \mathbf{u}'}{\partial t} &= -\nabla p + \frac{1}{\mu_0\rho}\overline{\mathbf{B}}\cdot\nabla\mathbf{b} \\ \frac{\partial \mathbf{b}'}{\partial t} &= \qquad\quad \overline{\mathbf{B}}\cdot\nabla\mathbf{u} \end{aligned} \right\} + \text{small(?) diffusion terms}. \tag{10.3}$$

To find the mean e.m.f. proportional to $\overline{\mathbf{B}}$, we can assume isotropy, so that $\alpha_{ij} = \alpha\delta_{ij}$. Thus we have (the dots denoting time derivatives)

$$\begin{aligned} \boldsymbol{\mathcal{E}} = \alpha\overline{\mathbf{B}} &= \overline{\mathbf{u}\times\mathbf{b}'} + \overline{\mathbf{u}'\times\mathbf{b}} \\ &\approx \int_0^{\tau_c} \left(\overline{\mathbf{u}\times\dot{\mathbf{b}}'} + \overline{\dot{\mathbf{u}}'\times\mathbf{b}}\right) dt \text{ where } \tau_c \text{ is a "correlation time"}. \end{aligned} \tag{10.4}$$

If τ_c is short relative to other timescales then we can use (10.3) to obtain

$$\alpha \approx -\frac{\tau_c}{3}\left(\overline{\mathbf{u}\cdot\nabla\times\mathbf{u}} - (\mu_0\rho)^{-1}\overline{\mathbf{b}\cdot\nabla\times\mathbf{b}}\right). \tag{10.5}$$

There are many assumptions made in this derivation, not least the one that equates correlation times for velocity and magnetic fields. Nonetheless the expression (10.5) does have the satisfying characteristic that if the "turbulence" takes the form of Alfvén waves, for which $\mathbf{u} = \pm\mathbf{b}/\sqrt{\mu_0\rho}$, then $\boldsymbol{\mathcal{E}}$ must vanish. Unfortunately the formula has been interpreted by many authors as giving a model of the effects of large imposed fields on α, with $\mathbf{u}$, $\mathbf{b}$ considered as the actual fields. In fact the formula can be justified only for *small* $|\overline{\mathbf{B}}|$, since equations (10.3) can then be linearized; and where the field

b has nothing to do with $\overline{\mathbf{B}}$ but is preexisting. Nonetheless it is useful as a guide to the initial growth rate of a large-scale field in the presence of MHD turbulence. It should be emphasised that the induction equation remains linear irrespective of the effects of the Lorentz force, and so the last term in (10.5) can *only* arise from magnetic fields that do *not* owe their existence to the imposed field $\overline{\mathbf{B}}$. This is not the situation considered by Moffatt (1978) and others.

Whether or not the above result remains true for large imposed fields, there remains the crucial question posed above: what is the form of the function $\mathcal{F}$ defined in (10.2), and what in particular is the crucial exponent a? In general terms we expect that $\mathcal{F}(X)$ decreases with X, and $\sim X^{-\beta}$ as $X \to \infty$, with $\beta \geq 1$. The existence of large-scale fields of significant amplitude suggests that a is small, while numerical calculations of idealized problems suggest that $a \sim 1$, which must lead to significant problems with the large-scale fields. In consequence these calculations have been criticized as inapplicable to real MHD turbulence. Nonetheless there are several theoretical reasons for supposing a significant, and the critics have not yet found a definitive solution to the difficulty.

The theoretical backing for a to be significant is provided by Gruzinov & Diamond (1994, 1995). They consider a situation in which magnetic and velocity fields are statistically stationary. This implies that the time derivative of the mean magnetic helicity $\overline{\mathbf{a} \cdot \mathbf{b}}$ vanishes, where **a** is the magnetic potential defined by $\mathbf{b} = \nabla \times \mathbf{a}$, $\nabla \cdot \mathbf{a} = 0$. The equations for **a** and **b** are

$$\left.\begin{aligned}
\frac{\partial \mathbf{a}}{\partial t} &= (\mathbf{u} \times \overline{\mathbf{B}}) + (\mathbf{u} \times \mathbf{b}) - \nabla \chi - \eta \nabla \times \mathbf{b} \\
\frac{\partial \mathbf{b}}{\partial t} &= \nabla \times (\mathbf{u} \times \overline{\mathbf{B}}) + \nabla \times (\mathbf{u} \times \mathbf{b}) - \nabla \times (\eta \nabla \times \mathbf{b})
\end{aligned}\right\} \tag{10.6}$$

where χ is the electrostatic potential. Setting $\dfrac{\partial}{\partial t}(\overline{\mathbf{a} \cdot \mathbf{b}}) = 0$, we obtain after some manipulation

$$\overline{\mathbf{B}} \cdot (\overline{\mathbf{u} \times \mathbf{b}}) = \overline{\mathbf{B}} \cdot \mathcal{E} = -\eta \overline{\mathbf{b} \cdot \nabla \times \mathbf{b}},$$

and so we have the exact result (not depending on any assumptions concerning small R_m or short correlation times)

$$\alpha = -|\overline{\mathbf{B}}|^{-2} \eta \, \overline{\mathbf{b} \cdot \nabla \times \mathbf{b}}. \tag{10.7}$$

It should be noted here that the field **b** is now the total small scale field; there is no approximation involving small $|\overline{\mathbf{B}}|$. Gruzinov & Diamond use (10.7) in combination with (10.5) to give a relation between α and $|\overline{\mathbf{B}}|$ of the form

$$\alpha = \alpha_0(1 + R_m|\overline{\mathbf{B}}|^2/B_E^2)^{-1},$$

where α_0 is the kinematic α-effect that holds when Lorentz forces are negligible. which suggests that $a = 1$. Although this result is very appealing, it must be recalled that the definitions of $\mathbf{b}$ in (10.5) and (10.7) are not obviously compatible. Further calculations establish that the part of $\mathcal{E}$ proportional to gradients of $\overline{\mathbf{B}}$ (the 'turbulent diffusivity') only depends on $|\overline{\mathbf{B}}^2|/B_E^2$.

The physical picture that backs up the theory has been elaborated by Cattaneo & Hughes (1996), and recently given support by Brandenburg (2001). The basic idea is simple. The dynamical effects of the magnetic field on the flow must be felt when the Lorentz forces become significant. In flows of astrophysical interest, $R_m \gg 1$ even on the small scales, and in this case $|\mathbf{b}| \gg |\overline{\mathbf{B}}|$. In fact when the fields are sufficiently weak the growth of small-scale field is limited by diffusion in regions of flow convergence, and so we expect $|\mathbf{b}| \sim R_m^{1/2}|\mathbf{B}|$ if the field is in sheets. When we have flux-tube-type structures, the amplification factor is larger but the dynamical effects smaller. In either case, when the small-scale field reaches the equipartition value we expect a significant change in the dynamo process. Thus the physical picture predicts $a \approx 1$. Although the small-scale field is highly intermittent the crucial mechanism of dynamo generation occurs precisely where the small-scale fields are being produced – and so such intermittency is unlikely to affect the value of a significantly.

These ideas have their origin in simpler studies in two dimensions (e.g. Vainshtein & Cattaneo 1992) investigating the effects of the Lorentz force on the diffusion rate of an imposed large scale field. Here there is no dynamo, but similar considerations suggest that the stretching properties of the flow are affected, leading to a decrease in the turbulent diffusivity, In that work it is argued that the conservation of the mean square magnetic potential in the absence of diffusion, together with the requirement that the turbulent diffusion have a value independent of η, requires the small-scale magnetic field to exist on diffusive length scales. There is a clear analogy in three dimensions with the conservation of magnetic helicity. This leads via (10.7) to the requirement of magnetic fields on diffusive scales in order that α not depend on the diffusivity. It is notable that there is no similar conservation law for mean square potential in three dimensions, and that the turbulent diffusivity in this case is affected much less by the imposed field (one can see that unknotted field lines can slip through highly conducting material in 3D without affecting the flow much). One would expect the helicity constraint to have an effect on this process; the situation remains unclear. The reduction in the α-effect occurs on this view because the Lorentz forces prevent the smallest scales of the magnetic field from reaching diffusion levels. In addition, when the magnetic Prandtl number is of order unity, as may be appropriate for the Sun, the MHD turbulence spectrum may contain a significant proportion of Alfvén waves, for which **u** and **b** are parallel and which thus give no contribution to $\mathcal{E}$. When the magnetic Prandtl number is very large, as may be the case in galaxies, then of course there are no Alfvén waves and the equilibration mechanism is different, perhaps leading to smaller values of a, as shown in recent work by Schekochihin, Cowley, Maron & Malyshkin (2002).

The idea that a is significant is given support from three very different numerical studies. The first (Brandenburg 2001) considers flow in a periodic domain, forced by a helical body force on a small scale. There is eventual growth of significant large-scale fields, which are force-free and can grow to large size free of dynamical constraints. While increasing R_m leads to more rapid initial growth, the time taken for final equilibration also increases. The α-effect is calculated by solving a short-time initial value problem, and by superposing a uniform mean field and calculating $\mathcal{E}$ directly. Both methods (see Figure 10.3) yield a significant dependence on R_m in the α-quenching formula, with $a \geq 1$. Brandenburg also finds that the turbulent diffusivity is quenched, but that the dependence on R_m is rather weaker, as suggested above. The remaining studies were carried out by Cattaneo & Hughes (1996) and Cattaneo, Hughes & Thelen (2002). In the first, a kinematic flow is forced that has the form of the so-called CP-flow of Galloway & Proctor (1992). A fully three-dimensional calculation is undertaken, starting from this velocity field with an imposed uniform field in the z-direction. Only that part of the α-effect which derives from fully three-dimensional, that is nonlinearly driven, flows is evaluated by direct calculation of $\mathcal{E}$ and the results show that $a \sim 1$ for the quenching properties. The magnitude of the turbulent fluctuations, however, scarcely changes with the imposed field. This last result was predicted previously by Cattaneo & Hughes (1996). In the paper of Cattaneo et al. the CP flow is again employed, but now solutions are sought in a long periodic box in the z-direction (the original flow being independent of z). The length of the box is chosen as 8 times the period of the most rapidly growing mode; the latter then plays the role of fluctuating field, while the mode with the same period as the box plays the role of the large-scale field. Two different case are considered. In one the initial condition has comparable energy in the small and large scales, while in the other the large-scale energy is initially much greater. The final

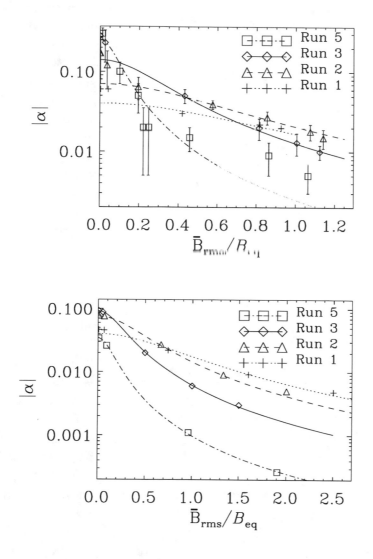

Fig. 10.3. Graphs of various runs from the paper of Brandenburg (2001), showing the reduction of α with increasing $\overline{\mathbf{B}}$. The top figure shows the result of solving a short-time initial value problem, and the bottom figure the value calculated from imposing a uniform field. The results are very similar. The lower curves correspond to greater values of R_m.

state appears to depend on these initial conditions. In the first case the nonlinear interactions between different wavenumbers force rapid growth of the large-scale field, although its natural growth rate is much less than that of the small-scale field, but growth stops when the large-scale field has

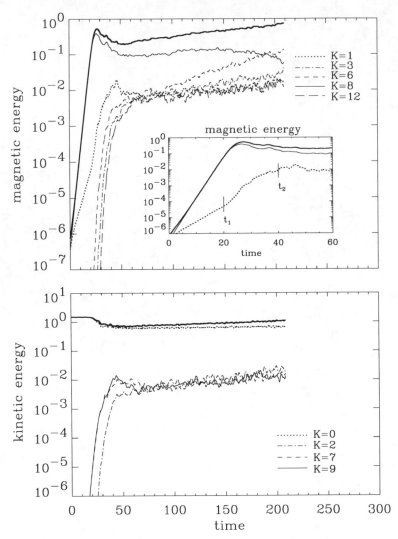

Fig. 10.4. Graphs of runs from the paper of Cattaneo et al. (2002). The different lines refer to modes of different wavenumber. The initial energy of the largest scale ($K = 1$) and the most unstable modes ($K = 8$) are comparable. Growth of the $K = 1$ mode is accelerated above its kinematic rate between times t_1 and t_2.

much lower energy than the short lengthscale mode. There seems to be a further adjustment on a much longer timescale. In the second case the large scale quickly equilibrates, leaving the other scales at lower values. Results are shown in Figures 10.4, 10.5. It turns out that the evolution of the large scale field can be discussed in terms of an α-effect. This is verified by looking at similar short box calculations and evaluating the α-effect as in the earlier

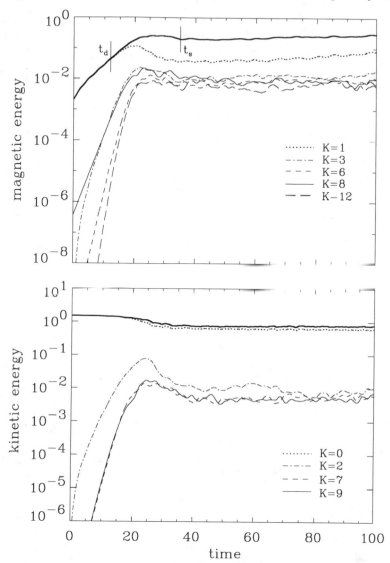

Fig. 10.5. As for the previous figure except that the energy in the largest scale is initially much greater than that in the other scales. The magnetic field becomes dynamically active at time t_d, and nonlinear saturation occurs at time t_s.

paper described above. The two methods give very similar results, justifying the interpretation. It is again found that the process of α-quenching depends strongly on R_m, as indeed does the initial value of α for weak imposed fields. (see Figure 10.6). From the results (Figure 10.7) we can see that α falls to diffusive values while the mean fields are well below equipartition values.

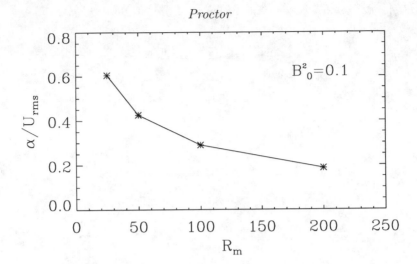

Fig. 10.6. The behaviour of the α coefficient for fixed $\overline{\mathbf{B}}$ as a function of R_m, in the calculation of Cattaneo et al. with comparable initial large and small-scale field. It can be seen that there is a strong reduction as R_m increases.

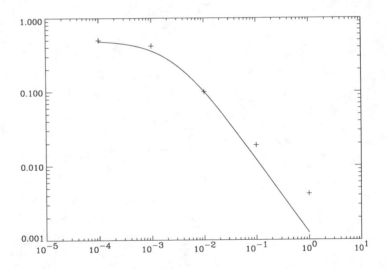

Fig. 10.7. The behaviour of the α coefficient (y-axis) for $R_m = 100$ as a function of $\overline{\mathbf{B}}^2$ (x-axis), in the same calculation as for the previous figure. It can be seen that there is a strong reduction in α when $R_m\overline{\mathbf{B}}^2$ is of order unity.

The conclusion of these studies, that α-quenching is very strong at large values of R_m, of course makes it difficult to see how large-scale fields could arise on other than irrelevant diffusive timescales. A possible chink in the reasoning has been identified by Blackman & Field (2000), who argue that the results depend strongly on the constraint that *small-scale* helicity (and not just total helicity) is conserved. Such conservation is natural in model experiments with periodic boundary conditions, but it may be that with more realistic boundary conditions the separate conservation of small-scale and large-scale magnetic helicity will be destroyed, allowing a decrease of small-scale helicity, which may affect the quenching process. Calculations so far have been inconclusive. Indeed Brandenburg & Dobler (2001), who carried out model calculations with forced helical flows with non-periodic boundary conditions, reported that the crucial value of a was still $\geq 1/2$, with peak fields actually reduced over the periodic case. The results were somewhat model dependent, however, in that while an imposed vertical field boundary condition lowered the mean field over the periodic case, the use of a "potential field" condition could actually increase it. However in the latter case the time taken to reach these larger values increased with R_m. In my view calculations with more realistic boundary conditions are worth pursuing further; nonetheless there are powerful local arguments, not dependent on any conservation law, which support the idea that the dynamo properties of a turbulent flow (which depend very subtly on its Lagrangian structure) are going to be strongly affected when the Lorentz forces become significant on the smallest scales.

10.5 Conclusion

In this short review I have tried to put forward the current state of play regarding the important question of the effects on and by large scale fields of small-scale MHD turbulence. The difficulty in the past has been the misleading prejudices induced by the study of models with small R_m on the smallest scales. Not only do these give a wrong picture of the nature of the α-effect, but they fail to take into account the fact that at large R_m the small-scale flow is likely to be a dynamo in its own right, with effects on dynamo generation for mean fields that are only now becoming apparent. There are many questions that need to be answered before a satisfactory theory can emerge that will account for the observed large-scale solar fields. An important aspect that has yet to receive full attention is the inhomogeneity of the process. This will inevitably lead to large scale fields on dynamic timescales, as foreshadowed by the calculations of Cattaneo et al. (2002).

Acknowledgements I am grateful for discussions over many years with my colleagues Nigel Weiss, Fausto Cattaneo and David Hughes.

References

Blackman, E.G. & Field, G.B. 2000, *ApJ*, **534**, 984

Brandenburg, A., 2001, *ApJ* **550**, 824

Brandenburg, A. & Dobler, W. 2001, *A&A*, **369**, 329

Cattaneo, F., 1999, *ApJ*, **515**, L39

Cattaneo, F. & Hughes, D.W. 1996, *Phys. Rev. E*, **54**, R4532

Cattaneo, F. & Hughes, D.W., 2001, *Astron. Geophys.*, **42**, 3.18

Cattaneo, F., Hughes, D.W. & Kim, E.-J. 1996, *PRL*, **76**, 2057

Cattaneo, F., Hughes, D.W. & Thelen, J.-C. 2002, *JFM*, **456**, 219

Cattaneo, F, Kim, E.-J., Proctor, M.R.E. & Tao, L. 1995, *PRL*, **75**, 1522

Childress,S. & Gilbert, A.D., 1995, *Stretch, Twist, Fold: The Fast Dynamo*, Springer

Galloway, D.J. & Proctor, M.R.E., 1992, *Nature*, **356**, 691

Gruzinov, A.V & Diamond, P.H. 1994, *PRL*, **72**, 1651

Gruzinov, A.V & Diamond, P.H. 1995, *Phys. Plasmas*, **2**, 1941

Knobloch, E., 1994, in Proctor, M.R.E. & Gilbert, A.D., eds, *Lectures on Solar and Planetary Dynamos* (Cambridge University Press), p. 331

Moffatt, H.K., 1978, *Magnetic Field Generation in Electrically Conducting Fluids*, Cambridge University Press.

Parker, E.N., 1955, *ApJ*, **122**, 293

Parker, E.N. 1993, *ApJ*, **408**, 707

Pouquet, A., Frisch, U. & Léorat, J., 1976, *JFM*, **77**, 321

Schekochihin, A., Cowley, S., Maron, J. & Malshkin, L., 2002, *Phys. Rev. E*, 65, 016305

Thelen, J.-C., 1997, *Acta Astron. et Geophys. Univ Comenianae*, **XIX**, 221

Title, A.M., 2000, *Phil. Trans. R. Soc. Lond.*, **358**, 657

Vainshtein, S.I. & Cattaneo, F. 1992, *ApJ*, **393**, 165

Weiss, N.O., 1994, in Proctor, M.R.E. & Gilbert, A.D., eds, *Lectures on Solar and Planetary Dynamos* (Cambridge University Press), p. 59

11

Dynamos in planets

CHRIS JONES

School of Mathematical Sciences
University of Exeter,
Exeter EX4 4QE, UK

Significant advances in our understanding of the geodynamo have been made over the last ten years. In this review, we consider the extent to which this knowledge can be used to understand the origin of the magnetic fields in other planets. Since there is much less observational data available, this requires a 'first principles' understanding of the physics of convection driven dynamos.

11.1 Introduction

The basic structure of the interior of the Earth has been worked out by seismologists. The iron core is divided at $r_{icb} = 1220$ km, the inner core boundary (ICB), into the solid, mainly iron, inner core below and the fluid outer core above. The exact composition of the outer core is not known, but the most plausible models suggest it is a mixture of liquid iron and various impurities, probably sulphur and oxygen (Alfè et al., 2000). The whole core is electrically conducting. Above the core-mantle boundary (CMB), at $r_{cmb} = 3485$ km, lies the rocky mantle. The electrical conductivity of the mantle is very small, except possibly very close to the CMB itself, where iron may have leaked into the mantle. The basic structure of the other terrestrial planets, in which we include the larger satellites, is believed to be similar to that of the Earth, but the size of the iron core varies considerably, and the division between the fluid outer core and the solid inner core, if it exists, has to be computed from theoretical models. The moon's core is probably only 5% of its mass, whereas Mercury's core accounts for nearly 70% of its mass. Except in the case of the Earth, where seismological information is available, the size of a planet's core is inferred from measurements of its density distribution, which in turn is derived from its gravitational field. This can be mapped accurately by space probes.

The geomagnetic field has been recorded at magnetic observatories since the time of Gauss, and excellent data are available for the recent field from magnetic satellite observations. Historical data are also available (e.g. Jackson, Jonkers & Walker, 2000), so that the behaviour of the field over the last few hundred years can be mapped with some accuracy. Although it is not possible to reconstruct the fluid flow beneath the CMB (even if the frozen flux assumption is made, flow locally parallel to the field cannot be detected), it is nevertheless possible to estimate the typical flow velocity at $3 \times 10^{-4} \, \mathrm{m\,s^{-1}}$, and to sketch the main features of the flow (Bloxham & Jackson, 1991). For example, the core fluid flow is primarily westward below the Atlantic, but appears to be significantly smaller below the Pacific, and there are gyres at high latitudes which may be connected with convection rolls (Longbottom et al., 1995). Velocities of this order give a convective turn-over time of order 10^3 years.

Paleomagnetic studies show that the geomagnetic field occasionally reverses polarity, and recently is has become clear (e.g. Gubbins, 1999) that excursions (major changes in the direction of the dipole component), and variations of the strength of the dipole (Valet & Meynardier, 1993), are far more common than full reversals. Data is available on the behaviour during a reversal (e.g. Hoffmann, 2000), which can be compared with theoretical reversal models. Reversal behaviour may be coupled to mantle convection, through an inhomogeneity in the heat flux at the CMB, which can affect core flow and hence the dynamo (e.g. Sarson et al., 1997).

Our knowledge of the physical properties of core material are derived from various sources: experiments such as the diamond anvil technique (Boehler, 1993) can suggest how material at high pressure and temperature behaves. More recently, *ab initio* quantum calculations (e.g. Vocadlo et al., 2000) are coming onstream. This work gives us estimates of the specific heat, the viscosity, and the thermal and electrical conductivities which are of importance in dynamo models. Braginsky & Roberts (1995) give the numerical values of many of the useful physical properties of the Earth's core, and Stevenson's comprehensive (1983) review has many of the corresponding properties in other planets. Another good source of physical data (with references) is Lodders & Fegley (1998).

This observational and experimental information is incorporated into geodynamo models. However, it is only recently that computer hardware powerful enough to solve relevant dynamo equations in reasonably realistic models has been developed. Unfortunately, it is still not possible to solve the equations numerically in the right parameter regime, but only in regimes in which the diffusivity has been artificially enhanced. As we see below, this is not necessarily a fatal objection, but it does mean that great care is needed before such simulations can be realistically applied to other planets.

Before the exploration of our solar system by space probes, our theoretical understanding of planetary interiors was insufficient to allow any worthwhile predictions of planetary magnetic fields. The rich diversity that was discovered came as a surprise. Now that far more information is available, we face the less challenging, but still very difficult, task of trying to fit the data into a coherent theoretical framework. There are many obstacles to this, some arising from lack of data about the physical conditions in planetary cores, and some from our lack of theoretical understanding of convective dynamos. The aim of this review is to explore what can be done, and where the major obstacles to progress lie.

11.2 Planetary magnetic fields

When considering the origin of their magnetic fields, it is natural to divide the planets into three groups. The terrestrial planetary dynamos are those with iron cores, which are either currently liquid, or have been liquid in the past. Such cores have an electrical conductivity σ of $\sim 4 \times 10^5 \, \mathrm{S \, m^{-1}}$, which corresponds to a magnetic diffusivity η of $\sim 2 \, \mathrm{m^2 \, s^{-1}}$. Here $\eta = 1/\sigma\mu$, and as all planetary cores have temperatures well above the Curie point, the permeability μ has its free space value. The second group are the giant planets, Jupiter and Saturn, which are sufficiently massive to have metallic hydrogen cores. The electrical conductivity here is uncertain, but recent data from shock pressure experiments (Nellis, 2000) suggest the magnetic diffusivity is $\eta \sim 1 \, \mathrm{m^2 \, s^{-1}}$ similar to that of the Earth, much higher than the theoretical estimate of Stevenson (1983) which was $\sim 4 \times 10^{-2} \, \mathrm{m^2 \, s^{-1}}$. The third group consists of the outer planets Uranus and Neptune, which are not large enough to have metallic hydrogen cores, but are believed to have cores with of ionic material (Stevenson, 1983, Holme & Bloxham, 1996) with a comparatively low electrical conductivity, corresponding to a magnetic diffusivity $\sim 10^2 \, \mathrm{m^2 \, s^{-1}}$. This considerable diversity in the physical conditions means we must be very wary of simplistic 'universal' laws, such as the magnetic Bode's law, which states that the magnetic dipole moment is proportional to the planet's angular momentum.

In Table 11.1, we list some of the properties of planetary magnetic fields. Magnetic fields are given in Tesla; 1 T $= 10^4$ gauss. The data for Table 11.1 were taken from Lodders & Fegley (1998), who give the references of the source papers. The dipole component of the field is always dominant far from the planet, and hence is the easiest to measure observationally. In the

Table 11.1. *Planetary magnetic fields*

	Dipole moment $\mathrm{T\,m^3}$	Planetary radius $10^6\,\mathrm{m}$	Core radius $10^6\,\mathrm{m}$	Max Field at CMB $10^{-4}\,\mathrm{T}$	Dipole inc. degrees	Rotation rate $10^{-5}\,\mathrm{s^{-1}}$
Mercury	4.3×10^{12}	2.438	1.9	0.014	< 10	0.124
Earth	1.6×10^{16}	6.371	3.48	7.6	11.5	7.3
Ganymede	1.4×10^{13}	2.634	0.48	2.5	~ 10	1.02
Jupiter	1.5×10^{20}	69.95	56	17	9.6	17.6
Saturn	4.2×10^{18}	58.30	32	2.5	0.8	16.4
Uranus	3.8×10^{17}	25.36	18	1.3	58.6	10.1
Neptune	2.0×10^{17}	24.62	20	0.52	47.0	10.8

case of the Earth, the dipole component is still the largest component even when the field is extrapolated down to the core-mantle boundary, but this is not necessarily true of other planetary magnetic fields.

The Earth and Ganymede, the largest of Jupiter's moons, are the only terrestrial planets where there is compelling evidence for a dynamo. Mercury is generally believed to have a dynamo, but its field is weak. It has a large iron core, extending to a radius of 1860 km (Spohn et al., 2001) compared with a total radius of 2440 km. Mercury is also a slow rotator, being locked in a 2:3 resonance with its orbital period.

The magnetic field of Ganymede was discovered by the Galileo mission. Its strength is similar to that of Io, the innermost of the Galilean satellites, but the ambient field of Jupiter at Io is of the same magnitude as the suggested internal field, so it is difficult to establish whether Io has a genuine internal field or not (Kivelson et al., 2001). Ganymede is further from Jupiter, where the jovian field is much weaker, so the internal origin of Ganymede's field is not in doubt. Ganymede was not expected to have a magnetic field, as it was thought to have cooled sufficiently to stop convecting. Tidal interactions may have reheated the planet after formation (Showman et al., 1997).

Rocks from both Mars and the Moon show strong remanent magnetism (see e.g. Stevenson, 2001; Runcorn, 1996). Both these bodies had ancient dynamos, which have now ceased to function. Both planets seem to have had fields of roughly similar strength to the Earth's current field. Venus has no current magnetic field, and the high temperature of its surface makes it unlikely that any remanent magnetism will survive.

The giant planets Jupiter and Saturn both have strong fields; Saturn's field is unusual for its high degree of axisymmetry; in Table 11.1 we note that the inclination of Saturn's dipole field is less than one degree from its rotation axis. A possible theoretical explanation has been given by Stevenson (1982). Jupiter is the only planet apart from the Earth for which we have information about the secular variation (Russell et al., 2001), that is the rate of change of the field. This is very important information, as it enables us to estimate the typical velocity in Jupiter's core, which is $\sim 2 \times 10^{-3}\,\mathrm{m\,s^{-1}}$. This velocity is five orders of magnitude less than the zonal flows in Jupiter's non-magnetic atmosphere, a striking demonstration of the importance of the magnetic field to the dynamics of convection.

The outer planets Uranus and Neptune have fields with exceptionally large inclinations between the magnetic dipole axis and the rotation axis. The fields also contain quadrupole and octupole modes of comparable value near the planetary surface. We argue below that these planets are probably in a different dynamical regime from that of other planetary dynamos.

11.3 Convective driving and thermal history

We shall suppose that all the known planetary dynamos are driven by convection, with the possible exception of a dynamo in Io where Kerswell & Malkus (1998) proposed tidal forcing as the driving. Stevenson (1983) describes a large number of alternative mechanisms proposed to account for planetary magnetism, but none can account for the large fields actually found.

The convection can be either thermally or compositionally driven. Compositional convection can occur when there is an inner core. The light elements mixed with iron in the outer core are released when iron is deposited on the inner core, and this impurity rich buoyant material can stir the core. The amount of buoyancy can be estimated in terms of the rate of growth of the inner core and the density jump between the outer and inner core. Thermal convection comes from various sources, the cooling of the planet, radioactivity in the core, and if there is an inner core, latent heat release at the ICB.

There are major uncertainties hampering the development of thermal history models both for the Earth and for other planets. The first is the uncertainty of the initial condition, that is how much heat the planets contained immediately after formation. The second uncertainty is how much radioactive material there is in the core. The third difficulty is that mantle convection helps determine the rate of cooling (more specifically the heat flux going through the CMB), and mantle convection is a very complex subject (see e.g. Schubert et al., 2001). To estimate the rate of cooling it is necessary to make some assumptions about the Nusselt number-Rayleigh number relation. The rheology of the mantle is uncertain; the viscosity is very temperature dependent, and the behaviour near the Earth's surface is dominated by solid subducting slabs, so it is not even clear that a fluid dynamic model is adequate.

In planetary cores, convection is efficient and the temperature gradient is close to adiabatic and the composition is close to uniform (outside of boundary layers). The adiabatic temperature gradient T_a is given by

$$T_a^{-1} \nabla T_a = \gamma \mathbf{g}/u_P^2 , \qquad \text{where} \qquad \gamma = \alpha u_P^2/c_p \qquad (11.1)$$

is the Grüneisen parameter, for which estimates are available (Merkel et al., 2000). Gravity $\mathbf{g}$ and the sound speed u_P are given in the preliminary reference Earth model PREM (Dziewonski & Anderson, 1981). Here α is the coefficient of thermal expansion and c_p is the specific heat at constant pressure. When solving equation (11.1), the temperature at the CMB must be supplied; in principle it can be found from mantle convection studies. The pressure is found from the hydrostatic equation $\nabla P_a = \rho_a \mathbf{g}$, the density ρ_a being given by PREM. The liquidus temperature at which freezing occurs increases with pressure as we go deeper into the core at a rate faster than the adiabatic temperature increases, so the inner core forms first at the centre of the planet. In principle, when the temperature and pressure are known, the freezing point and hence the location of the inner core is determined. In practice, the location of the Earth's inner core is known from seismology. This is fortunate, because the freezing point of iron is significantly depressed by the impurities in the outer core, and the exact amount of the depression is hard to estimate; for other planets, where seismology is not yet available, we have to rely on theoretical estimates to determine where the inner core lies. For the Earth, it is generally believed that the CMB temperature is at about $4\,000\,\mathrm{K}$ and equation (11.1) then gives the ICB temperature at about $5\,100\,\mathrm{K}$.

When the temperature structure is known, the next step is to use high pressure physics estimates of the thermal conductivity ($45\,\mathrm{W\,m^{-1}\,K^{-1}}$ is a typical value) to find the heat flux conducted down the adiabat. This is comparable to the convected flux. If the Nusselt number Nu is defined as the ratio of conducted to convected flux, then $Nu \sim 1$, very different from solar convection. Conduction down the adiabat generates entropy at a rate $\Sigma \sim 170\,\text{MW}\,\text{K}^{-1}$.

The energy balance is

$$Q_{\text{CMB}} = Q_{\text{ICB}} + Q_{\text{L}} + Q_{\text{S}} + Q_{\text{G}} + Q_{\text{R}} , \qquad (11.2)$$

which relates the heat flux Q_{CMB} coming out of the CMB to the small amount (0.3 TW) coming out of the inner core, Q_{ICB}. Q_{S} is the rate of core cooling, known from the time evolution of equation (11.1), provided T_{CMB} can be found from mantle convection studies. Q_{L} is the latent heat released at the ICB and Q_{G} is the gravitational energy liberated by the central condensation as the inner core grows. These can both be estimated in terms of the rate of growth of the inner core, which in turn depends on how the temperature structure given by equation (11.1) evolves with time. Q_{G} involves the density jump at the ICB. The Earth's inner core density $= 12730\,\text{kg}\,\text{m}^{-3}$ and the fluid outer core density $= 12160\,\text{kg}\,\text{m}^{-3}$ and the difference is the density jump across the core. However, not all this jump releases useful energy for the dynamo. There are two parts contributing (i) due to release of light material, and (ii) due to contraction on solidification. The estimates of Roberts et al. (2002), with the age of the inner core taken as 1.2 Gyrs, suggest that the useful part (i) gives $Q_{\text{G}} \sim 0.5$ TW. This age for the inner core is consistent with the Labrosse et al. (2001) value of 1 ± 0.5 Gyr. The cooling $Q_{\text{S}} \sim 2.3$ TW, and the latent heat $Q_{\text{L}} \sim 4.0$ TW. If Q_{R}, the radioactive term, is zero then we have $Q_{\text{CMB}} \sim 7.1$ TW. However, if there is radioactivity in the core, this estimate of the CMB heat flux could be a serious underestimate. The value of 7.1 TW for CMB heat flux apparently causes no great difficulty for mantle convection models, but the same is true for larger values, too.

The flux conducted down the adiabat near the CMB is around 6 TW using the above estimates, and because the latent heat is released at the ICB, the total heat flux exceeds the conducted flux everywhere, so there is convection throughout the core on this model. However, a rather small reduction in CMB heat flux would change this. If the CMB heat flux is less than 6 TW, the top of the core is subadiabatic. It would still convect through compositional convection, and would still be close to adiabatic, but one would expect convection there to be much less vigorous; this thermally stable layer is Braginsky's 'inverted ocean' (Braginsky, 1993).

Energy balance does not allow us to investigate the amount of dissipation Q_{D}, since the work done by buoyancy cancels the dissipation. We need to consider the entropy balance; following the discussion of Roberts et al.

(2002).

$$\mathcal{Q}_D = \frac{T_D}{T_{CMB}}[(\mathcal{Q}_{ICB} + \mathcal{Q}_L)(1 - \frac{T_{CMB}}{T_{ICB}})$$
$$+ (\mathcal{Q}_S + \mathcal{Q}_R)(1 - \frac{T_{CMB}}{\overline{T}}) + \mathcal{Q}_G - \Sigma T_{CMB}], \qquad (11.3)$$

recalling that Σ is the entropy production due to conduction down the adiabat. $\mathcal{Q}_D$ is almost entirely ohmic dissipation, viscous dissipation being orders of magnitude smaller in the core. T_D is the mean temperature at which the dissipation occurs (effectively where the dynamo operates most strongly), which clearly lies between T_{CMB} and T_{ICB}. $\overline{T}$ is the mean temperature of the outer core. Putting in numerical estimates (Roberts et al., 2002) gives

$$\mathcal{Q}_D = \frac{T_D}{T_{CMB}} (0.5\,\mathrm{TW} + \mathcal{Q}_G + 0.12\mathcal{Q}_R) , \qquad (11.4)$$

indicating that thermal and compositional convection both contribute roughly 0.5 TW, giving a total of around 1 TW to drive the dynamo. We are therefore aiming at finding a dynamo with about 1 TW of ohmic dissipation. This is consistent with the output of current dynamo models. There is still some uncertainty in the above heat flux estimates; for example Lister & Buffett (1995) estimated the conducted flux at the CMB as only 2.7 TW.

Early Earth

There is, however, a serious problem with all the above theory (Roberts et al., 2002). While it can explain the current geodynamo, what was happening before the inner core formed ~ 1.2 Gyr ago? According to the paleomagnetic evidence, the magnetic field dates back to at least 3.5 Gyr. It was suggested (Hale, 1987) that the field strength intensified 2.7 Gyr ago, possibly corresponding to the formation of the inner core. Before inner core formation, the latent heat and the gravitational energy sources are not available, only cooling. The power available for the geodynamo is then much reduced. Even more seriously, most, if not all, of the core would be stably stratified with the above estimate of the cooling term. It is not clear how a dynamo could be sustained under these circumstances.

If we assume the Earth was formed from material with solar abundances, there is a significant depletion of radioactive potassium, ^{40}K, in the mantle. This could either have been lost to space during the Earth's formation, which is the view favoured by most geochemists, or it could be trapped in the core. If it has been trapped in the core, then the CMB heat flux would be much greater, possibly even as much as 20 TW (Roberts et al., 2002), removing the difficulty with the early dynamo. Including radioactivity in the core also has the effect of altering the time at which the inner core formed, generally increasing the age of the inner core (Labrosse et al., 2001). Another possibility is that the primordial heat at formation was very large, and so the rate of cooling, Q_S, is much larger than our 2.3 TW estimate, especially during the time before the inner core formed.

11.4 Physical nature of convective dynamo solutions

A number of research groups have produced three-dimensional numerical solutions of the geodynamo equations, and these have been recently reviewed by Dormy et al. (2001), Jones (2000) and Busse (2000). We shall therefore focus on a few particular issues here.

The geodynamo equations are usually solved in the Boussinesq approximation, which are given in e.g. Jones (2000), although Glatzmaier & Roberts have also solved the anelastic equations, which allow for variations in the properties of the Earth's core (see e.g. Glatzmaier & Roberts, 1997). To avoid complications in what is already a formidable set of equations, only one source of convection (either thermal or compositional) is usually assumed.

The dimensionless parameters that occur in the equations are the Ekman number $E = \nu/2\Omega d^2$ (d is the core radius), the Roberts number $q = \kappa/\eta$, the Rayleigh number Ra and the Prandtl number $Pr = \nu/\kappa$. Here η is the magnetic diffusivity, ν is the kinematic viscosity, and κ is the thermal diffusivity.

In the inner core, the magnetic diffusion equation is solved, and appropriate continuity conditions are applied across the ICB (see e.g. Jones et al., 1995). For the mechanical boundary conditions at the ICB, Glatzmaier & Roberts (1996) used no-slip, while Kuang & Bloxham (1997) used stress-free, arguing that since viscosity is artificially enhanced in the models (see below), stress-free represents the physical situation better. The different assumptions for this boundary condition appear to make a significant difference to the nature of the solutions, but the detailed reasons for this are not yet apparent.

The main problem with geodynamo solutions is that it is not possible to solve the equations in the desired parameter regime. The molecular diffusion coefficients $\kappa \sim 2\times10^{-5}\,\mathrm{m^2\,s^{-1}}$ and $\nu \sim 10^{-6}\,\mathrm{m^2\,s^{-1}}$ lead to very small values of E and q which are numerically impossible to achieve. Even if it is argued that turbulent values of these diffusion coefficients are more appropriate (and then the question of whether isotropic or anisotropic diffusion is appropriate

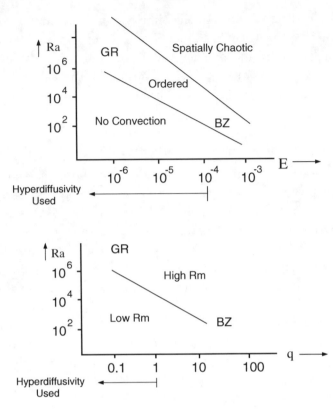

Fig. 11.1. The different regimes of parameter space explored by numerical models. Dynamo action is only possible with high R_m convection. BZ and GR locate typical solutions of the 'Busse-Zhang' and 'Glatzmaier-Roberts' type.

arises; Braginsky & Meytlis, 1990), $E \sim 10^{-9}$ which is still too small to deal with numerically.

In Figure 11.1 (Sarson, 2000), we show a schematic diagram indicating which parts of the parameter space have been explored (see also Busse, 2000). It is not possible to reduce the Ekman number much below 10^{-4} in spherical codes (for plane layer codes we can do better, see below). Hyperdiffusivity (see e.g. Zhang & Jones (1997) for an explanation of what this involves), which enhances diffusion in the latitudinal and azimuthal directions, but not the radial direction, has to be used to explore the low E regime, and this introduces further uncertainties. At, for example, $E \sim 10^{-3}$, dynamos are found at mildly supercritical Rayleigh number provided $q \gg 1$, the 'Busse-Zhang' regime. If $q \sim 1$ the magnetic Reynolds number is too small to give dynamo action. To correct this, we must increase the Rayleigh number, in order to increase the flow velocity. In principle, this should allow us to achieve dynamo action at lower q, but in practice raising Ra at fixed E makes the flow more chaotic and small scale, and no large scale dipole field results. We need to lower E as well as raise Ra in order to keep the flow sufficiently coherent to generate a dipole dominated field. This is the 'Glatzmaier-Roberts' regime (see also Kuang & Bloxham, 1997), but as noted above it can only be found by introducing hyperdiffusion, with its concomitant uncertainties.

There is, therefore, much less freedom to choose the parameters E and q than one would like. We still have to decide on what values of Pr and Ra to choose. Pr only affects the inertial term. For behaviour on a timescale of tens of years and greater, the inertial term is rather small, and its neglect can be formally justified by letting Pr be large. However, on molecular values at least, Pr is *small* not large. If the dynamo is in the correct low E regime, the inertial term will become less important as E is reduced and so the solutions will become *independent* of Prandtl number. Since dynamo codes are not yet run in the low E regime, it is not surprising that authors report significant Prandtl number dependence in their results. Dormy et al. (2001) note that since dynamo codes are not run in the correct regime, great care must be taken in interpreting the results, and in how the dimensionless variables are to be translated back into physical variables. Finally, how is the Rayleigh number to be chosen? This measures the ratio of the superadiabaticity to the diffusion coefficients, and there is no direct method of determining this. Instead, we choose the Rayleigh number so that the heat flux gives the correct value. We show below that this criterion gives us the the typical velocity of the flow, and this is similar to that of the 'westward drift' velocity.

Ohmic Dissipation

Since essentially all the magnetic energy generated by the dynamo ends up as ohmic dissipation, we can test whether our dynamo solutions have a total dissipation comparable with the available energy estimates given in the previous section.

Gubbins (1977) showed that if the field inside the core minimises the dissipation subject to the constraint that the field at the CMB is the observed field, this minimum dissipation is

$$Q_{\min} = \sum_{n=1}^{\infty} q_n \,, \qquad q_n = \frac{\eta r_{\mathrm{cmb}}}{\mu} \frac{(2n+1)(2n+3)}{n} R_n \,, \qquad (11.5)$$

where q_n is the dissipation from the spherical harmonics of order n, and

$$R_n = \left(\frac{r_{earth}}{r_{\text{cmb}}}\right)^{2n+4}(n+1)\sum_{m=0}^{n}\left[(g_n^m)^2+(h_n^m)^2\right] \qquad (11.6)$$

is the Mauersberger-Lowes spectrum extrapolated to the core surface (see e.g., Langel, 1987), and g_n^m and h_n^m are the usual Gauss coefficients (see e.g. Backus et al., 1996). The Mauersberger-Lowes spectrum R_n at the CMB is well-approximated for $n \geq 3$ by

$$R_n = 1.51 \times 10^{-8}\exp(-0.1n)\ \mathrm{T}^2\ , \qquad (11.7)$$

which leads to $\mathcal{Q}_{\min} \sim 44\,\mathrm{MW}$, with the peak dissipation occurring at around $n \sim 12$. This value of n is coincidentally at about the limit of what can be observed, as higher harmonics are obscured by crustal magnetism.

Dynamo models suggest that the actual dissipation $\mathcal{Q}_\mathrm{D} \gg \mathcal{Q}_{\min}$, so $44\,\mathrm{MW}$ is a gross underestimate. The dynamo is very inefficient, in the sense that the actual dissipation is orders of magnitude greater than the minimum necessary dissipation. The reasons are (i) most of the flux generated in dynamo models never leaves the core. The toroidal field generated is necessarily trapped in the core, but models show that only a small fraction of the poloidal field leaves the core to form the observed potential field. (ii) although the field escaping from the core is mostly dipolar in the models, the internal field has a much more complex structure than the very simple structure of the minimising field. So not only is there far more field in the core than is strictly necessary to generate the observed dipole field, its structure is also rather complex.

The upshot is that dynamo models do suggest that the dissipation is of the order of 1 TW, in agreement with the estimates of section 3. Any energy source which falls significantly short of this figure is insufficient. However, it is not yet possible to make very reliable estimates of the dissipation with the current generation of dynamo models, because the dissipation occurs mainly in the range $n \sim 10 - 40$ and this range is affected by hyperdiffusivity. An interesting theoretical question is what is the nature of the power spectrum in dynamo models. Formula (11.7) is empirical, and indeed power law spectra fit just as well. However, a recent simulation (Roberts & Glatzmaier, 2000) was fitted well by the formula

$$R_n = 1.51 \times 10^{-8}\exp(-0.055n)\ \mathrm{T}^2\ , \qquad (11.8)$$

suggesting that an exponential law may have some as yet unknown theoretical basis. The form of this power spectrum has important implications for dynamo theory, because it connects directly with an outstanding problem at the heart of geodynamo theory, the problem of finding an adequate energy source.

11.5 Dynamical regimes in planetary cores

Rotvig & Jones (2002) and Jones & Roberts (2000) have considered plane layer models to gain some understanding of the low E dynamical regime. In this geometry we can no longer compare results with geomagnetic studies, but there are significant computational advantages in Cartesian geometry (the non-existence of useful fast Legendre transforms is the root of the problem for spectral spherical codes). We can get E small enough to get into the correct dynamical regime, where the basic balance of terms is correct. This is signalled by the magnetic field satisfying Taylor's (1963) constraint. When this is achieved, the terms in the equation of motion are in MAC balance (Braginsky, 1967), that is viscous forces and inertial acceleration are negligible, while pressure force, Lorentz force and buoyancy force are all comparable with the Coriolis acceleration.

We therefore have (Starchenko & Jones, 2002)

$$2|\mathbf{\Omega} \times \mathbf{u}| \sim |\nabla p|/\rho \sim |\mathbf{j} \times \mathbf{B}|/\rho \sim g(\alpha T_{\mathrm{a}} S/c_p + \alpha_\xi \xi) , \qquad (11.9)$$

where S is the entropy fluctuation, ξ is the composition fluctuation, ρ is the density and α_ξ is the compositional expansion coefficient, with $\alpha_\xi \approx 0.6$ being typical for terrestrial cores.

This is a completely different balance from that in the solar convection zone, where mixing length balance occurs,

$$|\mathbf{u} \cdot \nabla \mathbf{u}| \sim U_*^2/l \sim g\alpha T_{\mathrm{a}} S/c_p , \qquad (11.10)$$

l being the mixing length and U_* being a typical velocity. In non-magnetic planetary atmospheres a geostrophic balance is common,

$$2\rho \mathbf{\Omega} \times \mathbf{u} \sim -\nabla p , \qquad (11.11)$$

with either the thermal and viscous terms being much smaller. In laboratory convection the motion is on short length scales (tall thin rolls) so that viscous forces can be significant through the particular geometry of the motion.

Taking S_* as a typical entropy fluctuation, and ignoring compositional terms as appropriate for Jupiter,

$$2\Omega U_* \sim g\alpha T_{\mathrm{a}} S_*/c_p . \qquad (11.12)$$

The heat flux equation gives

$$F \sim \rho T_a U_* S_* \sim \frac{Q}{4\pi r_{\text{cmb}}^2} \,. \tag{11.13}$$

Eliminating the entropy fluctuation S_*,

$$U_* \sim \left[\frac{g\alpha r_{\text{cmb}} Q}{M\Omega c_p} \right]^{\frac{1}{2}} \,. \tag{11.14}$$

Putting in the standard estimates for thermodynamic quantities, we obtain for Jupiter, $U_* \sim 2 \times 10^{-3}\,\text{m s}^{-1}$. For the Earth, we can form the mass flux equation analogous to the heat flux equation, and we obtain (see Starchenko & Jones, 2002 for details) $U_* \sim 2 \times 10^{-4}\,\text{m s}^{-1}$.

These estimates are in good agreement with velocities inferred from measurements of the secular variation (Bloxham & Jackson, 1991) for the Earth, Russell et al. (2001) for Jupiter. This agreement provides useful evidence that MAC balance does operate in the cores of these two planets. For example, if the mixing length balance, equation (11.10), is used in place of equation (11.9), the typical velocity is orders of magnitude too large.

The typical magnetic field can be estimated from

$$2\rho\Omega U_* \sim |\mathbf{j} \times \mathbf{B}| \sim B_*^2/\mu r_* \,, \tag{11.15}$$

where r_* is lengthscale of the variation of the field, $|\mathbf{B}|/|\nabla \times \mathbf{B}|$. Eliminating U_*,

$$B_* \sim \left[\frac{g\alpha\rho_*^2\mu^2 r_*^2 r_{\text{cmb}} Q\Omega}{Mc_p} \right]^{\frac{1}{4}} \,. \tag{11.16}$$

How do we choose r_*? Unfortunately, this is not at all clear. Studies of flux ropes (Galloway, Proctor & Weiss, 1978) suggest $r_* \sim R_m^{-1/2}$. Numerical simulations at $R_m \sim O(10^2)$ suggest $r_* \sim d/50$, where $d = r_{\text{cmb}} - r_{\text{icb}}$. Magnetic field saturates when the stretching properties of the flow are inhibited by Lorentz force. Dynamic simulations suggest that flux ropes thicken in the fully nonlinear regime: it might therefore be that r_* eventually becomes independent of R_m at large R_m. For the Earth, it is reasonable to take $r_* \sim d/50$ as suggested by the simulations with R_m close to its terrestrial value, and we obtain $B_* \sim 0.5 \times 10^{-2}$ T, about ten times the dipole field extrapolated to the CMB (Table 11.1), a reasonable value consistent with numerical models, which indicate that about 10% of the core field escapes through the CMB.

Interestingly, if we use $r_* \sim d/50$ for Jupiter, we get a core field of $B_* \sim 2 \times 10^{-2}$ T which is reasonable if about 10% of the core field escapes to form the observed dipole. If we assume that $r_* \sim R_m^{-1/2}$, this gives a smaller value of B_* (smaller than the observed field), because Jupiter has a higher magnetic Reynolds number.

For moderate R_m, geodynamo models typically give for the Elsasser number Λ

$$\Lambda = \frac{B_*^2}{2\mu\rho\Omega\eta} \sim 4 \Rightarrow r_* = 4dR_m^{-1} \sim d/50 \;, \qquad (11.17)$$

with a typically Earth-like value of $R_m \sim 200$. For Jupiter's metallic hydrogen core, we expect $R_m \sim 10^4$, and $\Lambda \sim 20$. A possible problem with this is that magnetic instabilities may occur when Λ is this big (e.g. Zhang, 1995). One possibility is that Jupiter's dynamo is located not deep inside the conducting core, but at the interface of the metallic hydrogen core and the molecular atmosphere. Since it is likely that the electrical conductivity goes smoothly to zero with distance from the centre (Kirk & Stevenson, 1987), there must be a zone where the Elsasser number and the magnetic Reynolds number assume Earth-like values, and this may be a promising location for the jovian dynamo.

Saturn may be driven by compositional as well as thermal convection, Stevenson (1982), and the uncertainty in the core energy fluxes means that typical velocities and field strengths are also uncertain. We can only compare with the field strength, and this suggests that Saturn, like Jupiter, is probably in MAC balance. A similar uncertainty holds for Ganymede; as mentioned above, the thermal history, and consequently the current core heat flux, is unknown.

Mercury's small size makes it likely that inner core nucleation started early, so that by now a large solid inner core probably exists (Stevenson et al., 1983). The sulphur (and other impurities) present depress the freezing point, and as the inner core grows, the relative fraction of impurity rises in the fluid left, so it is difficult to freeze the core entirely. The fluid outer core is therefore probably only a thin shell; the thickness of this shell depends on the (unknown) initial sulphur concentration, but values of ~ 100 km to ~ 500 km are plausible. The thermal stratification in the liquid iron core is likely to be stable (Stevenson et al., 1983) but gravitational energy $Q_G \sim 6 \times 10^9 (d/100)^2$ W, where d is the outer core thickness in km, is available from compositional convection (Stevenson, 1987). The rotation rate of Mercury is slow, and the core magnetic field is much weaker than that in other planets. These facts may well be related. If we take the shell thickness $d = r_{\text{cmb}} - r_{\text{icb}} \sim 100$ km, we obtain from MAC balance $U \sim 3 \times 10^{-3}$ m s^{-1} which implies a magnetic Reynolds number (based on length d) of over 100. To obtain the observed field strength, however, we need a rather small $r_* < 1$ km. It may be that Mercury's field has a relatively smaller dipole component than the Earth, so the core field may be more than ten times the escaping observed field. This could come about because the slower rotation and relatively stronger driving may impose less order on the flow (see Figure 11.1).

The planets Uranus and Neptune present a different problem. The heat flux coming out of Uranus is about 3×10^{14} W and for Neptune 3×10^{15} W. If we assume MAC balance, we have a typical velocity of 2×10^{-4} m s^{-1} for Uranus, and about 3 times larger for Neptune. With the large diffusivity $\eta \sim 10^2$ m^2 s^{-1}, we have for Uranus a magnetic Reynolds number of only about 40, which is probably too small to sustain dynamo action; Neptune is marginal. MAC balance is therefore probably not possible in these planets. Holme & Bloxham (1996) also point out that the ohmic dissipation associated with the observed fields would be larger than the total heat flux if the core field is significantly larger than the observed field.

The absence of a dynamo in Venus is also of interest. Even with its slow rotation rate, Coriolis acceleration should still be important in the core, and dynamo action should result. Stevenson et al. (1983) suggested that slower cooling due to the high surface temperature meant that Venus' inner core had not yet formed, depriving it of much of its energy source. However, the very slow rotation must be important, because the Earth apparently maintained a dynamo before its inner core formed.

11.6 Conclusions

Although numerical convection driven geodynamo models have not yet reached the parameter regime relevant to the Earth, they are probably now not far off entering the MAC balance regime. The analysis of section 5 suggests that provided the heat flux, the rotation rate and the thermodynamic quantities are specified correctly, the typical velocity, and hence the magnetic Reynolds number, will then be predicted correctly. The magnetic field strength is harder to get right, because magnetic saturation is not so well understood, and is probably model dependent. A possible way forward is to try to improve our understanding of the magnetic energy spectrum and how it relates to ohmic dissipation.

The problem of planetary magnetic fields is closely linked to the thermal history of the planets, because this determines the amount of available thermal and gravitational energy. It also determines the point at which the core becomes stably stratified, which is most likely why Mars and the Moon have ceased to be dynamos. A serious problem is posed by recent thermal history calculations, which indicate that the solid inner core is only around 1.2 Gyr old. Current models of the Earth's thermal history do not seem to provide enough heat to drive the dynamo before this time, despite evidence for a geomagnetic field dating back 3.5 Gyr. Either more primordial heat is required, or significant radioactivity was present in the core 1.2 Gyr ago.

The uncertainties about the thermal history of the planets are even greater, and so it is not realistic to hope for a theory that can predict planetary magnetic fields solely from the physical data. Instead, we must try to piece together the thermal history of the planets using the data provided by magnetic field observations and our gradually improving understanding of dynamo theory. Our ability to build dynamo models is now getting to the point where applying the knowledge obtained from geodynamo studies to other planets, and perhaps even to extra-solar planets, could be fruitful.

References

Alfè, D., Gillan, M. J. & Price, G. D., 2000, *Nature*, **405**, 172.

Backus, G., Parker, R. & Constable, C., 1996, *Foundations of Geomagnetism*, Cambridge University Press.

Bloxham, J. & Jackson, A., 1991, *Rev. Geophys.*, **29**, 97.

Boehler, R., 1993, *Nature*, **363**, 534.

Braginsky, S. I., 1967, *Geomagn. Aeron.*, **7**, 1050.

Braginsky, S. I. & Meytlis, V. P., 1990, *GAFD*, **55**, 71.

Braginsky, S. I., 1993, *J. Geomagn. Geoelectr.*, **45**, 1517.

Braginsky, S. & Roberts, P. H., 1995, *GAFD*, **79**, 1.

Busse, F. H., 2000, *Ann. Rev. Fluid Mech.*, **32**, 383.

Dormy, E., Valet, J-P. & Courtillot, V., 2000, *Geochem. Geophys. Geosyst*, **1**, paper number 2000GC000062.

Dziewonski, A. M. & Anderson, D. L., 1981, *Phys. Earth Planet. Inter.*, **25**, 297.

Galloway, D. J., Proctor, M. R. E. & Weiss, N. O., 1978, *JFM*, **87**, 243.

Glatzmaier, G. A. & Roberts, P. H., 1996, *Science*, **274**, 1887.

Glatzmaier, G. A. & Roberts, P. H., 1997, *Contemp. Phys.*, **38**, 269.

Gubbins, D., 1977, *J. Geophys.*, **43**, 453.

Gubbins, D., 1999, *Geophys. J. Int.*, **137**, F1.

Hale, C. J., 1987, *Nature*, **329**, 233.

Hoffman, K. A., 2000, *Phil. Trans. R. Soc. Lond. A*, **358**, 1181.

Holme, R. & Bloxham, J., 1996, *J. Geophys. Res. - Space Physics*, **101**, 2177.

Jackson, A., Jonkers, A. R. T. & Walker, M., 2000, *Phil. Trans. R. Soc. Lond. A*, **358**, 957.

Jones, C. A., 2000, *Phil. Trans. R. Soc. Lond. A.*, **358**, 873.

Jones, C. A., Longbottom, A. W. & Hollerbach, R., 1995, *Phys. Earth Planet. Inter.*, **92**, 119.

Jones, C. A. & Roberts, P. H., 2000, *JFM*, **404**, 311.

Kerswell, R. R. & Malkus, W. V. R., 1998, *Geophys. Res. Lett.*, **25**, 603.

Kirk, R. L. & Stevenson, D. J., 1987, *ApJ*, **316**, 836.

Kivelson, M. G., Khurana, K. K., Russell, C. T., Joy, S. P., Volwerk, M., Walker,

R. J., Zimmer, C., Linker, J. A., 2001, *J. Geophys. Res. - Space Physics*, **106**, 26121.

Kuang, W. & Bloxham, J., 1997, *Nature*, **389**, 371.

Labrosse, S., Poirier, J-P. & Lemouel, J-L., 2001, *Earth Planet. Sci. Lett.*, **190**, 111.

Langel, R. A., 1987, in Jacobs, J. A., ed., *Geomagnetism* 1, Academic Press, New York, p. 249.

Lister, J. R. & Buffett, B. A., 1995, *Phys. Earth Planet. Interiors*, **91**, 17.

Lodders, K. & Fegley, B., 1998, *The Planetary Scientist's Companion*. Oxford University Press.

Longbottom, A. W., Jones, C. A. & Hollerbach, R. 1995, *GAFD*, **80**, 205.

Merkel, S., Goncharov, A. F., Mao, H.-K., Gillet, Ph. & Hemley, R. J., 2000, *Science*, **288**, 1626.

Nellis, W.J., 2000, *Planet. Space Sci.*, **48**, 671.

Roberts, P. H. & Glatzmaier, G. A., 2000, *Phil. Trans. R. Soc. Lond. A.*, **358**, 1109.

Roberts, P. H., Jones, C. A. & Calderwood, A., 2002, in Soward, A. M., Jones, C. A. & Zhang, K., eds, *Earth's Core and Lower Mantle*, Taylor and Francis, in the press.

Rotvig, J. & Jones, C. A., 2002, *Phys. Rev. E.*, in press.

Runcorn, S. K., 1996, *Geochim. Cosmochim. Acta*, **60**, 1205.

Russell, C. T., Yu. Z. J. & Kivelson, M. G., 2001, *Geophys. Res. Lett.*, **28**, 1911.

Sarson, G. R., 2000, *Phil. Trans. R. Soc. Lond. A.*, **358**, 921.

Sarson, G. R., Jones, C. A. & Longbottom, A. W., 1997, *Phys. Earth Planet. Interiors*, **101**, 13.

Schubert, G., Turcotte, D. P., & Olson, P., 2001, *Mantle Convection*. Cambridge University Press.

Showman, A. P., Stevenson, D. J. & Malhotra, R., 1997, *Icarus*, **129**, 367.

Spohn, T., Ball, A. J., Seiferlin, K., Conzelmann, V., Hagermann, A., Komle, N. I. & Kargl, G., 2001, *Planet. Space Sci.*, **49**, 1571.

Starchenko, S. & Jones, C. A., 2002, *Icarus*, **157**, 426.

Stevenson, D. J., 1982, *GAFD*, **21**, 113.

Stevenson, D. J., 1983, *Rep. Prog. Phys.*, **46**, 555.

Stevenson, D. J., 1987, *Earth Planet. Sci. Lett.*, **82**, 114.

Stevenson, D. J., 2001, *Nature*, **412**, 214.

Stevenson, D. J., Spohn, T. & Schubert, G., 1983, *Icarus*, **54**, 466.

Taylor, J. B., 1963, *Proc. R. Soc. Lond. A*, **274**, 274.

Valet, J. P. & Meynardier, L., 1993, *Nature*, **366**, 234.

Vocadlo, L., Brodholt, J., Alfè, D., Gillan, M. J. & Price, G. D., 2000, *Phys. Earth Planet. Interiors*, **117**, 123.

Zhang, K., 1995, *Proc. R. Soc. Lond. A*, **448**, 245.

Zhang, K. & Jones, C. A., 1997, *Geophys. Res. Lett.*, **24**, 2869.

III Physics and structure of stellar interiors

12

Solar constraints on the equation of state

WERNER DÄPPEN

*Department of Physics and Astronomy,
University of Southern California,
Los Angeles, CA 90089-1342, USA*

*Helioseismology has become a very successful diagnosis of the equation of
state of the plasma of the solar interior. Although the gas in the solar interior is only weakly coupled and weakly degenerate, the great observational
accuracy of the helioseismological measurements puts strong constraints on
the nonideal part of the equation of state. The helioseismic verification of
major nonideal effects in the equation of state of solar matter has become
well established. The dominant contribution is the Coulomb pressure, con-
ventionally described in the Debye-Hückel approximation. However, in the
last years, the increased precision of the helioseismic diagnosis has brought
significant observational progress beyond the Debye-Hückel approximation.
The helioseismic detection of a signature of relativistic electrons was a striking example. Very recently, effects of the excited states of the atoms and ions
of heavy elements were discovered, which have a promising potential both for
statistical mechanics and solar physics, in particular, the helioseismic determination of the heavy-element abundance.*

12.1 Introduction

Precise measurements of solar oscillation frequencies provide data for accurate inversions for the sound speed in the solar interior. Except in the
very outer layers, the stratification of the convection zone is almost adiabatic and the Reynolds stresses are negligible. The sound-speed profile is
governed principally by the specific entropy, the chemical composition and
the equation of state, and it is therefore essentially independent of the uncertainties in the radiative opacities. The inversions thus reveal, via minute
variations in the adiabatic exponent of the solar plasma, physical processes
that have just a small influence on the equation of state.

Although a simple ideal-gas model of the plasma of the solar interior was adequate before helioseismology, helioseismic equation-of-state analyses require more sophisticated physical models. The need to go beyond the ideal-gas approximation for helioseismic applications had been recognized in the early 1980s (e.g. Berthomieu et al. 1980, Ulrich 1982, Noels et al. 1984). With the better data available towards the end of the 1980s, a clearer picture began to emerge. Christensen-Dalsgaard et al. (1988) demonstrated that it was essential to include the leading Coulomb correction. The Coulomb correction is due to the sum of all pair interactions between charged particles (electrons, nuclei and compound ions); together they lead to a screening of the Coulomb potential and to a negative pressure correction with respect to the ideal-gas value. The leading-order Debye-Hückel (DH) theory is a good approximation for solar conditions.

Helioseismic equation-of-state studies typically use solar models based on sophisticated new equations of state, in particular, the ones underlying the two ongoing major opacity recomputation efforts. One of these efforts is the international Opacity Project (OP; see the books by Seaton 1995, Berrington 1997); it contains the so-called Mihalas-Hummer-Däppen equation of state (Hummer & Mihalas 1988; Mihalas, Däppen & Hummer 1988; Däppen et al. 1988; hereinafter MHD) and it deals with *heuristic* concepts about the modification of atoms and ions in a plasma. The other effort is being pursued at Lawrence Livermore National Laboratory by the OPAL group (Iglesias & Rogers 1996; Rogers, Swenson & Iglesias 1996); its equation of state is based on a detailed *systematic* method to include density effects in a plasma.

Specific reviews address the helioseismic determination of the equation of state (e.g., Christensen-Dalsgaard & Däppen 1992; Baturin et al. 2000). The article by Christensen-Dalsgaard et al. (2000) contains a significant part dedicated to the helioseismic equation-of-state diagnosis, too. Finally, there is an article (Däppen & Guzik 2000), which contains a practical compilation of available stellar equations of state (and opacities), including their detailed specifications, and information about how to obtain them.

Although approximate asymptotic techniques (see Christensen-Dalsgaard et al. 1985; Gough 1993) exist to invert solar oscillation frequencies for the internal sound speed, for an accurate analysis of the observations, a fully-fledged, non-asymptotic numerical treatment of the oscillations is mandatory (see Gough et al. 1996). Figure 12.1 is a typical result of such a numerical inversion (Basu & Christensen-Dalsgaard 1997). It shows the relative difference (in the sense sun – model) between the squared sound speed obtained from inversion of oscillation data and that of a two standard solar models. The two solar models used are identical in all respects except for their equation of state, MHD (circles) and OPAL (triangles), respectively.

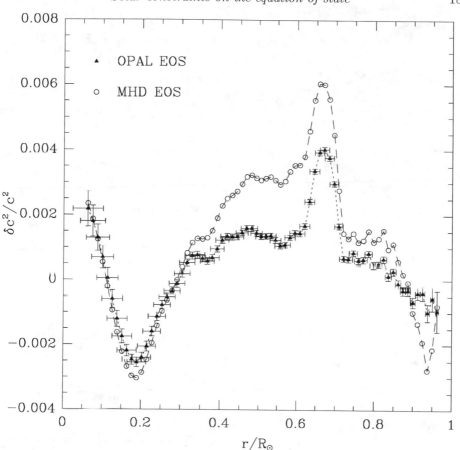

Fig. 12.1. Difference between squared sound speed from inversion and two solar models. Figure by S. Basu.

For the present purpose, we can consider inversion results such as Figure 12.1 as the *data* of helioseismology, disregarding the procedure through which they were actually obtained from solar oscillation frequencies. It follows from Figure 12.1 that in most parts of the sun the OPAL equation of state is a better approximation to reality than MHD, but OPAL needs to be improved as well. In the outermost layers of the sun, however, the general trend might be reversed (Section 12.2.3).

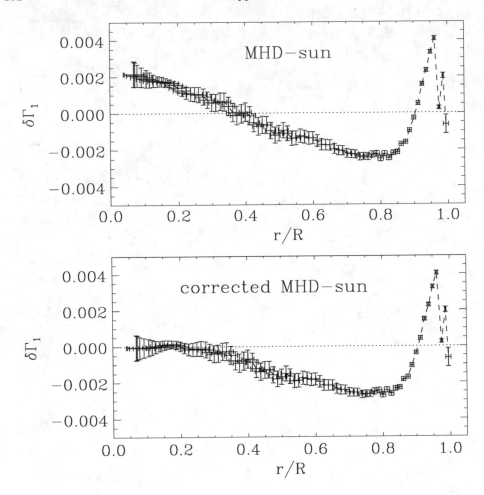

Fig. 12.2. Difference between γ_1 (here denoted Γ_1) of a solar model and observation, for models with nonrelativistic electrons (top) and relativistic electrons (bottom). Figures by J. Elliot.

12.2 Equation of state issues

12.2.1 *Coulomb correction*

As mentioned above, the most important result from earlier helioseismic equation-of-state analyses was that it is essential to add the leading Coulomb correction (Debye-Hückel term) to ideal-gas thermodynamics (Christensen-Dalsgaard et al. 1988, 1996). The relative Coulomb pressure correction peaks in the outer part of the convection zone (about –8 per cent) and in the solar core (about –1 per cent). Both MHD and OPAL contain the Debye-Hückel correction; the good agreement seen in Figure 12.1 would be an order of magnitude less without it (Christensen-Dalsgaard et al. 1996). The discrepancy between theory and present observations is clearly much smaller than the Debye-Hückel correction.

12.2.2 Relativistic electrons

The strong constraints from helioseismology revealed the influence of *relativistic* electrons. The original versions of MHD and OPAL had not included relativistic electrons (although both did include degeneracy). In a recent helioscismic inversion for the adiabatic gradient $\gamma_1 = (\partial \ln p / \partial \ln \rho)_s$ (s being specific entropy), Elliot and Kosovichev (1998) found a discrepancy between, on the one hand, the observed structure of the sun, and, on the other hand, models using the OPAL or MHD equation of state.

The top panel of Figure 12.2 shows this discrepancy for MHD. The relevant deviation occurs in the central 30% parts of the sun. (A corresponding figure for OPAL would look essentially the same.) A *relativistic* treatment of the degenerate electrons in the solar model (bottom panel) removes the discrepancy nicely. As a result, both MHD (Gong et al. 2001b; Gong et al. 2001c) and OPAL (Rogers, *private communication*) were since upgraded to include relativistic electrons.

12.2.3 Effect of excited states in hydrogen and helium

Another effect beyond the Debye-Hückel correction is the signature of the internal partition functions. Nayfonov and Däppen (1998) discovered a "wiggle" in the thermodynamic quantities, located in the hydrogen and helium ionization zones. This effect, due to excited states, has probably already been observed in the sun, because new observations (Basu et al. 1999) suggest that in the top 2% of the solar radius, MHD models can give a more accurate match with the data than OPAL models. Since it turns out that in this region, the discrepancy between MHD and OPAL is essentially reflected by the aforementioned wiggle (Nayfonov & Däppen 1998), the result of the inversion (Basu et al. 1999) could mean a validation of an MHD-like treatment (Hummer & Mihalas 1988) of exited states.

The main result of Basu et al. (1999) is shown in Figure 12.3. It is the result from an inversion of observed solar oscillation frequencies for the *intrinsic* γ_1 difference between the sun and a solar model. The intrinsic difference is that part of the γ_1 difference which is due to the difference in the equation of state itself; there is a further component to the γ_1 difference caused by the change to the structure of the solar model resulting from

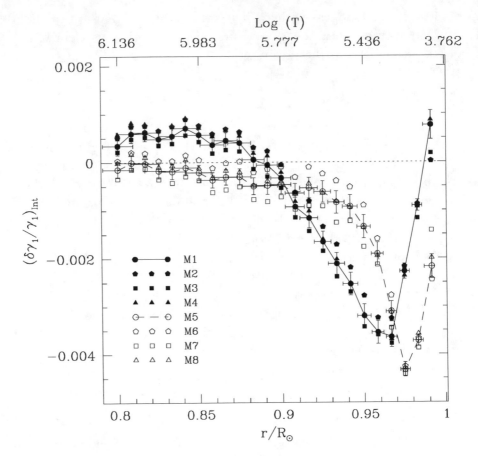

Fig. 12.3. Intrinsic difference between γ_1 obtained from an inversion of helioseismological data (Basu et al. 1999), and γ_1 of 4 MHD models (M1-4: filled points) and 4 OPAL models (M5-8: empty points), respectively. All results are in the sense "sun – model" (for more detail, see Basu et al. 1999).

the difference in the equation of state (Basu & Christensen-Dalsgaard 1997). The error bars shown in Figure 12.3 are based on combined errors of the inversion method and observational errors.

First, Figure 12.3 reveals the accuracy of present-day helioseismology, which has attained almost 10^{-4} for γ_1. Second, Figure 12.3 reveals the influence of the equation of state. The method is by an analysis of the difference between the solar values obtained from inversions and the ones computed in reference models (standard solar models). Since in Figure 12.3 two sets of reference models (MHD and OPAL) are compared with the real solar data, whereas all other model ingredients are kept the same, the figure not only reflects the difference between the solar models and the sun but also the difference between the MHD and OPAL models themselves.

Figure 12.3 should not be over-interpreted, however, because present uncertainties in the inversion of the upper layers of the sun (e.g., turbulent pressure, magnetic fields, nonlocal thermodynamic effect due to radiation, uncertainties in the chemical composition) preclude so far a definitive interpretation, and further clarifying work is in progress. In the slightly deeper regions (below a depth of about 3% of the solar radius) the findings of the study (Basu et al. 1999) are more reliable, and they confirm the findings of Figure 12.1, that is, overall OPAL is a better equation of state than MHD.

However, should the results in the top 2% of the sun remain in favor of MHD, they would demonstrate the significance of the different implementations of many-body interactions in the two formalisms. In principle, since density decreases in the upper part, OPAL, by its nature of a systematic expansion, inevitably becomes itself more accurate; but MHD might, by its heuristic approach (and by luck!), have incorporated even finer, higher-order effects. Since helioseismology gives localized information, it is natural that the various equations of state have their preferred regions in the sun. One should, however, resist the temptation to produce a "combined" solar equation of state, with different pieces for different parts of the sun. Such a hybrid solution is fraught with danger, because patching equations of state together can introduce spurious effects (Däppen et al. 1993). It is better to seek an improvement of individual equations of state, such as MHD and OPAL, in parallel and independently, guided by the progress of helioseismology.

12.2.4 Heavy elements

In solar modeling, an adequate treatment of the heavy elements and their excited states is important. The treatment is subject to the stringent requirements of helioseismology. In a recent analysis (Gong et al. 2001a), the contribution of various heavy elements in a set of thermodynamic quantities was examined.

To show specifically the contribution of each individual heavy element, Gong et al. (2001a) have calculated solar models with particular mixtures. In each case, hydrogen and helium abundances were fixed with mass fractions $X = 0.70$ and $Y = 0.28$; the remaining 2% heavy-element contribution was topped off by only one element, carbon, nitrogen, oxygen and neon, respectively. Gong et al. (2001a) then compared these models with a pure hydrogen-helium mixture (such a mixture is obtained by replacing the two per cent reserved for heavy elements with additional helium). The expecta-

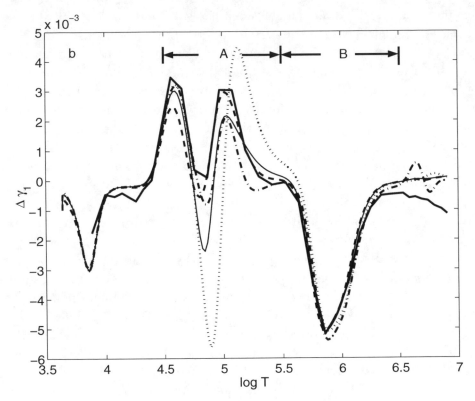

Fig. 12.4. Difference in thermodynamic quantities between the H-He-C mixture and the H-He mixture for various equations of state. Differences are in the sense $[X(Model)_{H-He-C} - X(Model)_{H-He}]$. Line styles are: Dashed line: MHD; Thick Solid line: OPAL; Dashed-dotted line: CEFF; Dotted line: SIREFF; Thin Solid line: MHD$_{GS}$. See Gong et al. (2001a) for more details.

tion is that the biggest deviations between these special models, and the one with the complete heavy-element mixture, will result (i) from the change in the total number of particles per unit volume and (ii) from the different ionization potentials of the respective elements. Because the solar plasma is only slightly non-ideal, the leading pressure term is still given by the ideal-gas equation $pV = Nk_BT$, with p standing for pressure, N the number of particles, and k_B the Boltzmann constant. Because of their higher mass, the number of heavy-element atoms is obviously smaller than that of helium atoms representing the same mass fraction. However, against this reduction in total number there is an offset due to the ionization of their larger number of electrons which becomes stronger at higher temperatures. The resulting net change of the total number of particles is therefore a combination of these two effects. Details can be found in Gong et al. (2001a).

A further comparison done by Gong et al. (2001a) had the purpose to disentangle even further the contribution of the mere presence of each heavy element from the more subtle influence of different physical formalisms. In Figure 12.4 various H-He-C models (with mass fractions 0.70:0.28:0.02) are compared with the analogous H-He models (0.70:0.30) for a set of different equations of state, respectively. In addition to OPAL and MHD, three other equations of state were considered. First, MHD_{GS}, which is MHD but with partition functions truncated to ground states; second CEFF, which is the Eggleton, Faulkner & Flannery (EFF) (1973) formalism with an added Coulomb term in the Debye-Hückel approximation (Christensen-Dalsgaard & Däppen 1992); third, SIREFF, another EFF sibling (Guzik & Swenson 1997), which includes hydrogen molecules and mimics OPAL in some respects.

Comparing H-He-C with H-He models allows one to eliminate most of the difference due to the treatment of hydrogen and helium in each individual equation of state. The comparison therefore isolates the behavior of carbon in each of our equations of state. (The slight roughness of the OPAL curve in Figure 12.4 is due to the fact that we have almost reached the limit interpolation accuracy of the available OPAL tables.)

From Figure 12.4, which shows $\Delta\gamma_1$ graph between H-He-C and H-He, it follows that γ_1 reveals the biggest differences between the various equations of state in the temperature range of $4.5 < \log T < 5.5$ (named region "A" in the figure). Common features can be identified however, when $\log T > 5.5$ and $\log T < 4.5$.

The bump in $\Delta\gamma_1$ in the region "B" of Figure 12.4 is totally independent of the equation of state used. Gong et al. (2001a) showed that the bump feature is likely due to the ionization of carbon at that temperature in general, and importantly, quite independent of details in the equation of state. The strength and robustness of this profile and its relative independence on the equation of state qualify it ideally for a helioseismic heavy-element abundance determination. This is true because the profile does not only appear for carbon, but also for all other heavy elements which exhibit a similar feature independent of the equation of state.

12.3 Resolution power of helioseismology

Admittedly, the aforementioned fine effects of heavy elements and excited states are small: they influence γ_1 by a few 10^{-3}. However, as shown before, present-day helioseismology has already sufficient resolving power to study

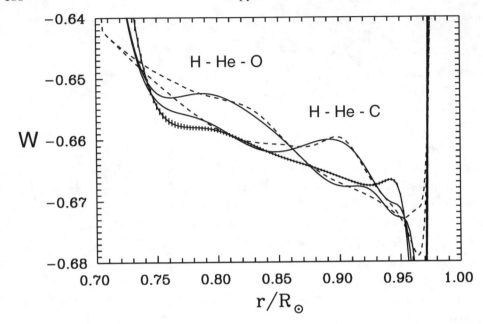

Fig. 12.5. The quantity W=(dc²/dr)/g(r) obtained from an inversion of solar p-mode frequencies and three solar models with different heavy-element abundances (Baturin et al. 2000). All 3 models include Debye-Hückel screening. See text for further details.

them. This is further demonstrated in Figure 12.5 taken from the work of Baturin et al. (2000). The figure shows the result of an *asymptotic* inversion introduced by Gough (1984) for the helioseismic helium abundance determination. The inversion is for the so-called quantity W, which is a function of solar structure

$$W = \frac{dc^2}{dr} \frac{1}{g(r)} .$$

(12.1)

Here, $g(r)$ is the local gravitational acceleration at position r in the sun. The quantity W is useful because of its equality with a purely thermodynamic quantity if the stratification is assumed to be perfectly adiabatic (otherwise, the equality is violated by the amount of the non-adiabaticity). For adiabatic stratification, the following relation holds (Gough 1984)

$$W = \frac{1 - \gamma_1 - \gamma_{1,\rho}}{1 - \gamma_{1,c^2}} ,$$

(12.2)

with the derivatives

$$\gamma_{1,\rho} = \left(\frac{\partial \ln \gamma_1}{\partial \ln \rho}\right)_{c^2}, \gamma_{1,c^2} = \left(\frac{\partial \ln \gamma_1}{\partial \ln c^2}\right)_{\rho}. \tag{12.3}$$

In Figure 12.5, the curve computed from inversion of observational data is marked with vertical bars (estimated inversion error). The pair of solid and dashed lines that follows the observational W most closely is from a model with the usual solar composition (dashed line: W computed directly from the solar model; solid line: result of the same inversion procedure as for the observational curve, but with artificial mode frequencies from the solar model). The two other pairs of solid and dashed lines ("H-He-C" and "H-He-O") are from models with a heavy-element composition of pure carbon and oxygen, respectively, of an amount equal to the total solar heavy-element abundance. Solid and dashed lines have the same meaning as before. In contrast to Figure 12.3, the inversions of Figure 12.5 are absolute (that is, not with respect to a reference model).

It is not surprising that since the quantity W involves *derivatives* of γ_1, it is even more sensitive to equation-of-state effects than γ_1 itself. That this is indeed the case is clearly reflected in Figure 12.5, where the two models, which are, after all, very close to each other, lead to values of W which differ by an amount far larger than the accuracy of the inversion (indicated by error bars). Figure 12.5 convincingly demonstrates that even small and subtle effects, such as those due to excited states of atoms and ions of heavy elements, are already well within the reach of present-day observational accuracy.

12.4 Conclusions

The increasing precision of the helioseismic diagnosis has brought significant observational progress beyond the Debye-Hückel approximation. For instance, evidence for relativistic effects were found, despite the fact that, even in the solar center, electrons are largely nonrelativistic. Other detectable subtle plasma effects concern the excited states of bound systems, hydrogen especially, but also other atoms and ions. Since helioseismology gives localized information, it is natural that the various equations of state have their preferred regions in the sun. As predicted with remarkable foresight by Gough (1984), this delocalization allows us to obtain information about the physical theories in one part of the sun, to be used in other locations for purposes such as the determination of the chemical composition. In large parts of the solar convection zone, the thermodynamic quantities are very sensitive to the detailed physical treatment of heavy elements, in particular, regarding the excited states of all atoms and ions of the heavy elements. Some of these heavy-element features depend strongly both on the given species (atom or ion) and its detailed internal partition function, whereas other features only depend on the presence of the species itself, not on details such as the internal partition function. The latter features are obviously well suited for a helioseismic abundance determination, while the former features present a unique opportunity to use the sun as a laboratory to test the validity of physical theories of partial ionization in a relatively dense and hot plasma.

Acknowledgments I thank Jørgen Christensen–Dalsgaard for helpful discussions and for providing solar models. This work was supported by the grant AST-9987391 of the National Science Foundation.

References

Basu, S. & Christensen-Dalsgaard, J. 1997, *A&A*, **322**, L5

Basu, S., Däppen, W. & Nayfonov, A. 1999, *ApJ*, **518**, 985

Baturin, V. A., Däppen, W., Gough, D. O. & Vorontsov, S. V. 2000, *MNRAS*, **316**, 71

Berrington, K. A. 1997, *The Opacity Project*, vol. II, Institute of Physics Publishing. Bristol

Berthomieu, G., Cooper, A. J., Gough, D. O., Osaki, Y., Provost, J. & Rocca, A. 1980, in Hill, H. A., Dziembowski, W., eds., *Lecture Notes in Physics* **125**, Springer, Heidelberg, p. 307

Christensen-Dalsgaard, J. & Däppen, W. 1992, *Astron. Astrophys. Rev.*, **4**, 267

Christensen-Dalsgaard, J., Gough, D. O. & Toomre, J. 1985, *Science*, **229**, 923

Christensen-Dalsgaard, J., Däppen, W. & Lebreton, L. 1988, *Nature*, **336**, 634

Christensen-Dalsgaard, J., Däppen, W., Dziembowski, W. A. & Guzik, J. A. 2000, in İbanoğlu, C., ed., *Variable stars as essential astrophysical tools*, Kluwer, Dordrecht, p. 59

Christensen-Dalsgaard, J., Däppen, W., and the GONG Team 1996, *Science*, **272**, 1286

Däppen, W. & Guzik, J. A. 2000, in İbanoğlu, C., ed., *Variable stars as essential astrophysical tools*, Kluwer, Dordrecht, p. 177

Däppen, W., Mihalas, D., Hummer, D. G. & Mihalas, B. W. 1988, *ApJ*, **332**, 261

Däppen, W., Gough, D. O., Kosovichev, A. G. & Rhodes, E. J., Jr. 1993, in Weiss, W., Baglin, A., eds., *Proc. IAU Symp. 137, Inside the Stars*, ASP Conf. Ser. **40** Astron. Soc. Pac., San Francisco, p. 304

Eggleton, P. P., Faulkner, J. & Flannery, B. P. 1973, *A&A*, **23**, 325

Elliott, J. R. & Kosovichev, A. G. 1998, *ApJ*, **500**, L199

Gong, Z.-G., Däppen, W. & Nayfonov, A. 2001a, *ApJ*, **563**, 419

Gong, Z.-G., Däppen, W. & Zejda, L. 2001b, *ApJ*, **546**, 1178

Gong, Z.-G., Zejda, L., Däppen, W. & Aparicio, J. M. 2001c, *Comp. Phys. Commun.*, **136**, 294

Gough, D. O. 1984, *Mem. Soc. Astron. Ital.*, **55**, 13

Gough, D. O. 1993, in Zahn, J.-P., Zinn-Justin, J., eds., *Astrophysical Fluid Dynamics*, North-Holland, Amsterdam, p. 399

Gough, D. O., Kosovichev, A. G., Toomre, J., and the GONG Team 1996, *Science*, **272**, 1296

Guzik, J. A. & Swenson, F. J. 1997, *ApJ*, **491**, 967

Hummer, D. G. & Mihalas, D. 1988, *ApJ*, **331**, 794

Iglesias, C. A. and Rogers, F. J. 1996, *ApJ*, **464**, 943

Mihalas, D., Däppen, W. & Hummer, D. G. 1988, *ApJ*, **331**, 815

Nayfonov, A. & Däppen, W. 1998, *ApJ*, **499**, 489

Noels, A., Scuflaire, R. & Gabriel, M. 1984, *A&A*, **130**, 389

Rogers, F. J., Swenson, F. J. & Iglesias, C. A. 1996, *ApJ*, **456**, 902

Seaton, M. J. 1995, *The Opacity Project* Vol. I, Institute of Physics Publishing. Bristol

Ulrich, R. K. 1982, *ApJ*, **258**, 404

13

³He transport and the solar neutrino problem

CHRIS JORDINSON

Institute of Astronomy, University of Cambridge,
Madingley Road, Cambridge CB3 0HA, UK

³He transport in the solar core has been suggested as a solution to the solar neutrino problem. I investigate the consequences of imposing a flow on the solar core and show that it is unlikely that a flow could exist that would re produce the best-fit astrophysical solution to the experimental neutrino fluxes from Homestake, SAGE, GALLEX and SuperKamiokande.

13.1 Introduction

Before the announcement of the results from the Sudbury Neutrino Observatory (Ahmad et al. 2001), the measurements of the fluxes of neutrinos coming from nuclear reactions in the core of the sun were inconsistent with solar models. It has been argued that a so-called standard solar model can never be consistent with the experimental fluxes, and this has been used as an argument for the necessity of flavour transitions. However, non-standard solar models where ³He is burnt out of equilibrium have been suggested as astrophysical solutions to the neutrino problem (e.g. Dilke & Gough, 1972 and Gough, 1991), and Cumming & Haxton (1996) showed how a solar model with a redistribution of ³He in the core could overcome the problems of a standard solar model.

It was argued (Bahcall et al., 1997), using simple one-dimensional models with a mixed core, that Cumming and Haxton's model was inconsistent with helioseismology. As helioseismology had measured only a horizontal average of quantities in the solar interior, and as the suggested mechanism is fundamentally at least two-dimensional, the mechanism cannot be ruled out until a more realistic, two-dimensional model has been produced. In this paper I attempt to reproduce Cumming and Haxton's effect on the neutrino fluxes in a two-dimensional solar model with a flow imposed on the core to advect ³He. I am unable to find a flow that reproduces their effect

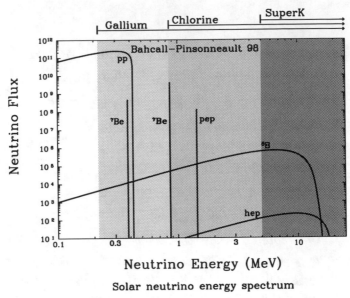

Fig. 13.1. The energy spectrum of neutrinos produced by various reactions and the energy thresholds of the different type of detectors. Taken from Bahcall (1996).

and conclude that it is not possible owing to the large variation in the ^{3}He equilibration timescale throughout the core.

13.2 Neutrinos and the neutrino problem

In the solar core, neutrinos are produced by several different nuclear reactions; the ones that provide the largest contribution to the fluxes detected at the Earth all come from the pp cycle of reactions:

$$p(p, \beta^+\nu)D(p, \gamma)^3He(^3He, 2p)^4He \qquad \text{ppI}$$

$$p(e^-p, \nu)\,D \qquad ^3He(^4He, \gamma)^7Be(e^-, \nu)^7Li(p,^4He)^4He \qquad \text{ppII}$$

$$^7Be(p, \gamma)^8B(\beta^+\nu)^8Be^*(^4He)^4He \qquad \text{ppIII}$$

The 3 main neutrino-producing reactions (the $p+p$, $^7Be+e^-$ and ^8B decay reactions) produce neutrinos with different energy spectra, and the reactions used to detect the neutrinos at the Earth have different energy thresholds, as shown in Figure 13.1. Based upon a standard solar model, the predicted contributions of the reactions to the experimentally measured fluxes are shown in Figure 13.2.

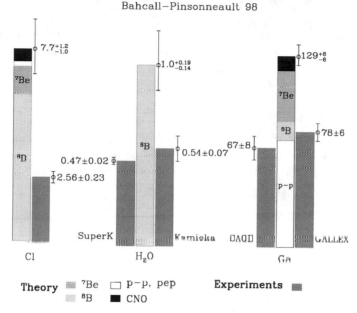

Fig. 13.2. The neutrino fluxes measured by the different experiments and the predicted contributions to them from various different reactions in a standard solar model. Taken from Bahcall (2000).

Possibly the most striking feature of this figure is that the predicted flux of ^{8}B neutrinos is much too high. If this is reduced so as to agree with the water experiments (SuperK, Kamioka), there is then no room left for the ^{7}Be neutrinos in the chlorine experiment (Homestake) or the gallium experiments (SAGE, GALLEX). If we assume that no neutrino oscillations are taking place, then the best-fit "astrophysical" solution to the experimental data would require the neutrino fluxes from the different reactions to be

$$\phi(\mathrm{pp}) \sim \phi^{\mathrm{SSM}}(\mathrm{pp})$$
$$\phi(^7\mathrm{Be}) \sim 0$$
$$\phi(^8\mathrm{B}) \sim 0.4\phi^{\mathrm{SSM}}(^8\mathrm{B}),$$

i.e. it needs to reduce

$$R_1 \equiv \frac{\mathrm{ppII} + \mathrm{ppIII}}{\mathrm{ppI}} \qquad \text{and} \qquad R_2 \equiv \frac{\mathrm{ppII}}{\mathrm{ppIII}}.$$

If we assume that the sun is in thermal equilibrium, we must leave

$$\mathrm{ppI} + 0.956\mathrm{ppII} + 0.508\mathrm{ppIII}$$

unchanged, as this is proportional to the luminosity due to the pp cycle (which makes up most of the sun's luminosity).

The reaction rates r are very sensitive functions of temperature T. In the range of temperatures appropriate for the sun's core, the reaction rates can be approximated by power laws, the indices of which (and the order of magnitude of the time a reaction takes) are as follows:

Reaction	Timescale (y)	$\dfrac{\partial \log(r)}{\partial \log(T)}$
$p(p, \beta^+\nu)D(p, \gamma)^3He$	10^{10}	4
$^3He(^3He, 2p)^4He$	10^5	16
$^3He(^4He, \gamma)^7Be$	10^6	17
$^7Be(e^-, \nu)^7Li(p, ^4He)^4He$	10^{-1}	$-1/2$
$^7Be(p, \gamma)^8B$	10^2	13
$^8B(\beta^+\nu)^8Be^*(^4He)^4He$	10^{-8}	0

Considering the ^{3}He-destroying reactions, we find that

$$\text{ppI} \propto X_3^2 T^{16} \qquad \text{and} \qquad \text{ppII} + \text{ppIII} \propto X_3(1 - X)T^{17},$$

and from the ^{7}Be-destroying reactions,

$$\text{ppII} \propto X_7(1 + X)T^{-\frac{1}{2}} \qquad \text{and} \qquad \text{ppIII} \propto X_7 X T^{13};$$

here X, X_3 and X_7 are the abundances by mass of H, ^{3}He, and ^{7}Be, respectively, It follows that the two ratios we wish to lower are

$$R_1 \propto (1 - X)X_3^{-1}T \qquad \text{and} \qquad R_2 \propto (1 + X^{-1})T^{-13.5}. \tag{13.1}$$

In a standard solar model without motion in the core, species are destroyed in the same place that they are created. Approximately 80% of the ^{3}He is destroyed the ppI reaction, so its abundance is roughly determined by balancing its creation by the p + p reaction with its destruction by the ^{3}He + ^{3}He reaction, giving

$$X^2 T^4 \propto X_3^2 T^{16} \Rightarrow X_3 \propto X T^{-6},$$

and so the two ratios become

$$R_1 \propto (X^{-1} - 1)T^7 \qquad \text{and} \qquad R_2 \propto (1 + X^{-1})T^{-13.5}.$$

Clearly within the framework of a standard solar model there is no way of lowering both of these ratios by changing the core temperature, as reducing the central temperature lowers R_1 but increases R_2. Altering the hydrogen

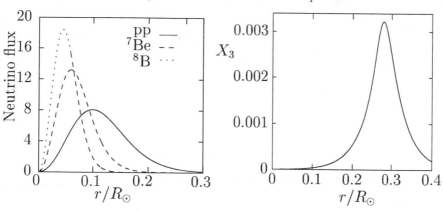

Fig. 13.3. Standard solar neutrino fluxes (left) and ^{3}He profile (right).

abundance sufficiently would be inconsistent with the amount of helium expected from big bang-nucleosynthesis.

13.3 Cumming and Haxton's model

Cumming and Haxton (1996) proposed a mechanism by which both ratios could be reduced, reproducing something like the best-fit astrophysical solution. To get around the restrictive temperature dependencies above, they suggested burning ^{3}He out of equilibrium by transporting it around the core. Unlike Dilke and Gough (1972), where ^{3}He is burnt out of equilibrium by periodic mixing events, they considered a steady-state model.

The contribution to the neutrino fluxes as a function of radius, and the ^{3}He abundance for a standard solar model are given in Figure 13.3. The ppII and ppIII reactions occur predominantly at very small radii, where the ^{3}He abundance is very low. The ratio of the two rates (R_2) increases with radius, owing to the decrease in temperature.

Cumming and Haxton assumed that ^{3}He was transported quickly from larger radii, where its equilibrium abundance is high but the temperature is too low for the ppII and ppIII reactions to occur, to smaller radii, where the temperature is high enough to suppress ppII terminations in favour of ppIII terminations. Ratio R_1 is reduced, as X_3 has been increased, and ratio R_2 is reduced, as the ^{3}He is being burnt at low radii and therefore high temperature (e.g. equation 13.1). They adjusted the ^{3}He profile artificially whilst maintaining global equilibrium. Their modified ^{3}He profile is shown in Figure 13.4 and a schematic of their suggested flow is shown in Figure 13.5.

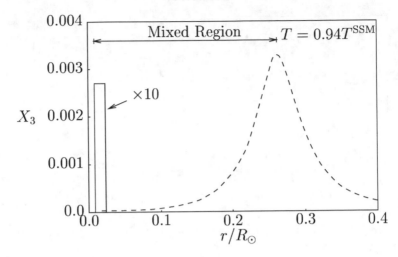

Fig. 13.4. Modified ³He profile (from Cumming and Haxton, 1996).

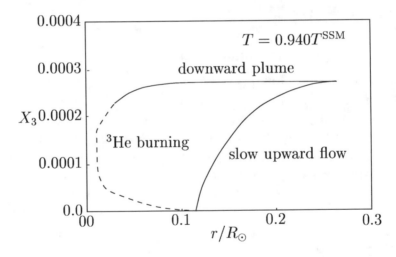

Fig. 13.5. Schematic of flow (from Cumming and Haxton, 1996).

13.4 Modelling the flow

To see if such a scheme could really work, I modelled the kinematic effect of imposing a steady flow on the solar core. The timescale of the flow is of the order of the timescale for ³He destruction in the core, i.e. about 10^7 years. This is much less than the hydrogen burning timescale, so the hydrogen abundance should be approximately constant along streamlines. For simplicity I assume that the hydrogen abundance is constant within the region mixed by the flow, and equal to the SSM abundance elsewhere. The

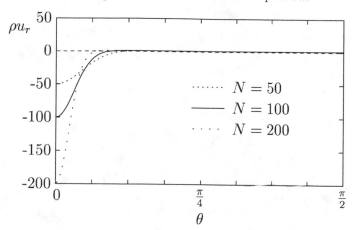

Fig. 13.6. Latitudinal dependence of the vertical mass flux, normalized to one at the equator, as a function of colatitude.

flow timescale is much longer than the lifetime of ^{7}Be in the core (which is less than a second), so I assume that the ^{7}Be abundance is equal, everywhere, to the local equilibrium abundance. The flow timescale is also many orders of magnitude larger than the dynamical time for the solar core, so I assume hydrostatic equilibrium. I am not considering how the flow is driven, only its effect on the neutrino fluxes.

Since we want to reproduce a valid model for the sun, I matched my model to the radius, mass, luminosity and temperature of a standard solar model at the base of the convection zone. I assume that things are approximately spherically symmetrical in the convection zone, so I set the temperature to be uniform at its base and I match the non-spherically-symmetric parts of the gravitational field to a vacuum solution.

I describe my flow velocity $\mathbf{u}$ by a stream function ψ:

$$\rho\mathbf{u} = \nabla \times \left(\frac{\psi(r,\theta)}{r\sin\theta} \hat{\phi} \right),$$

where r is distance to the centre, θ is colatitude and $\hat{\phi}$ is a unit vector in the longitude direction; also ρ is density. Note that the continuity equation is satisfied implicitly. The flow needs to have a sharp downflow to carry ^{3}He into the core without it getting burnt too soon, and a slow upflow to allow the ^{3}He to be replenished. For the form of my stream function I choose

$$\psi(r,\theta) = \Psi(r) \left(\cos^{N+1}\theta - \cos\theta \right)$$

with N even, which gives the following form for the vertical mass flux:

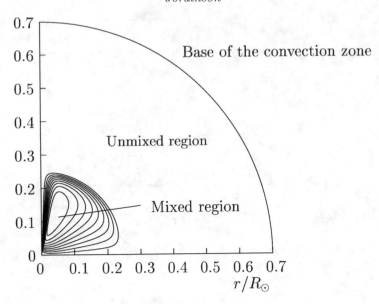

Fig. 13.7. Cross section of the flow.

$$\rho u_r = \frac{\Psi(r)}{r^2}\left(1 - (N+1)\cos^N\theta\right).$$

The latitudinal dependence is shown in Figure 13.6. There is a sharp down-flow at the pole and a slower upflow elsewhere. N is the ratio of the maximum downflux to the maximum upflux, and also determines the solid angle occupied by the downflow. The following parameters can be adjusted to create different flows:

- N, the ratio of the downward flux at the pole to the upward flux at the equator.
- $\Psi(r)$, determining the speed of the flow and the widths of the regions where it turns around.
- The radius of the mixed region.

An example flow, with $N = 100$, is given in Figure 13.7.

13.5 The equations

The equations that I solve are the equations for hydrostatic equilibrium:

$$\nabla p = \rho\mathbf{g},$$

advection, creation and destruction of ^{3}He:

$$\mathbf{u}\cdot\nabla X_3 = M_3(r_{\text{pp}}X^2 - 2r_{33}X_3^2 - r_{34}X_3Y),$$

and energy generation and transport:

$$\rho\mathbf{u}\cdot\nabla\epsilon - \frac{p}{\rho}\mathbf{u}\cdot\nabla\rho = \rho\varepsilon + \nabla\cdot\left(\frac{4acT^3}{3\kappa\rho}\nabla T\right),$$

Poisson's equation:

$$\nabla\cdot\mathbf{g} = 4\pi G\rho,$$

and the ideal gas equation of state

$$p = \frac{\mathcal{R}\rho T}{\mu} \qquad \text{where} \qquad \mu^{-1} = \frac{3}{4} + \frac{5}{4}X,$$

as I assume that the gas is a fully ionized mixture of hydrogen and helium. In these equations, p and μ are the pressure and mean molecular mass of the gas, and ϵ is its internal energy per unit mass ($= \frac{3}{2}\frac{\mathcal{R}}{\mu}T$ for a perfect gas). The acceleration due to gravity is $\mathbf{g}$. The mass of a ^{3}He nucleus is denoted M_3 and the abundance of ^{4}He is Y. The rates of the $p + p$, ^{3}He$+^3$He and ^{3}He$+^4$He reactions are denoted r_{pp}, r_{33} and r_{34}, and the rate of energy generation per unit mass (which is assumed to be just from the pp chain reactions) is denoted by ε. The opacity is κ. $G, \mathcal{R}, a$ and c are the gravitational constant, the ideal gas constant, the first radiation constant and the speed of light.

To solve the resulting set of partial differential equations, I expand my variables as the sum of products of functions of radius and functions of colatitude angle θ, e.g.

$$\rho(r,\theta) = \sum_{i=0}^{M}\rho_i(r)f_i(\cos^N\theta)$$

etc. The functions f_i are polynomials chosen so that $f_i(\cos^N\theta)$ are orthonormal over the surface of a sphere, so

$$f_0 = 1,$$
$$f_1 = (1+2N)^{\frac{1}{2}}\left[\left(1+\frac{1}{N}\right)\cos^N\theta - \frac{1}{N}\right],$$
$$f_n = \sum_{i=0}^{n}\cos^{Ni}\theta(-1)^{n-i}\frac{(1+2nN)^{\frac{1}{2}}}{i!(n-i)!}\prod_{k=i}^{n+i-1}\left(k+\frac{1}{N}\right)$$

This choice of functions allows me to resolve structure near to the pole whilst truncating the expansion at a few terms. The expansion gives rise to coupled ordinary differential equations which I solve by relaxation, starting from a

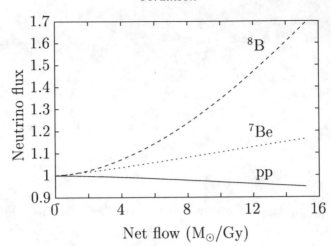

Fig. 13.8. Normalized neutrino fluxes against flow speed.

standard solar model, slowly homogenizing the hydrogen abundance in the core, and then increasing the flow speed.

13.6 Results

I experimented with different geometries and speeds of flow, but consistently found that as the flow speed increased, the fluxes of ^{7}Be and ^{8}B neutrinos also increased, although the ratio of ^{7}Be to ^{8}B neutrinos did go in the desired direction. Figure 13.8 shows typical results for how the neutrino fluxes depended on the speed of the flow, using a flow with $N = 100$.

The resulting ^{3}He profiles along streamlines are shown for two representative $N = 100$ flows in Figures 13.9 and 13.10. The flow speed for the "slow" flow (Figure 13.9) was chosen so that the flow has only a slight effect on the ^{3}He profile. The fluid rises slowly at most latitudes, with the abundance of ^{3}He always being slightly below its local equilibrium value. The fluid is then dragged down (more quickly than the upflow but still quite slowly) and the ^{3}He abundance takes on values slightly above the local equilibrium value.

In the "fast" flow (Figure 13.10) the flow has significant effect on the ^{3}He abundance, making it substantially below the local equilibrium value on the upflow and then dragging it down so fast that it essentially does not burn until deeper into the core. One thing to notice is the scales on the vertical (X_3) axes in the two graphs. The maximum abundance reached in the fast flow has gone down by an order of magnitude from that in the slow flow; the flow is too fast for the abundance to reach its equilibrium values at the

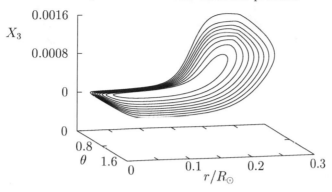

Fig. 13.9. The ^{3}He abundance along streamlines for a slow (0.2M$_\odot$/Gy) flow.

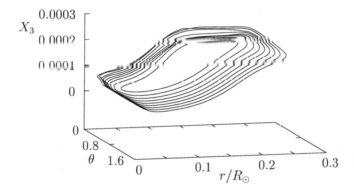

Fig. 13.10. The ^{3}He abundance along streamlines for a fast (20M$_\odot$/Gy) flow.

larger radii and so the maximum abundance is closer to the equilibrium abundance at a much lower radius.

To recover the effect Cumming and Haxton suggested, we would have to make the fluid spend longer in the outer regions, allowing ^{3}He to build up, and less time in the inner regions so that the ^{3}He abundance there can be kept high. From continuity arguments, the average amount of time the fluid can spend within a given region is proportional to the mass it contains, so it is the ratio of the masses of the two regions that determines the ratio of the times spent in each. If we look at the timescale for ^{3}He equilibration as a function of radius we see a difference of many orders of magnitude between the centre and outer parts of the core. This difference will be exaggerated when the hydrogen abundance of the core is homogenized (as in a SSM the rate of production of ^{3}He in the core is further decreased by a drop in the hydrogen abundance). However, the mass distribution throughout the sun shows that there is not this many-orders-of-magnitude difference in the

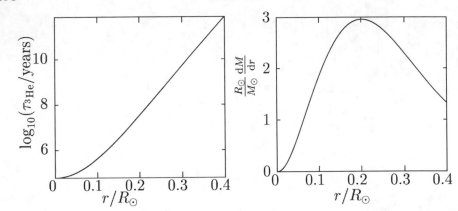

Fig. 13.11. ^{3}He equilibrium timescale (left) and mass distribution (right) of a standard model.

amount of mass in the different regions. The mass distribution and the ^{3}He timescale are show in Figure 13.11.

13.7 Conclusions

The timescales for ^{3}He to reach equilibrium are too different in the ^{7}Be producing regions and the ^{3}He rich regions. A flow slow enough to allow ^{3}He to accumulate at large radii spends too long in the hotter ^{7}Be producing region at lower radii so ^{3}He reaches its equilibrium abundance there.

The overall effect of imposing a flow is to lower the mean ^{3}He abundance, favouring ppII over ppI terminations and increasing the ^{7}Be and ^{8}B neutrino fluxes. I have not been able to reproduce the important features of Cumming & Haxton's solution to the solar neutrino problem by ^{3}He transport.

References

Ahmad, Q. R., Allen, R. C., Andersen, T. C. et al., 2001, *PRL*, **87**, 071301

Bahcall, J. N., 1996, *ApJ*, **467**, 475

Bahcall, J. N., 2000, in Diwan, M. V., Jung, C. K., eds, AIP Conf. Proc. 533, *Next Generation Nucleon Decay and Neutrino Detector (NNN99), SUNY*, p. 91

Bahcall, J. N., Pinsonneault, M. H., Basu, S. & Christensen-Dalsgaard, J., 1997, *PRL*, **78**, 171

Cumming, A. & Haxton, W. C., 1996, *PRL*, **77**, 4286

Dilke, F. W. W. & Gough D. O., 1972, *Nat*, **240**, 262

Gough, D. O., 1991, *Ann NY Acad Sci*, **647**, 199

14

Mixing in stellar radiation zones

JEAN-PAUL ZAHN

Observatoire de Paris, 92195 Meudon, France

Stars undergo some mild mixing in their radiation zones, which is due to a thermally driven large scale circulation, and presumably to turbulence caused by shear instabilities. It is the rotation of the star which is responsible for these motions, and therefore the transport of angular momentum must be described in time and space when modeling stellar evolution. We review the present state of the problem and discuss briefly the open questions.

14.1 The observational evidence

At first sight, there seems to be no mixing at all in stellar radiation zones, since a thoroughly mixed and therefore homogeneous star would not evolve to the red giant stage. This is why such mixing is ignored in the standard modeling of stellar evolution. However there are several signs that at least some partial mixing occurs in radiative interiors, and that this may have an impact on the later phases of stellar evolution.

Let us start by reviewing briefly the observational evidence pointing to such mixing.

- Models of built by pretending that there is no mixing in the radiation zones do not agree well with the observed global properties of stars, such as their luminosity and radius (or effective temperature). This is apparent when comparing theoretical isochrones with their observed counterpart in the Hertzsprung-Russel diagram, for stars with more than about 2 solar masses. The situation improves if one allows for some extra mixing beyond the convective core.

- It is well known that Li is depleted with age in main-sequence stars of effective temperature less than $6\,200\,\mathrm{K}$, although the base of their convection zone has a temperature which is too low for this element to be destroyed through nuclear burning (e.g., Herbig 1965; Boesgaard & Tripicco 1986). Another transport mechanism is thus required to carry Li down to high enough temperature.

- Further up in the Hertzsprung-Russel diagram, many O and B stars display an overabundance of ^{4}He and ^{14}N at their surface, which proves that these elements have been transported there from the stellar core, where they have been produced (Lyumbimkov 1984; Herrero et al. 1992; Gies & Lambert 1992). This anomaly seems to be correlated with the rotation rate.

- A similar effect is seen in giant stars, where one observes a ratio of ^{12}C over ^{13}C of 6 to 12, compared to that of about 90 measured in the solar system (e.g., Day et al. 1973). This cannot be accounted for just by convective dredge-up.

- Another, albeit indirect proof of partial mixing is the very existence of "normal" stars, among those which have negligible surface convection zones ($T_{\text{eff}} > 8\,000\,\text{K}$). Without such mixing, radiative levitation and gravitational settling would act selectively on the various chemical elements, and all stars would exhibit strong differences in their surface composition. This fact was pointed out already by Eddington (1926), and recent work by Richer et al. (1999) illustrates how even a weak level of turbulence smoothes out such composition differences.

- Finally, another indirect proof is provided by the variation of sound speed with depth in the Sun, which is now determined with high precision by helioseismology: below the convection zone the observed sound speed profile departs significantly from that predicted by the standard models, even when including microscopic diffusion, and this can only be explained by invoking some extra mixing.

14.2 Possible causes of mixing

What is responsible for the mixing in stellar radiation zones? Several mechanisms have been proposed, which we shall now examine.

14.2.1 Convective overshoot and penetration

The most important cause of mixing in stellar interiors is thermal convection. The convective turnover time scales as (cf. Zahn 1989)

$$t_{\text{conv}} \approx \left[\frac{MR^2}{L}\right]^{1/3} , \tag{14.1}$$

with M, R and L being the mass, radius and luminosity of the star. That time is of the order of one year in the solar convection zone and in the convective core of massive main-sequence stars, and therefore those regions can be considered as thoroughly mixed.

But this mixing is not confined to the unstable, superadiabatic domain. It is well known, from laboratory experiments and from the observation of geophysical fluids, that the convective motions penetrate or overshoot into the adjacent stable regions. The exact amount of extra mixing due to that process is still not reliably known. Several prescriptions have been proposed – for a recent review, we refer the reader to Zahn (2002). But usually stellar evolution codes just parameterize the extent of this mixing as a fraction of the local pressure scale-height (or of the size of the convective core), with the free parameter being adjusted such as to yield a satisfactory agreement between models and observations. Admittedly, this is an uncomfortable situation, which will only improve thanks to high resolution numerical simulations, as now being performed by Brummell, Clune & Toomre (2002). These simulations are very costly, and it is out of the question to implement them in stellar structure codes. But they can be used to validate the closure assumptions made in turbulence models, such as that built by Canuto & Dubovikov (1998), which may then be used to model penetrative convection with much more modest computer resources, as done by Kupka & Montgomery (2002) for A type stars.

It is important to note that since the mixing due to overshoot and penetration occurs only in the vicinity of the convective regions, it cannot account for the appearance at the surface of elements which have been synthesized in the core, as is observed in massive stars. That property requires mixing in the bulk of the radiation zone, of the kind we shall consider next.

14.2.2 Meridional circulation

In the radiation zone of a rotating star, the isothermal surfaces are distorted by the centrifugal force, with the result that the radiative flux is no longer divergence-free (Von Zeipel 1924). This loss of thermal equilibrium induces a large-scale axisymmetric circulation, which was first sketched out by Eddington (1925) and Vogt (1925). A complete description was given later by Sweet (1950), who calculated the relevant time-scale (since named Eddington-Sweet time):

$$t_{\mathrm{ES}} = t_{\mathrm{KH}} \frac{GM}{\Omega^2 R^3}, \quad \text{with } t_{\mathrm{KH}} = \frac{GM^2}{RL} \tag{14.2}$$

being the global thermal adjustment time (or Kelvin-Helmholtz time), Ω the angular velocity and G the gravitational constant. Sweet's result suggested that rapidly rotating stars should be well mixed by this circulation, and that they would be prevented from becoming red giants. But soon Mestel (1953) showed that the inhomogeneities produced through the burning of hydrogen in the core would interfere with the circulation and could eventually stop it (see also Mestel 2003).

A similar circulation may also be induced at the boundary of a radiation zone, by a differentially rotating convection zone. This is observed in the Sun, where acoustic sounding has shown that the rotation rate changes from latitude-dependent in the convection zone, to almost uniform in the radiative interior below, with the transition occurring in a thin boundary layer, the tachocline. In the absence of other stresses, such as due to a magnetic field or to waves, the circulation in the tachocline has an overturn time given by (Spiegel & Zahn 1992)

$$t_{\text{tac}} = \frac{r^2}{K} \frac{N^2}{\Omega^2} \left(\frac{\Delta\Omega}{\Omega}\right)^{-1} \left(\frac{\Delta r}{r}\right)^4, \qquad (14.3)$$

with Δr being the thickness of the tachocline, $\Delta\Omega$ the applied differential rotation, K the thermal conductivity, and N the buoyancy frequency, which in a chemically homogeneous region is given by

$$N^2 = \frac{g\delta}{H_P}(\nabla_{\text{ad}} - \nabla). \qquad (14.4)$$

We have used the classical notations: g for the gravity, H_P for the pressure scale-height, $\nabla = d\ln T/d\ln P$ for the gradient of temperature T with respect to pressure P; also $\nabla_{\text{ad}} = (\partial\ln T/\partial\ln P)_{\text{ad}}$, the derivative being for an adiabatic change, and $\delta = -(\partial\ln\rho/\partial\ln T)_P$, where ρ is density.

14.2.3 Turbulence caused by differential rotation

In general, the rotation regime achieved by the interplay of the circulation discussed above and the applied torques is not uniform, and the shear of differential rotation is liable to various instabilities, which will generate turbulence and therefore mixing. For a complete review of these instabilities, we refer the reader to Pinsonneault (1997). Here we shall consider only those which apparently play a major role, namely the shear instabilities.

14.2.3.1 Turbulence produced by the vertical shear

Let us first consider the instability produced by the vertical shear, $\Omega(r)$. This instability is very likely to occur, because the Reynolds number characterizing such shearing flows in stars is very high, due to the large sizes involved. Depending on the velocity profile, the instability may be linear, or of finite amplitude. In the absence of stratification, turbulence can be sustained whenever the Reynolds number $Re = wl/\nu$ is larger than about $Re_c = 40$, as it has been discussed by Schatzman, Zahn & Morel (2000). (The Reynolds number Re is expressed here in terms of the velocity w and the size l of the largest turbulent eddies, and ν is the kinematic viscosity.) However a radiation zone is stably stratified, and that stratification acts to hinder the shear instability. In an ideal fluid, i.e., without thermal dissipation, the instability occurs only if

$$\frac{N^2}{(\mathrm{d}V_h/\mathrm{d}z)^2} \leq Ri_c, \tag{14.5}$$

where V_h is the horizontal velocity, z the vertical coordinate, and N was introduced in equation (14.4). This condition is known as the *Richardson criterion*, and the critical Richardson number Ri_c is of order unity.

In a stellar radiation zone, this criterion is modified because the perturbations are no longer adiabatic, due to radiative leakage. When the radiative diffusivity K exceeds the turbulent diffusivity $D_v = wl$, the instability criterion takes the form (Townsend 1958; Dudis 1974; Zahn 1974; Lignières et al. 1999)

$$\frac{N^2}{(\mathrm{d}V_h/\mathrm{d}z)^2} \left(\frac{wl}{K}\right) \leq Ri_c. \tag{14.6}$$

From the largest eddies which fulfill this condition, one can deduce an estimate for the turbulent diffusivity acting in the vertical direction in the radiation zone of a star:

$$D_v = wl = Ri_c \frac{K}{N^2} \sin^2\theta \left(\frac{\mathrm{d}\Omega}{\mathrm{d}\ln r}\right)^2, \tag{14.7}$$

provided that $D_v \geq Re_c \nu = 40\nu$.

The instability criterion (14.6) holds in regions of uniform composition, where the stability is enforced only by the temperature gradient; when the molecular weight μ varies with depth, it seems at first sight that one should replace this criterion by the original one, equation (14.5), where now the buoyancy frequency is dominated by the gradient of molecular weight:

$$N^2 \approx N_\mu^2 = \frac{g\varphi}{H_P} \frac{\mathrm{d}\ln\mu}{\mathrm{d}\ln P},$$

with $\varphi = (\partial \ln\rho/\partial \ln\mu)_{P,T}$. As Meynet & Maeder (1997) pointed out, this severe condition would prevent any mixing in early-type main-sequence stars, contrary to what is observed. We shall see below how the stabilizing effect of μ-gradients can be weakened too.

14.2.3.2 Turbulence produced by the horizontal shear

Likewise, the horizontal shear $\Omega(\theta)$ will also generate turbulence, most likely through a finite-amplitude instability, because most plausible rotation profiles are linearly stable. What type of turbulence then will occur is still a matter of debate. Its vertical component will be constrained through the stratification, according to condition (14.6), and therefore it is likely that this turbulence will be anisotropic, with much stronger transport in the horizontal than in the vertical direction. In other words, the turbulent diffusivity is a tensor, with its horizontal component D_h much larger than the vertical one D_v.

Such anisotropic turbulence will interfere with the advective transport due to the meridional circulation, and it will turn it into a diffusion, as it was shown by Chaboyer & Zahn (1992). Assuming that the vertical velocity of the circulation is given by $u_r(r,\theta) = U(r)P_2(\cos\theta)$, the resulting diffusivity will be

$$D_{\text{eff}} = \frac{1}{30}\frac{(rU)^2}{D_h},\tag{14.8}$$

provided that $rU \geq D_h$. Unfortunately, a reliable prescription for the horizontal diffusivity D_h is still lacking, and when modeling this effect one has to make use of a free parameter.

Another property of such anisotropic turbulence is that, by smoothing out chemical inhomogeneities on level surfaces, it reduces the stabilizing effect of the vertical μ-gradient. The Richardson criterion for vertical shear instability then takes the form (Talon & Zahn 1997)

$$\frac{wl}{K+D_h}N_T^2 + \frac{wl}{D_h}N_\mu^2 \leq Ri_c\left(\frac{dV_h}{dz}\right)^2,\tag{14.9}$$

where N_T^2 stands for the thermal part of the buoyancy frequency given by equation (14.4). From this criterion one deduces, as before, the vertical component of the turbulent viscosity:

$$D_v = Ri_c\left[\frac{N_T^2}{K+D_h} + \frac{N_\mu^2}{D_h}\right]^{-1}\sin^2\theta\left(\frac{d\Omega}{d\ln r}\right)^2.\tag{14.10}$$

A similar prescription has been proposed by Maeder (1997).

14.3 Rotational mixing

The three causes of mixing which have just been discussed are intimately linked with the (differential) rotation. Therefore, when modeling the evolution of a star including these mixing processes, it is necessary to calculate also the evolution of its rotation rate $\Omega(r, \theta)$. The latter changes with time because angular momentum is advected by the large-scale circulation. At first sight, the problem looks very simple to handle, just dealing with laminar flows. However, as already mentioned, the differential rotation which is generated by this advection generates turbulence, which in turn transports angular momentum. In addition, torques may be applied to the star, either externally due to mass loss or accretion, or to tides exerted by a companion or an accretion disk, or else internally due to magnetic stresses, or even to waves, as we shall see. To take all these effects into account, one has thus to solve the transport equation for the angular momentum:

$$\frac{\partial}{\partial t} \left[\rho r^2 \sin^2 \theta \, \Omega \right] + \nabla \cdot \left[\rho \mathbf{u} r^2 \sin^2 \theta \, \Omega \right] = \text{applied torques.} \qquad (14.11)$$

The problem may be significantly simplified if the horizontal shear produces the anisotropic turbulence mentioned above. This turbulence then acts to reduce its cause, namely the differential rotation in latitude, and one can neglect the variations of Ω in latitude. To lowest order Ω is then a function of r only, a regime which may be called "shellular" rotation; it makes it easier to calculate the meridional velocity, which separates into $u_r(r, \theta) = U(r) P_2(\cos \theta)$. For a detailed account of how this modelization is implemented, we refer to Zahn (1992) and to Maeder & Zahn (1998).

14.3.1 Rotational mixing of type I

Let us first examine the simplest case, where the torques exerted by waves or magnetic fields are negligible compared to the other effects. Angular momentum is thus transported by the same processes which are responsible for mixing, namely meridional circulation and turbulent diffusion. The angular velocity then obeys the following transport equation, obtained by averaging equation (14.11) over θ:

$$\frac{\partial}{\partial t} \left[\rho r^2 \Omega \right] = \frac{1}{5 r^2} \frac{\partial}{\partial r} \left[\rho U r^2 \Omega \right] + \frac{1}{r^2} \frac{\partial}{\partial r} \left[\rho \nu_{\mathrm{v}} r^4 \frac{\partial \Omega}{\partial r} \right], \qquad (14.12)$$

with $\nu_{\mathrm{v}} \approx D_{\mathrm{v}}$ given by equation (14.10). Note the presence of the advection term on the RHS: depending on the sign of U, angular momentum may be transported up the gradient of Ω, which is never the case when the

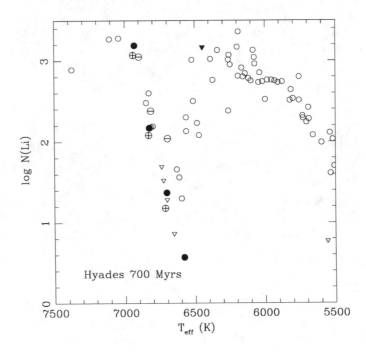

Fig. 14.1. Models reproducing the Li gap in the Hyades; on the blue side with rotational mixing of type I, where angular momentum is transported by meridional circulation and turbulence (filled dots, dots with plus or minus), and on the red side with rotational mixing of type II, where uniform rotation is imposed (filled triangle). The models have been calculated by Talon & Charbonnel (1998); observational data (open circles, and open triangles for upper limits) are from Boesgaard & Tripicco (1986) and Thorburn et al. (1993). (Reprinted with permission from Talon & Charbonnel, Astronomy & Astrophysics, **335**, 959, 1998.)

effect of meridional circulation is modeled as a diffusive process, as done by Pinsonneault et al. (1989) and Chaboyer et al. (1995).

The circulation is governed mainly by the outer boundary condition. When there is no or little loss of angular momentum from the star, the circulation settles into a regime where a weak inward flux of angular momentum compensates the turbulent diffusion directed upwards. On the other hand, when the star loses angular momentum, the circulation adjust itself such as to transport that angular momentum towards the surface (Zahn 1992). We thus expect more mixing to occur in the latter case.

Massive main-sequence stars belong to the first category, and rotational mixing has been quite successful in improving their modeling. The theoretical isochrones fit well the observed ones, even without introducing convective overshoot from the core, and such rotational mixing accounts for the observed enhancement of He and N at the surface of early-type stars. This was first illustrated with the evolution of a 9 $M_\odot$ star by Talon et al. (1997), and then generalized to all massive stars by Meynet & Maeder (2000). One important property of these models is that, because their effect is weakened by the anisotropic turbulence, the μ-gradients do not stop the circulation as was predicted by Mestel (1953). Combined with a suitable prescription for the mass loss, this type of mixing also yields the observed proportion of blue and red giants. Finally, such mixing accounts perfectly for the destruction of Li on the blue side of the Li gap (Talon & Charbonnel 1999), as shown in Figure 14.1; in their calculations they took the same value for the adjustable parameter characterizing the anisotropic turbulence than that which was used to model the 9 $M_\odot$ star by Talon et al. (1997). This type of mixing also provides a plausible interpretation of the peculiar abundances of subgiants (Palacios & Charbonnel, private communication).

The second case applies to solar-type stars, which are spun down through a magnetized wind. But for these stars the models built with that type of rotational mixing fail several observational tests. For instance, the models predict that the mixing in the radiation zone is correlated with the loss of angular momentum. This may be verified through the surface abundance of ^{7}Li: this fragile element is destroyed at some distance of the base of the convection zone, and it can only get there if that region is the seat of extra mixing. Indeed, in some clusters there is an anti-correlation between the Li abundance and the rotation rate. But the crucial test is the behavior of tidally-locked binaries: there the angular momentum lost by the wind is drawn from the orbital motion, and it need not to be extracted from the deep interior of the stars (Zahn 1994). This seemed to be supported by the observations of two of such binaries in the Hyades: according to Thorburn et al. (1993), they displayed a higher Li abundance than single stars of the same mass. But recently Balachandran (2002) re-examined these stars, and she could not detect the presence of Li. Another problem is that this type of mixing would deplete ^{9}Be by a factor of 3 in the Sun, whereas a careful analysis of the solar spectrum shows that Be has been very little destroyed (Balachandran & Bell 1998). Finally, models built according to equation (14.12) keep a fast rotating core (Matias & Zahn 1998), which is ruled out by helioseismology. Models which ascribe the transport of angular momentum to a diffusive process only lead to the same conclusion (Pinsonneault et al. 1989; Chaboyer et al. 1995).

We must therefore conclude that another, more powerful process is responsible for the transport of angular momentum in solar-type stars. Before we discuss the possible mechanisms, let us just assume that they are efficient enough to enforce uniform rotation.

14.3.2 Rotational mixing of type II

What we mean by this denomination is that the chemical elements are still transported by the meridional circulation and the turbulence caused by the residual differential rotation, but that angular momentum is carried by another process, which establishes an almost uniform rotation profile in the radiation zone.

With that assumption, Talon & Charbonnel (1999) have shown that one can explain why the Li abundance takes its original value on the red side and in the vicinity of the Li gap (Figure 14.1). The same assumption was made by Théado & Vauclair (2001) to model the effect of composition gradients in halo stars, in order to explain the Li plateau in these stars.

These first attempts are promising, but they bypass the question which must now be answered, namely what causes the extraction of angular momentum and enforces uniform rotation in solar-type stars. Two physical processes have been suggested: magnetic torques and gravity waves. At this point, it is hard to decide which is the most efficient, as we will see in the next section. But before we shall examine another cause of mixing, which occurs just below the convection zone of solar-type stars.

14.3.3 Tachocline mixing

As mentioned already, the differential rotation which is imposed on the top of the radiation zone of solar-type stars drives a circulation which will cause some mixing there. In the absence of other processes, the transition layer between the two regimes, of differential rotation above and uniform rotation below, would tend to spread into the radiation zone by thermal diffusion, and in the present Sun it would extend down to about 1/3 of the solar radius. But helioseismology tells us that this layer is thinner than 5% of the radius.

Two mechanisms have been proposed to stop the spread of the tachocline. One is anisotropic turbulence produced by the shear of the differential rotation, if it acts indeed to reduce the shear and tends to restore uniform rotation (Spiegel & Zahn 1992). Models of the solar tachocline have been built according to this scheme by Brun et al. (1999) and Brun et al. (2002). These models depend on two parameters, one defining the buoyancy frequency and hence the degree of stability of this region, and the other which fixes the strength of the horizontal turbulence. One constraint is provided by the observed thickness of the tachocline in the present Sun, which leaves one adjustable parameter, say the buoyancy frequency, to be chosen within reasonable limits. Models built in this way display temperature and composition profiles which are compatible with helioseismic inversions. Furthermore, they predict a destruction in time of ^{7}Li which can account for the observed one, while leaving ^{9}Be intact.

The spread of the tachocline may also be halted by a fossil magnetic field, as suggested by Gough & McIntyre (1998). Recent calculations performed by P. Garaud substantiate this mechanism (cf. Garaud 2001, 2003), although it is the high viscosity she had to employ for numerical reasons, associated with non-penetrative boundaries, which drives there the circulation. Work is in progress to render more realistic these models of the magnetic tachocline.

14.4 Open questions

In spite of some successes which have been highlighted above, the problem of the mixing in stellar radiation zones can hardly be considered as solved. More work is needed to refine the models, and to derive reliable prescriptions for the turbulent transport. Let us review some of the open questions.

14.4.1 Does turbulence caused by a horizontal shear act to reduce that shear?

We have seen that some models invoke anisotropic turbulence produced by the latitudinal shear $\Omega(\theta)$ to reduce the differential rotation. This property has been challenged by McIntyre (2003), who points to geophysical situations where turbulence rather tends to smooth the potential vorticity. In the Couette-Taylor experiment, however, the laminar profile for which vorticity is constant along the radius turns into a turbulent one for high enough Reynolds number, where vorticity is no longer constant (cf. Richard & Zahn 1999). Admittedly, turbulence is rather isotropic in the laboratory experiment, and it is not obvious that the result would be the same in a strongly stratified fluid, in which turbulence would be anisotropic. But it is difficult to avoid the conclusion that turbulence acts in a way to suppress or to weaken its cause.

14.4.2 How does a poloidal field avoid imprinting the differential rotation of the convection zone into the radiation zone?

Magnetic torques are a promising explanation for the uniform rotation of the solar interior. One would have to invoke a fossil field, because an alternating dynamo field, as in the Sun, would remain confined at the base of convection zone, due to the skin effect caused by ohmic dissipation (see Garaud 2001). But to establish uniform rotation in the radiation zone, such a fossil field must a have a special topology. The torque exerted by the Lorentz force acts so as to render the angular velocity uniform along the field lines of the meridional field (Ferraro's law). If this meridional field threads into the convection zone, like a dipolar field for instance, it will imprint the differential rotation of that zone into the radiation zone, as illustrated by the calculations performed by McGregor & Charbonneau (1999). Thus a quasi-uniform rotation, as observed in the Sun, can only be enforced if the meridional field does not penetrate into the convection zone, or if it does so only over a small interval of latitude. This seems quite possible, but it has not been proved yet by fully consistent calculations, where the advection of the field by the thermally driven circulation is taken in account.

14.4.3 Can waves extract angular momentum from the solar interior?

Another possible way of transporting angular momentum is by waves emitted at the base of the convection zone: internal waves, for which buoyancy is the restoring force, and gravito-inertial waves, where the Coriolis force also plays a role. These waves have a sufficiently low frequency, and a short enough wavelength, that they are damped through thermal diffusion in the radiation zone, thus depositing there the momentum they carry. This mechanism was first considered by Kumar & Quataert (1997) and Zahn et al. (1997), but their treatment was incorrect, as pointed out by Gough & McIntyre (1998) and Ringot (1998).

In fact, the actual properties of wave transport turns out to be much more complicated than anticipated. Depending on the spatial and temporal frequencies of the waves, the deposit of angular momentum takes place just below the convection zone, or much deeper down.

Let us first examine the behavior of waves which are dissipated in the vicinity of the convection zone. Prograde waves carry positive angular momentum, retrograde waves negative angular momentum. When they travel in a medium which is rotating faster then the region where they have been emitted, their frequency is Doppler-shifted, leading to higher dissipation for the prograde waves than for the retrograde waves. For this reason the angular velocity tends to increase where it was already higher, and its slope steepens until the shear becomes unstable. That turbulent layer then merges with the convection zone. But in the meanwhile the retrograde waves have deposited negative angular momentum somewhat further down, thus building there another shear layer flowing in opposite direction, which now takes the place of the former one. And the cycle continues. A similar phenomenon is observed in the Earth atmosphere, where it is called quasi-biennial oscillation (cf. McIntyre 2002).

The question arises how this oscillating layer located just below the convection zone will filter the waves of longer wavelength. If the oscillation were perfectly symmetrical in time, its effect would be the same on the prograde and on the retrograde waves which traverse it. But if there is even a slight increase of Ω with depth in that layer, the prograde waves will be more dissipated, which allows the retrograde waves to extract angular momentum from the deep interior. This has been established recently through numerical simulations performed by Talon et al. (2002), using a rather crude approach where the Coriolis force has been neglected and where the waves have been averaged over all latitudes.

Future developments will tell which of these processes dominates the angular momentum transport in the radiative interior of solar-type stars. Let us hope that we will get the answer when we celebrate the 70th birthday of our dear Douglas Gough!

References

Balachandran, S. C., 2002, in Rickmann, H., ed., *Mixing and diffusion in Stars; Joint Discussion 5 at 24th IAU Gen. Ass., Highlights of Astronomy*, **12**, Astr. Soc. Pac., San Francisco, p. 276.

Balachandran, S. C. & Bell, R. A., 1998, *Nature*, **392**, 791.

Boesgaard, A. M. & Tripicco, M. J., 1986, *Ap. J.*, **302**, L49.

Brummell, N., Clune, T. & Toomre, J., 2002, *ApJ*, **570**, 825.

Brun, A. S., Turck-Chièze, S., & Zahn, J.-P., 1999, *ApJ*, **525**, 1032.

Brun, A. S., Antia, H. M., Chitre, S. M. & Zahn, J.-P., 2002, *A&A*, **391**, 725.

Canuto, V. M. & Dubovikov, M., 1998, *ApJ*, **493**, 834.

Chaboyer, B., Demarque, P. & Pinsonneault, M. H., 1995, *ApJ* **441**, 865.

Chaboyer, B. & Zahn, J.-P., 1992, *A&A*, **253**, 173.

Day, R. W., Lambert, D. L., & Sneden, C., 1973, *ApJ*, **185**, 213.

Dudis, J. J., 1974, *JFM*, **64**, 65.

Eddington, A. S., 1925, *Observatory*, **48**, 78.

Eddington, A. S., 1926, *The Internal Constitution of the Stars* (repr. 1959 Dover; New York)

Garaud, P., 2001, *Ph. D. thesis, Cambridge Univ.*

Garaud, P., 2003, in this volume.

Gies, D. R. & Lambert, D. L., 1992, *ApJ*, **387**, 673.

Gough, D. O. & McIntyre, M. E., 1998, *Nature*, **394**, 755.

Herbig, G. H., 1965, *ApJ*, **141**, 588.

Herrero, A., Kudritzki, R. P., Vilchez, J. M., Kunze, D., Butler, K., & Haser, S., 1992, *A&A*, **261**, 209.

Kumar, P. & Quataert, E. J., 1997, *ApJ*, **475**, L143.

Kupka, F. & Montgomery, M. H., 2002, *MNRAS*, **330**, L6.

Lignières, F., Califano, F. & Mangeney, A. 1999, *A&A*, **349**, 1027.

Lyubimkov, L. S., 1984, *Astrophysics*, **20**, 255.

Maeder, A., 1997, *A&A*, **321**, 134.

Maeder, A. & Zahn, J., 1998, *A&A*, **334**, 1000.

MacGregor, K. B. & Charbonneau, P., 1999, *ApJ*, **519**, 911.

Matias, J. & Zahn, J.-P. 1998, in Provost, J. & Schmider, F.-X., eds, *Sounding Solar and Stellar Interiors, Proc. IAU Symp. 181, poster vol.*, Université de Nice, p. 103.

McIntyre, M. E., 2002, in Pearce, R. P., ed., *Meteorology at the Millennium*, London, Academic Press and Royal Meteorol. Soc., p. 283.

McIntyre, M. E., 2003, in this volume.

Mestel, L., 1953, *MNRAS*, **113**, 716.

Mestel, L., 2003, in this volume.

Meynet, G. & Maeder, A., 1997, *A&A*, **321**, 465.

Meynet, G. & Maeder, A., 2000, *A&A*, **361**, 101.

Pinsonneault, M., 1997, *ARAA*, **35**, 557.

Pinsonneault, M., Kawaler, S. D., Sofia, S. & Demarque, P., 1989, *ApJ*, **338**, 424.

Richard, D. & Zahn, J.-P., 1999, *A&A*, **347**, 734.

Richer, J., Michaud, G., & Turcotte, S., 2000, *ApJ*, **529**, 338.

Ringot, O. 1998, *A&A*, **335**, L89.

Schatzman, E., Zahn, J.-P. & Morel, P., 2000, *A&A*, **364**, 876.

Spiegel, E. A. & Zahn, J.-P., 1992, *A&A*, **265**, 106.

Sweet, P. A., 1924, *MNRAS*, **110**, 548.

Talon, S. & Charbonnel, C., 1998, *A&A*, **335**, 959.

Talon, S., Kumar, P. & Zahn, J.-P., 2002, *ApJ*, **574**, 175.

Talon, S. & Zahn, J.-P., 1997, *A&A*, **317**, 749.

Talon, S., Zahn, J.-P., Maeder, A. & Meynet, G. 1997, *A&A*, **322**, 209.

Théado, S. & Vauclair, S., 2001, *A&A*, **375**, 70.

Thorburn, J. A., Hobbs, L. M., Deliyannis, C. P., & Pinsonneault, M. H. 1993, *ApJ*, **415**, 150.

Townsend, A. A., 1958, *JFM*, **4**, 361.

Vogt, H., 1925, *Astron. Nachrichten*, **223**, 229.

Von Zeipel, H., 1924, *MNRAS*, **84**, 665.

Zahn, J.-P., 1974, in Ledoux, P., Noels, A. & Rodgers, A. W., eds, *Stellar Instability and Evolution, Proc. IAU Symp. 59*, Reidel, Dordrecht, p. 185.

Zahn, J.-P., 1989 *A&A*, **220**, 112.

Zahn, J.-P., 1992, *A&A*, **265**, 115.

Zahn, J.-P., 1994, *A&A*, **288**, 829.

Zahn, J.-P., 2002, in Aerts, C., Bedding, T. R. & Christensen-Dalsgaard, J., eds, *Radial and Nonradial Pulsations as Probes of Stellar Physics, Proc. IAU Symp. 185*, Astr. Soc. Pac. Conf. Ser., vol. **367**, 58.

Zahn, J.-P., Talon, S., & Matias, J., 1997, *A&A*, **322**, 320.

Element settling and rotation-induced mixing in slowly rotating stars

SYLVIE VAUCLAIR

Laboratoire d'Astrophysique, Observatoire Midi-Pyrénées, 31400-Toulouse, France

The element settling which occurs inside stars, due to the combined effect of gravity and thermal gradient (both downwards), radiative transfer (upwards) and concentration gradients, leads to abundance variations which cannot be neglected in computations of stellar structure. This process is now generally introduced as a "standard process" in stellar evolution codes. The new difficulty is to explain why, in some cases, element settling does not proceed at all as expected. Macroscopic motions, like rotation-induced mixing, may increase the settling time scales, but then it introduces in radiative regions extra mixing with consequences which are not always observed as predicted. We have recently developed a new approach for treating rotation-induced mixing in which we include the feedback effect of the settling-induced μ-gradients (Vauclair 1999, Théado & Vauclair 2001, 2002). This effect, which was not included in previous computations, leads to first order terms in the meridional circulation velocity. It results in a mixing process, just below the convective zone, quite different from that induced by normal circulation. For the first time, we have evidence of a mixing region which is precisely confined and directly modulated by the settling itself. This will have interesting consequences for the computations of abundance variations in stars.

15.1 Introduction

Although element settling inside stars was already recognized as a fundamental process at the very beginning of the computation of stellar structure and evolution (Eddington 1926), it has long been forgotten by stellar astrophysicists. In the 1970s, Michaud and collaborators (Michaud 1970, Michaud et al. 1976) suggested that the abundance anomalies observed in the so-called Ap and Am stars were due to the competition between gravitational settling and radiative acceleration. From then on, more and more precise computations showed that this was indeed the most important effect which led to the anomalies, although it had to be modulated by several kinds of macroscopic motions (Vauclair et al. 1978a,b).

The region in the HR diagram where stars can suffer large settling effects is limited for hot stars by rapid mass loss and for cool stars by deep convection zones (Vauclair & Vauclair 1982). For solar-type stars, the settling time scale below the convective zone is larger than the stellar lifetime. For this reason, when Douglas Gough asked me, in the early 1980s, whether I thought that element settling could be important in the Sun, I answered, "No!", because the helium depletion in the Sun should not be more than about 20%. "Ha ha", he replied, "but this is important!" He was right, as usual. Later on, helioseismology proved that helium and heavier elements do settle below the solar convective zone, even if the induced gradient has to be smoothed down by some mixing (also needed to account for the lithium nuclear burning) (Christensen-Dalsgaard, Proffitt & Thompson 1993, Richard et al. 1996, Christensen-Dalsgaard et al. 1996, Gough et al. 1996, Brun et al. 1998).

Element settling in stars is now generally introduced in the computation of stellar models as a "standard process". The new difficulty is to explain why, in some cases, it does not seem to occur as expected. This is the case, for example, in population II stars where the lithium abundance is remarkably constant for main-sequence stars with effective temperatures between 5500 and 6500 K. Element settling alone would lead to a decrease of lithium for the hottest stars, which is not observed, and the mixing which could prevent it would lead to a large lithium dispersion, which is not observed either (Michaud, Fontaine & Beaudet 1984, Vauclair & Charbonnel 1995, 1998, Pinsonneault et al. 1999). This last problem may however be overcome if, in the computation of rotation-induced mixing, one takes into account the influence of the gradient in mean molecular weight μ (hereafter μ-gradient) induced by helium settling (Vauclair 1999, Théado & Vauclair 2001, 2002).

15.2 Element settling in stellar radiative zones

Stellar radiative regions, in which no macroscopic motions take place, are subject to element diffusion induced by the pressure and temperature gradients and by radiative acceleration: this process is believed to be the reason for the large abundance variations observed in main-sequence stars, horizontal branch stars and white dwarfs (Vauclair & Vauclair, 1982).

Inside convective regions, the rapid macroscopic motions mix the gas components and force their abundance to remain homogeneous. The chemical composition observed in the external regions of cool stars is thus affected by the settling which occurs below the outer convective zones. As the settling time scales vary in the first approximation like the inverse of the density, the expected variations are smaller for cooler stars, which have deeper convective zones. While some elements can see their abundances vary by several orders of magnitude in the hottest Ap stars, the variations in the Sun are not larger than a few times 10%, too small to be measured by spectroscopy.

When mild macroscopic motions occur in the stellar radiative regions, they slow down the settling but do not prevent it completely unless an equilibrium concentration gradient is reached (Schatzman 1969, 1977). Evidence of such a competition between element settling and macroscopic motions appears in many stellar cases. The observed chemical composition is then used to constrain the hydrodynamical processes which occur inside the stars. Here I will discuss two cases: that of the Sun and that of main-sequence halo stars.

15.2.1 The solar case

Thanks to helioseismology, we know the sound speed inside the Sun with a precision of about 0.1% and can see evidence of the occurrence of helium settling as predicted by diffusion computations. Solar models computed in the old "standard" way, in which the element settling is totally neglected, do not agree with the inversion of the seismic modes. This result has been obtained by many authors, in different ways (see Gough et al. (1996) and references therein). There is a characteristic discrepancy of order 0.5%, just below the convective zone, between the sound speed computed in the models and that of the seismic Sun. Introducing element settling improves considerably the consistency with the seismic Sun, but some discrepancies do remain, particularly below the convective zone where a peak appears in the sound-speed difference. The reason for this peak is probably the steepness of the μ-gradient induced by pure microscopic diffusion (Richard et al. 1996). The helium profiles obtained directly from helioseismology (Basu 1997, Antia & Chitre 1998) indeed show a helium gradient below the convective zone which is smoother than the gradient obtained with pure settling (Figure 15.1).

Such motions are also needed to reproduce the observed abundances of light elements, especially lithium. The abundance determinations in the solar photosphere show that lithium has been depleted by a factor of about 140 compared to the protosolar value, while beryllium is normal (Balachandran & Bell 1998). Furthermore, observations of the ^{3}He/^{4}He ratio in the solar wind and in the lunar rocks (Geiss 1993, Geiss & Gloecker 1998) show that this ratio may not have increased by more than about 10% during the

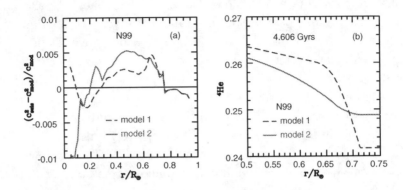

Fig. 15.1. (a) Relative differences between the square of the sound speeds computed in recent Toulouse solar models (including reaction rates from NACRE [Nuclear Astrophysics Compilation of REaction rates]) and those obtained from seismic inversions by Basu 1997, 1998; dashed line: pure element settling; solid line: element settling and rotation-induced mixing. (b) Helium gradients below the convective zone for the same cases. The helium gradient obtained when rotation-induced mixing is included is closer to that deduced from helioseismology than the one with pure settling, which is too steep. However, some discrepancy remains in the sound speed.

last 3 Gyr in the Sun. While the occurence of some mild mixing below the solar convective zone is needed to explain the lithium depletion and helps reconcile the models with the helioseismological constraints, the ^{3}He/^{4}He observations put a strict constraint on its efficiency. The effect of μ-gradients on the mixing processes has to be invoked to explain these observations in a consistent way.

15.2.2 The lithium plateau in halo stars

The lithium plateau observed in halo stars has long appeared as a paradox in the general context of the lithium abundance behaviour in the outer layers of stars. First, the plateau is flat; secondly, the lithium abundance dispersion is extremely small. This seems in contradiction with the large lithium variations observed in younger stars. It is also difficult to understand theoretically: as lithium nuclei are destroyed by nuclear reactions at a relatively low temperature (about 2.5 million degrees), the occurrence of macroscopic motions in the outer layers of stars easily leads to lithium depletion at the surface. On the other hand, if no macroscopic motions occur in the stellar

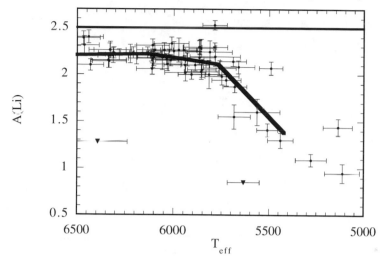

Fig 15.2. Lithium abundances in population II stars as a function of effective temperature; the ordinate $A(Li)$ represents the logarithm of the lithium abundance in the scale $\log H = 12$; the observational points and errors are from Bonifacio & Molaro (1997); the solid curve represents the average computational results obtained by Théado & Vauclair 2001, for an original lithium abundance $A(Li) = 2.5$.

gas, lithium is subject to microscopic diffusion which, in the case of halo stars, should also lead to depletion.

Several ideas have been proposed to account for the lithium behaviour in halo stars. The most promising possibilities are rotation-induced mixing, which could reduce lithium in the same way for all the stars (Vauclair 1988, Pinsonneault et al. 1992, 1999), and mass-loss, which could oppose the lithium settling (Vauclair & Charbonnel 1995, 1998). In both cases however, the parameters should be tightly adjusted to prevent any dispersion in the final results.

Vauclair (1999) and Théado & Vauclair (2001) pointed out that the effect of the μ-gradients induced by element settling was not introduced in previous computations of rotationally-induced mixing. This can change the mixing efficiency and help in understanding the observations, as discussed below (Figure 15.2).

15.3 Rotation-induced mixing in the presence of gravitationally-induced μ-gradients

The meridional circulation velocity in stars, in the presence of μ-gradients, is the sum of two terms, one due to the classical thermal imbalance (so-called

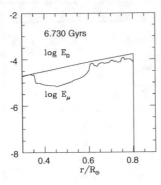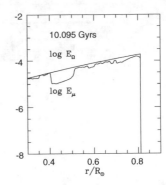

Fig. 15.3. Evolution of μ-induced currents with time inside a star. The graphs show the variations with depth of both $|E_\Omega|$ and $|E_\mu|$ in a $0.70 M_\odot$ halo star with $[Fe/H] = -2$ and $V_{\rm rot} = 5\,{\rm km\,s}^{-1}$ for two ages: 6.730 and 10.095 Gyrs. The region where the two currents cancel each other develops downwards with time, beginning just below the convective zone.

Ω-currents) and the other one due to the induced horizontal μ-gradients (μ-induced currents, or μ-currents in short). In the most general cases, μ-currents are opposite to Ω-currents (Mestel 1953, Zahn 1992, Maeder & Zahn 1998, Mestel 2003).

When element settling occurs below the stellar outer convective zone, a small helium gradient builds up, even in the presence of circulation. Vauclair (1999) and Théado & Vauclair (2001) showed that the resulting μ-gradients rapidly become large enough to create μ-currents of the same order as Ω-currents (Figure 15.3). Then the behaviour of the mixing processes is dramatically modified.

15.3.1 Computations of Ω and μ-currents

The vertical meridional circulation velocity in stars may be written

$$u_r = \frac{\nabla_{\rm ad}}{\nabla_{\rm ad} - \nabla + \nabla_\mu} \frac{\varepsilon_\Omega}{g} \,, \tag{15.1}$$

where g is the local gravitational acceleration, $\nabla_{\rm ad}$ and ∇ are respectively the usual adiabatic and actual logarithmic gradients $({\rm d}\ln T/{\rm d}\ln P)$ of temperature T with respect to pressure P, and ∇_μ the mean molecular weight gradient $({\rm d}\ln\mu/{\rm d}\ln P)$. The factor ε_Ω is defined below (equation 15.4). Writing u_r as

$$u_r = U_r\, P_2\,(\cos\theta) \,, \tag{15.2}$$

where P_2 is a Legendre polynomial and θ is co-latitude, the horizontal meridional velocity is given by

$$u_\theta = -\frac{1}{2\rho r} \frac{\mathrm{d}}{\mathrm{d}r} (\rho r^2 U_r) \sin\theta \cos\theta , \qquad (15.3)$$

ρ being the density. The expression for ε_Ω is then obtained as a function of the Ω and μ-currents:

$$\varepsilon_\Omega = \left(\frac{L}{M}\right) (E_\Omega + E_\mu) P_2(\cos\theta) , \qquad (15.4)$$

with

$$E_\Omega = \frac{8}{3} \left(\frac{\Omega^2 r^3}{GM}\right) \left(1 - \frac{\Omega^2}{2\pi G\bar{\rho}}\right) , \qquad (15.5)$$

$$E_\mu = \frac{\rho_m}{\bar{\rho}} \left\{ \frac{r}{3} \frac{d}{dr} \left[\left(H_T \frac{d\Lambda}{dr}\right) - (\chi_\mu + \chi_T + 1)\Lambda \right] - \frac{2H_T\Lambda}{r} \right\} . \qquad (15.6)$$

Here L and M are the luminosity and mass of the star; $\bar{\rho}$ represents the density average on the level surface ($\simeq \rho$) while ρ_m is the mean density inside the sphere of radius r; H_T is the temperature scale height; Λ represents the horizontal relative μ fluctuations; and χ_μ and χ_T represent the derivatives

$$\chi_\mu = \left(\frac{\partial \ln \chi}{\partial \ln \mu}\right)_{P,T} ; \quad \chi_T = \left(\frac{\partial \ln \chi}{\partial \ln T}\right)_{P,\mu} . \qquad (15.7)$$

In these expressions, the deviations from perfect gas law have been neglected, as have the energy production terms which are completely negligible in the regions of the star where the process takes place. We have also assumed differential rotation to be negligible, as observed in the radiative interior of the Sun from helioseismic studies. The corresponding condition on Ω is

$$\left|\frac{\partial \ln \Omega}{\partial \ln r}\right| < \frac{\Omega^2 r^3}{GM} . \qquad (15.8)$$

15.3.2 *Self-regulating process*

The hydrodynamical situation inside the star when the two currents become of the same order of magnitude is difficult to treat numerically, as the meridional circulation velocity nearly vanishes while two opposite "big numbers" cancel each other. A 2D numerical simulation of this process has been done in a static stellar model and gives interesting results, which will be extensively described by Théado & Vauclair (2002).

In the critical situation, the horizontal μ-gradient Λ remains everywhere close to the critical value Λ_c, except just below the convective zone. As microscopic diffusion does proceed, Λ is forced to remain small in the boundary layer, because the chemical composition is homogeneous inside the convective zone. As a consequence, the vertical velocity u_r strongly increases with radius as it goes from regions where $|E_\mu| \simeq |E_\Omega|$ to regions where $E_\mu \simeq 0$. The horizontal flow, which is derived from the mass continuity equation, is related to the induced negative divergence of the vertical flux:

$$\mathrm{div}(u_r \rho r^2) = -u_\vartheta \rho r \ . \tag{15.9}$$

Contrary to "normal" circulation, the direction of u_ϑ is reversed: it now goes from the downgoing flow towards the upgoing one while it normally proceeds the other way round. A loop forms below the convective zone so that the matter which goes down at the equator is brought back up to the convective zone in the upwards flow. The horizontal velocity in this boundary region is much larger than the vertical velocity because of the large variation of $(u_r \rho r^2)$ (equation 15.9). Its order of magnitude is given by

$$u_\vartheta H_\Lambda \sim u_r r \ , \tag{15.10}$$

where H_Λ is the diffusion scale height, needed for Λ to increase from zero to Λ_c. From numerical simulation, H_Λ is of order $r/100$, so that the horizontal velocities are about 100 times larger than the vertical ones. Below this region, the mixing velocities are very small in all the layers where the two currents are of the same order of magnitude.

15.4 Conclusion

The effect of the diffusion-induced μ-gradients on the meridional circulation, which were not introduced in previous computations, seems of fundamental importance in computing abundance variations in stellar surfaces. In the past, this effect was overcome by some parametrized inertial term introduced to avoid numerical instabilities. The difficulty comes from the fact that two large terms nearly cancel, which is difficult to treat numerically in the process of stellar evolution. The preliminary results presented here have been obtained with some assumptions. Two cases have been treated:

1) For lithium in halo stars, computations have been done in a stellar evolution code, with the assumption that, as soon as $|E_\mu| \simeq |E_\Omega|$, the horizontal and vertical μ-gradients adjust themselves to keep close to the critical regime (Théado & Vauclair 2001). We have shown that, under this assumption, the lithium plateau is well reproduced.

2) To test the process more precisely, we have done a 2D simulation in a static model, without any such assumption. We find that, as soon as $|E_\mu| \simeq |E_\Omega|$, a new loop takes place just below the convective zone, while the μ-gradients remain close to the critical values.

In the near future, the consequences of this process will be tested for the Sun and solar-type stars. The basic interest is that, for the first time, we have evidence of a mixing process which is directly modulated by the clement settling.

References

Antia, H.M. & Chitre, S.M., 1998, *A&A* **339**, 239

Balachandran, S.C. & Bell, R.A., 1997, *American Astron. Soc. Meeting* **191**, 7408

Basu, S., 1997, *MNRAS* **288**, 572

Basu, S., 1998, *MNRAS* **298**, 719

Bonifacio, P. & Molaro, P., 1997, *MNRAS* **285**, 847

Brun, A. S., Turck-Chièze, S. & Morel, P., 1998, *ApJ* **506**, 113

Christensen-Dalsgaard, J., Proffitt, C.R. & Thompson, M.J., 1993, *ApJ* **408**, L75

Christensen-Dalsgaard, J., Dappen, W., et al., 1996, *Science* **272**, 1286

Eddington, A.S., 1926, *The internal constitution of the stars*, Cambridge Univ. Press, Cambridge

Geiss, J. 1993, in Prantzos, N., Vangioni-Flam, E. & Cassé, M., eds, *Origin and Evolution of the Elements*, (Cambridge Univ. Press), p. 90

Geiss, J. & Gloecker, G., 1998, *Space Sci. Rev.* **84** , 239

Gough, D. O., Kosovichev, A. G., Toomre, J., et al., 1996, *Science* **272**, 1296

Maeder, A. & Zahn, J.-P., 1998, *A&A* **334**, 1000

Mestel, L., 1953, *MNRAS* **113**, 716

Mestel, L., 2003, in this volume

Michaud, G., 1970, *ApJ* **160**, 641

Michaud, G., Charland, Y., Vauclair, S. & Vauclair, G., 1976, *ApJ* **210**, 447

Michaud, G., Fontaine, G. & Beaudet, G., 1984, *ApJ* **282**, 206

Pinsonneault, M. H., Deliyannis, C. P. & Demarque, P., 1992, *ApJS* **78**, 179

Pinsonneault, M. H., Walker, T. P., Steigman, G. & Narayanan, V. K., 1999, *ApJ* **527**, 180

Richard, O., Vauclair, S., Charbonnel, C. & Dziembowski, W.A., 1996, *A&A* **312**, 1000

Schatzman, E., 1969, *A&A* **3**, 331

Schatzman, E., 1977, *A&A* **56**, 211

Théado, S. & Vauclair, S., 2001, *A&A* **375**, 70

Théado, S. & Vauclair, S., 2002, *ApJ*, in press

Vauclair, G., Vauclair, S. & Michaud, G., 1978a, *ApJ* **223**, 920

Vauclair, S., 1988, *ApJ* **335**, 971

Vauclair, S., 1999, *A&A* **351**, 973

Vauclair, S. & Charbonnel, C., 1995, *A&A* **295**, 715

Vauclair, S. & Charbonnel, C., 1998, *ApJ* **502**, 372

Vauclair, S. & Vauclair, G., 1982, *ARAA* **20**, 37

Vauclair, S., Vauclair, G., Schatzman, E. & Michaud, G., 1978b, *ApJ* **223**, 567

Zahn, J.-P., 1992, *A&A* **265**, 115

IV Helio- and asteroseismology

16

Solar structure and the neutrino problem

HIROMOTO SHIBAHASHI

Department of Astronomy, University of Tokyo, Tokyo 113-0033, Japan

There has been a long-standing discrepancy between the number of neutrinos expected from the sun and the number we actually detect. One possible interpretation for this was that our theoretical solar model was wrong. However, recent progress of helioseismology has shown that the real sun is very close to the latest solar models. On the other hand, very recent experiments of neutrino detection provided us evidence for neutrino oscillation. I discuss what we should do and what we can do in this situation for the neutrino physics from the astrophysical side.

16.1 Historical review: the solar neutrino problem

The energy source of sunshine (and shining of stars in general) is now thought to be nuclear fusion. To get direct evidence that nuclear reactions are really occurring in the sun is, however, a very challenging task. It takes $\sim 10^4$ years for photons generated by nuclear fusion near the solar centre to reach the solar surface, because the photons interact so frequently with matter in the sun. Hence, the photons by which we can see the sun right now do not tell us the physical state of the present solar core. On the other hand, since neutrinos interact little with matter, unlike photons, and travel at the speed of light, the neutrinos generated by nuclear reactions in the sun reach the earth only eight minutes after they are generated. The only possible way to confirm that nuclear reactions are indeed occurring in the sun is, therefore, to detect the neutrinos produced by nuclear reactions and to measure the neutrino fluxes from the sun.

The pioneering experiment to detect solar neutrinos, started by Raymond Davis Jr. at the Homestake mine, has, however, surprisingly shown that the capture rate of solar neutrinos is substantially smaller than the theoretically expected number. This discrepancy is known as the solar neutrino problem,

and has been one of the big problems concerning astrophysics and particle
physics (for reviews, see, e.g. Bahcall & Davis 1982, Bahcall 1989). The discrepancy implies either (i) something is wrong with the solar modeling, (ii)
there is an unknown behaviour of neutrinos or (iii) the experiments are incorrect. Since various aspects of the theory of stellar structure and evolution,
on which most of our understanding of the universe is based, are calibrated
with the sun, the solar neutrino problem and the resultant questions about
solar structure have been one of the big problems in astrophysics.

Neutrinos are generated in the sun mainly through three different nuclear processes in the pp-chain, — that is, the pp-neutrinos produced in
the pp-I branch, the ^{7}Be-neutrinos produced in the pp-II branch, and the
^{8}B-neutrinos produced in the pp-III branch:

pp-I: ^{1}H(^{1}H,e$^+\nu$)^{2}H(^{1}H,γ) ^{3}He(^{3}He,2^{1}H)^{4}He
pp-II: ^{3}He(^{4}He,γ)^{7}Be(e$^-$,ν) ^{7}Li(^{1}H,γ)^{8}Be$\rightarrow$2^{4}He
pp-III: ^{7}Be(^{1}H,γ)^{8}B(,e$^+\nu$) ^{8}Be$\rightarrow$2^{4}He

where, e.g., ^{2}H(^{1}H,γ)^{3}He means ^{3}He forms from ^{2}H by the capture of ^{1}H
followed by emission of a γ-ray. Since the main energy source is, we believe,
the pp-I branch, the pp-neutrino flux is much higher than the neutrino
fluxes associated with the pp-II and pp-III branches. However, since the
neutrino capture rate is dependent not only on the flux but also on the
energy spectrum, it is important to note that the energy spectra of the
neutrinos produced by these processes differ. The energy spectra of the
pp-neutrinos and the ^{8}B-neutrinos are continuous; the former is lower than
0.42 MeV while the latter extends up to 15 MeV. The ^{7}Be-neutrinos show
two line spectra at 0.861 MeV and 0.783 MeV.

Since the energy threshold of the Homestake experiment, which counts
^{37}Ar atoms produced by the neutrino capture of ^{37}Cl, is 0.814 MeV, the
Homestake experiment is sensitive mainly to neutrinos generated by the nuclear reaction of ^{8}B in the pp-III branch, which is highly sensitive to the
temperature near the solar centre. As the pp-III branch is not the main
source for the solar photon luminosity, it seemed that some slight reduction of the temperature near the solar centre would be able to explain the
discrepancy. Indeed, various attempts have been made along this line, but
none of them has succeeded in solving the solar neutrino problem without
any contradiction in the micro-physics or various observational data of the
sun. As a consequence, stable evolution of the sun, which was generally believed, was questioned. Dilke & Gough (1972) then suggested that the sun
might be unstable against some low-order gravity modes at certain epochs
of its main sequence evolution and that the deep interior could be mixed

intermittently due to the instability, resulting in a temporary depression of the solar neutrino flux. The occurrence of the suggested instability was soon confirmed by a global stability analysis of relevant modes for a realistic solar evolutionary sequence. Although it has not yet been established whether this instability does result in mixing, standard evolution of the sun has been questioned.

Three experiments started in late 1980s to detect solar neutrinos, — Kamiokande (and its successor Super-Kamiokande), GALLEX (and its successor GNO) and SAGE. All these new experiments show a deficit of neutrinos from the sun, and hence the third possibility listed above is proven to be highly unlikely. The energy threshold of Kamiokande, in which Cerenkov light emitted by electrons in water as a result of elastic scattering due to neutrinos is detected, is $\sim 7.5\,\mathrm{MeV}$, hence Kamiokande is sensitive only to the ^{8}B-neutrinos. The Kamiokande result, which showed the capture rate is about a half of the theoretical prediction, was once regarded as a confirmation of the Homestake result. On the other hand, the energy threshold of GALLEX and SAGE, which count radioactive ^{71}Ge atoms produced by neutrino capture of ^{71}Ga, is $0.233\,\mathrm{MeV}$ and this low threshold energy makes possible the detection of the pp-neutrinos generated by the pp-I branch, which is believed to be the main source of the solar photon luminosity, in addition to the higher energy ^{8}B-neutrinos and ^{7}Be-neutrinos. Both of these two experiments showed the same result; the capture rate is much lower than the expected value and is compatible with the predicted value of the pp-neutrinos but not with the grand sum including ^{7}Be- and ^{8}B-neutrinos.

It was noticed after careful comparison that the Kamiokande result is not consistent with the Homestake result. Kamiokande is sensitive only to ^{8}B-neutrinos while Homestake is sensitive not only to ^{8}B-neutrinos, but also to ^{7}Be-neutrinos, hence the Homestake measurement should be in excess of that predicted from the measured flux obtained at Kamiokande. The real data show, however, that this is not the case. This problem is often referred to as the beryllium problem, and it implies that the solar neutrino problem is caused by our lack of knowledge in neutrino physics rather than in astrophysics.

16.2 Historical review: helioseismology

In 1960 it was noticed that the solar surface was almost entirely covered with vertically oscillatory elements (Leighton et al. 1962). This oscillatory motion was called the "five-minute" oscillation, as its period was near five minutes. Later, these oscillations were identified as the superposition of eigenmodes of the sun — that is, many of the nonradial p-modes of the sun are excited and observed as five-minute oscillations. Excitation of the observed eigenmodes is now interpreted as a consequence of global resonance of the sun with acoustic noise generated by turbulent motion in the solar convection zone. Soon attention was paid to the possibility of using a large number of modes as diagnostics of the internal structure of the sun. One of the important characteristics of nonradial p-mode oscillations is that the oscillatory motion is well-trapped between a certain depth (which is dependent on the ratio of the frequency and the spherical degree of the mode) and a layer near the surface. The identified eigenmodes at the early phase of helioseismology were those trapped in the solar convection zone.

The first successful output of helioseismology was inference of the depth of the solar convection zone. The observed power spectrum is usually plotted on a two-dimensional diagram, often called the (k, ω)-diagram, on which the spherical degrees and the frequencies of the modes are plotted as the x- and y-axes, respectively. The power of the oscillations is concentrated in several distinct ridges. But the observed ridges appeared systematically below the theoretical eigenfrequencies. Roughly speaking, the eigenfrequencies of p-modes are determined by the travel time of the sound wave in the p-mode cavity. Hence the fact that the eigenfrequencies of the model are higher than those of the real sun means that the temperature in the convective zone of the model is higher than that of the real sun. Since the surface temperature is fixed, this in turn means that the estimated convective energy transport used in the model is less efficient than in the real sun. As a consequence, it was concluded that the solar convective region should be deeper than expected (Berthomieu et al. 1980). Stimulated by this fruitful output, helioseismology was then regarded as a new diagnostic tool of the invisible solar internal structure and was thus expected to be a useful key to solving the solar neutrino problem (see reviews, e.g. by Deubner & Gough 1984, Gough & Toomre 1991).

Detection of the oscillation signals in the five-minute period range in integrated sunlight over the whole disk by the Birmingham group and by the Nice group made this hope particularly realistic. The small patches of the oscillation patterns of high degree modes cancel each other in integrated sunlight, hence the whole-disk measurements are sensitive only to modes with low spherical degrees such as $l \simeq 0 - 4$. The detected oscillations were identified as a superposition of p-modes in this range of spherical degree l, which penetrate into the deep interior. Soon, various types of excellent narrow-band filters were invented and two-dimensional disk-image observations with them became popular, and p-modes with a wide range of spherical degree l were identified. With the development of the ground-based observation network (GONG) and the launch of a satellite (SOHO) dedicated to helioseismology, the eigenfrequencies of p-modes with a wide range of spherical degrees $l = 0 - 1000$ in the frequency range of $\sim 1.5 - 5.5$ mHz are now systematically measured with relative accuracies of the order of 10^{-5} (Hill et al. 1996, Kosovichev et al. 1997).

With a wide range of the combination of the eigenfrequencies and spherical degrees, we can now diagnose the invisible solar interior independently of the solar neutrino detection experiments. The p-mode pulsation characteristics of a star are determined by the sound-speed and the density profiles in the star. The equation governing linear adiabatic stellar pulsation is symbolically written in the form

$$\omega^2 \boldsymbol{\xi} = \mathcal{L}(\boldsymbol{\xi}; c(r), \rho(r)), \tag{16.1}$$

where ω and $\boldsymbol{\xi}$ denote the eigenfrequency and the (normalized) eigenfunction, respectively, and $c(r)$ and $\rho(r)$ denote the sound-speed and the density profiles, respectively, as functions of the distance r to the centre. It is known that the eigenfrequencies obey a variational principle. Hence, if we take a good reference model of the sun and compare its eigenfrequencies with those of the real sun, the differences in the eigenfrequencies between the model and the real sun are expressed in terms of the differences in the sound-speed and the density profiles between the model and the real sun:

$$\delta\omega^2 = \int \boldsymbol{\xi}^* \cdot \left(\frac{\partial \mathcal{L}}{\partial \ln c} \frac{\delta c}{c} + \frac{\partial \mathcal{L}}{\partial \ln \rho} \frac{\delta \rho}{\rho} \right) dm. \tag{16.2}$$

Equation (16.2) can be regarded as an integral equation with two unknown functions $\delta c(r)$ and $\delta\rho(r)$.

Various sophisticated methods of solving this type of integral equations have been invented, and the sound-speed and density profiles in the sun have been well determined (e.g. Gough & Thompson 1991). The transition of the temperature gradient from radiative to adiabatic at the base of the convection zone leads to a kink in the sound-speed profile. From this information, we now know precisely the location of the base of the convection zone (Christensen-Dalsgaard et al. 1991, Basu & Antia 1997).

As for solar modeling, standard evolutionary models of the sun have been improved with updated input physics. The relative difference in the sound-speed profile between the latest evolutionary models of the sun and the real sun is only of the order of 0.1%, which is surprisingly small from a general point of view. In this sense, it might be said that the latest evolutionary models are in good agreement with the helioseismic data. Indeed, the small-

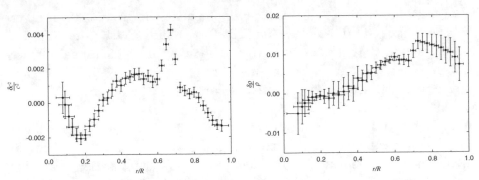

Fig. 16.1. The relative differences in the squared sound-speed profile (left) and in the density profile (right) between the helioseismically inversion and the evolutionary solar model by Christensen-Dalsgaard et al. (1996). The former is based on the frequency data obtained through 360 days by MDI on SOHO. Note that the relative difference is small, but it is larger than the statistical error bars.

ness of the relative difference in the sound-speed profile is often quoted as evidence for justification of the evolutionary models of the sun. However, it should be stressed here that the statistical error in the sound-speed inversion is substantially smaller than the difference between the models and the real sun. Hence, we should say that there still remain discrepancies between the latest evolutionary models and the helioseismic data.

16.3 Neutrino oscillation: MSW effect

A promising candidate for the particle physics solution of the solar neutrino problem is the MSW effect (Wolfenstein 1978, Mikheyev & Smirnov 1986), by which the apparent reduction of the solar neutrinos is explained by transformation of the electron-neutrinos generated in the solar core to those of other kinds (tau-neutrinos and muon-neutrinos) as a consequence of interaction with electrons in the sun. According to the standard theory of particle physics, neutrinos are massless particles. If the mass is non-zero, however, it is expected that the mass-eigenstate of neutrinos is not identical with the flavour-eigenstate — that is, the apparent states of electron-neutrino, muon-neutrino, and tau-neutrino should be described as three different mixed states of the three mass-eigenstates. How many electron-neutrinos generated in the sun are changed to other types of neutrinos is dependent on the energy of the neutrinos, the electron density profile in the sun, and the characteristics of neutrinos, which are the differences in the squared mass among the three mass-eigenstates and 'the mixing angles'. The energy spectrum of the generated neutrinos and the electron density profile are well defined by the solar internal structure. However, the mass differences and the mixing angles are not predicted theoretically at this stage and they are treated as free parameters. Since the pp-neutrinos, ^{7}Be-neutrinos, and ^{8}B-neutrinos produced in the sun have different energy spectra, the reduction rates due to the MSW effect are different among these three kinds of neutrinos for a given set of the neutrino parameters. By comparing each of the four observations (Homestake, Super-Kamiokande, GALLEX, and SAGE) with the corresponding theoretical prediction, we can limit the allowed regions in the parameter space. The problem is then to find the common regions for these allowed regions. The change from one flavour-eigenstate to another may frequently occur, being dependent on the parameters; this then is the phenomenon called neutrino oscillation. The physical cause of neutrino oscillation is evidently the non-zero mass of neutrinos. Hence, if the MSW effect is the solution of the solar neutrino problem, its impact is strong in particle physics as well as in other fields of astrophysics, such as cosmology.

Neutrino oscillation was hypothetical for a long time, but its reality was recently justified at Super-Kamiokande (Fukuda et al. 1998) by the careful observation of neutrinos produced by cosmic rays in the terrestrial atmosphere. Each pion produced by the collision of cosmic-ray protons with atoms in the atmosphere decays into a muon and a muon-neutrino ($\pi^{\pm} \rightarrow \mu^{\pm} + \nu_{\mu}(\bar{\nu}_{\mu})$). The former decays further into three particles — an electron, a muon-neutrino, and an electron-neutrino ($\mu^{\pm} \rightarrow e^{\pm} + \bar{\nu}_{\mu}(\nu_{\mu}) + \nu_e(\bar{\nu}_e)$). As a consequence, a single pion produces two muon-neutrinos and one electron-neutrino, hence the expected ratio of muon-neutrinos to electron-neutrinos, $(\nu_{\mu} + \bar{\nu}_{\mu})/(\nu_e + \bar{\nu}_e)$, is exactly 2. However, the observations show that this ratio is substantially smaller than 2 (about 60% of the theoretical prediction). Furthermore, the Super-Kamiokande experiments show clearly that this ratio is strongly dependent on the zenith angle of the incoming neutrinos. Since the cosmic-ray distribution is almost isotropic, the ratio is expected to be independent of the zenith-angle. The observed zenith-angle dependence contradicts this simple picture, and is in rather good agreement with the neutrino oscillation hypothesis. The anomalous ratio and the zenith-angle dependence of the atmospheric neutrinos are now recognized as clear evidence for neutrino oscillations (Fukuda et al. 1998).

Although the neutrino oscillation found in the atmospheric neutrinos cannot be applied to the solar neutrino problem directly, as the energy ranges of the solar neutrinos and of the atmospheric neutrinos are different, it seems unreasonable to reject the possibility of neutrino oscillations only in the case of the solar neutrinos. It may be fair to say that the recent general trend is, therefore, to shift the responsibility of the solar neutrino problem onto the MSW effect by assuming non-zero mass of neutrinos and to determine the possible range of the neutrino parameters.

16.4 SNO and Super-Kamiokande

The argument that the Super-Kamiokande (SK) experiment (based on the water-Cerenkov method) is sensitive only to solar ^{8}B-neutrinos might have been misleading in the sense that readers may have thought that SK was sensitive only to electron-neutrinos. The truth is that SK is also sensitive to muon-neutrinos and tau-neutrinos, but the sensitivity to these types of neutrinos is only $\sim 15\%$ of that to electron-neutrinos;

$$\phi_{\text{SK}} \simeq \phi(\nu_e) + 0.15\phi(\nu_{\mu,\tau}). \qquad (16.3)$$

While SK tries to catch Cerenkov light emitted by electrons running in light water, H_2O, as a consequence of elastic scattering by neutrinos, the Sudbury Neutrino Observatory (SNO) uses heavy water, D_2O, to measure neutrino fluxes via the different ways in which neutrinos interact with the heavy water. One of the neutrino reactions in heavy water is the "charged current reaction" (CC), in which, as the electron-neutrino approaches the deuterium nucleus, a W boson is exchanged and this changes the neutron in deuterium to a proton and the neutrino to an electron that emits Cerenkov light. Hence the neutrino flux measured in this way is that of electron-neutrinos only;

$$\phi_{\text{SNO}}^{\text{CC}} = \phi(\nu_e). \qquad (16.4)$$

Another type reaction is the "neutral current reaction," in which the deuterium nucleus is broken by neutrino hitting, and gamma-rays are emitted when the liberated neutron is captured by another nucleus. The third reaction is elastic scattering of electrons by neutrinos, and this is the same mechanism as that used in the SK experiments.

If we accept the MSW hypothesis described in the previous section, the flux of electron-neutrinos generated in the solar core is the total of the electron-neutrino flux and the muon- and the tau-neutrino fluxes measured at the earth. The SNO data of elastic scattering are less accurate than the SK data, as the total volume of water used in the SNO experiment is less than that used in SK and the integration time has not been long enough, hence the event number at SNO has been less than that at SK. Therefore, by combining the SK data with the SNO CC data, we can estimate with a reasonable accuracy the original ^{8}B-neutrino flux;

$$\phi_\odot(\nu_e) = \phi(\nu_e) + \phi(\nu_{\mu,\tau}) \simeq \phi_{\mathrm{SNO}}^{\mathrm{CC}} + (\phi_{\mathrm{SK}} - \phi_{\mathrm{SNO}}^{\mathrm{CC}})/0.15. \qquad (16.5)$$

The result is in good agreement with the theoretically expected value based on the evolutionary model of the sun, and this news was sensationally released as evidence for neutrino oscillation (Ahmad et al. 2001).

16.5 Recipe for construction of an evolutionary solar model

As described in previous sections, the solution of the solar neutrino problem is now highly likely to lie not with the sun, but with the neutrinos. The next important task is to determine the neutrino parameters. In this situation, what can astrophysicists do concerning the neutrino problem? It should be noted here that we need a good solar model to derive the neutrino parameters since they are obtained by comparing the observed neutrino fluxes with the theoretical expectation. It should also be noted that the deviation in the sound-speed profile and in the density profile of the evolutionary solar models from the helioseismically determined profiles is larger than the errors in helioseismology. Therefore, it is expected that we can construct a solar model, by using helioseismic information efficiently, which matches the seismic data better than the evolutionary models.

Let us review how to construct the standard evolutionary solar models. The basic assumptions are

- the model is in hydrostatic equilibrium,
- the model is in thermal equilibrium,
- chemically homogeneous at zero age,
- no mass loss and no mass accretion during evolution,
- the updated microphysics is used in calculations of nuclear reaction, opacity, diffusion, and convection,
- abundance changes are caused only by nuclear reactions and diffusion,
- the present age of the sun is $t_\odot = 4.5 \times 10^9$ yr,
- $M = 1\,M_\odot$, $R = 1\,R_\odot$, and $L = 1\,L_\odot$ at $t = t_\odot$, and
- the initial abundance ratio is chosen so that the surface abundances at $t = t_\odot$ match the present abundances observed at the photosphere.

Among them, the first assumption and the values of the present solar mass, radius, and luminosity should be definitely accepted. The real sun must be hydrostatic, otherwise the sun would shrink or expand within its dynamical timescale which is of the order of one hour. As for the mass, radius, and luminosity, they are very precisely measured compared with the other astrophysical quantities. In comparison with these, the other assumptions have less experimental support. For example, stable evolution of the sun has been questioned as described in Section 16.1.

The luminosity and the surface abundances of the model are mainly dependent on the initial abundances of hydrogen X_0 and of heavy elements Z_0. The radius of the model is strongly dependent on the treatment of convection. However, the theory of convection is still incomplete and we have to introduce a free parameter to express the efficiency of convective energy transport, — usually the mixing length, l. In summary, three parameters (X_0, Z_0, l) are introduced and they are chosen so that the last two requirements listed above are satisfied.

16.6 Recipe for construction of a seismic solar model

In order to estimate the neutrino fluxes from the sun, we need the temperature and the chemical-composition profiles in the sun. This requires more than the sound-speed profile $c(r)$ and the density profile $\rho(r)$, both of which are primarily determined from helioseismology. In the following, let us depart from the standard construction of a solar model and try to reconstruct a solar model by using only the experimentally well-measured quantities. These quantities are the mass $M_\odot$, the radius $R_\odot$, the photon luminosity $L_\odot$, and the sound-speed and the density profiles obtained from helioseismology. Whether the sun is in thermal equilibrium is uncertain, since even if the real sun is not in thermal equilibrium it takes about 10^7 yr for it to recover its equilibrium state. The justification for this assumption can be made by the solar neutrino flux measurement. If the ^{8}B-neutrino flux based on the constructed seismic solar model is in good agreement with the flux estimated with equation (16.5), this assumption may be justified. In summary, the assumptions in constructing a seismic solar model are as follows:

- the mass is $M_\odot$,
- the radius is $R_\odot$,
- the photon luminosity is $L_\odot$,
- the sound-speed profile $c(r)$ is that obtained from helioseismology,
- the density profile $\rho(r)$ is that obtained from helioseismology,
- the model is in hydrostatic equilibrium,
- the model is in thermal balance,
- the updated microphysics is adopted concerning the nuclear reactions, opacity, and the equation of state, and

- the envelope is chemically homogeneous, and Z/X there matches the spectroscopically determined value.

The basic equations for constructing a model with the above assumptions are formally identical with those used in constructing evolutionary models. The only difference is that the independent variable used in the evolution calculation is m which is a Lagrangian coordinate while the Eulerian coordinate r is adopted in the present case:

$$\mathrm{d}m/\mathrm{d}r = 4\pi r^2 \rho, \tag{16.6}$$

$$\mathrm{d}P/\mathrm{d}r = -Gm\rho/r^2, \tag{16.7}$$

$$\mathrm{d}L_r/\mathrm{d}r = 4\pi r^2 \rho \varepsilon, \tag{16.8}$$

$$\mathrm{d}T/\mathrm{d}r = \begin{cases} -3\kappa\rho L_r/(16\pi a c T^3 r^2) & \text{if radiative} \\ (1 - 1/\Gamma_2) T\,\mathrm{d}\ln P/\mathrm{d}r & \text{if convective} \end{cases} \tag{16.9}$$

with the boundary conditions $m = 0$ and $L_r = 0$ at $r = 0$ and $m = M_\odot$ and $L_r = L_\odot$ at $r = R_\odot$, where the symbols have their usual meanings (P: pressure, T: temperature, m: mass inside the radius r, L_r: luminosity at the radius r, ε: rate of nuclear energy generation, κ: opacity, a: Stefan constant, c: the speed of light, G: gravitational constant, Γ_2: adiabatic exponent).

In addition to these basic equations, we need the equation of state, the equations for the opacity and for the nuclear reaction rates,

$$\rho = \rho(P, T, X_i), \tag{16.10}$$

$$\kappa = \kappa(P, T, X_i), \tag{16.11}$$

$$\varepsilon = \varepsilon(P, T, X_i), \tag{16.12}$$

which link the thermal quantities and the chemical abundances.

It may be instructive to remember that, in constructing evolutionary models, the chemical composition profiles are given at each time step by following their temporal evolution

$$\partial X_i(t, m)/\partial t = (\partial X_i/\partial t)_{\text{nuclear}} + (\partial X_i/\partial t)_{\text{diffusion}} \tag{16.13}$$

assuming chemical homogeneity at zero age, $\partial X_i(t = 0, m)/\partial m = 0$, and then the basic equations can be solved as a closed system. A suspected instability and the resultant mixing may introduce an additional term in the right-hand side of equation (16.13), but such a term is ignored in the standard evolution scenario.

If we distinguish only hydrogen and helium separately as X and Y, respectively, and treat all other elements collectively as heavy elements, Z, then the sound-speed and the density can be regarded as functions of the chemical composition, X and Z, and any two other thermodynamical quantities such as P and T:

$$c = c(P, T, X, Z) \qquad (16.14)$$

$$\rho = \rho(P, T, X, Z). \qquad (16.15)$$

It should be noted here that, as demonstrated in Section 16.2, the sound-speed profile and the density profile in the solar interior have already been determined from helioseismology. Hence equations (16.14) and (16.15) inversely relate the hydrogen abundance, X, and the heavy element abundance, Z, at a given r to the pressure, the temperature, the sound-speed, and the density; $X = X(P, T, c_{\mathrm{inv}}, \rho_{\mathrm{inv}})$ and $Z = Z(P, T, c_{\mathrm{inv}}, \rho_{\mathrm{inv}})$, where $c_{\mathrm{inv}}(r)$ and $\rho_{\mathrm{inv}}(r)$ denote the seismically determined sound-speed and density profiles, respectively. The opacity and the nuclear reaction rates are, in turn, given in terms of $(P, T, c_{\mathrm{inv}}, \rho_{\mathrm{inv}})$ by equations (16.11) and (16.12), respectively. Thus all the variables appearing in the right-hand side of equations (16.6) – (16.9) can be expressed in terms of the variables in the left-hand side, and hence these equations are solvable. Note that in this way we obtain directly a model of the present-day sun. Note also that we do not need to make assumptions about the chemical composition profiles in the sun, but obtain the X, Y, and Z profiles as a part of the solutions of equations (16.6) – (16.9).

The depth of the base of the convection zone, r_{conv}, is well determined from helioseismology. Takata & Shibahashi (1998) shifted the outer boundary from $r = R_\odot$ to $r = r_{\mathrm{conv}}$ and required that the radiative temperature gradient matches the adiabatic temperature gradient there. The outer boundary conditions are then $L_r = L_\odot$ and $\nabla_{\mathrm{ad}} = 3\kappa L_r P/(16\pi acGmT^4)$ at $r = r_{\mathrm{conv}}$. This cutting off of the convective outer 30% of the sun has little effect on the solar interior where the neutrinos are generated. By setting the outer boundary at the base of the convection zone and treating only the radiative core, we do not need to worry about the treatment of convection.

It should be noted that the sound-speed profile is more accurately determined helioseismically than the density profile, as seen in Figure 16.1. Indeed, in early 1990s, attention was paid to the sound-speed profile rather than to the density profile. In those days, Shibahashi (1993) outlined a recipe for constructing a solar model with the constraint of the seismically determined sound-speed profile and with the assumption of a homogeneous

Z profile and simplified microphysics. Takata & Shibahashi (1998) polished this recipe and succeeded in constructing a realistic solar model by imposing a constraint of the seismically determined sound-speed profile $c_{\text{inv}}(r)$ and with the assumption of a homogeneous Z profile in advance.

16.7 Seismic solar model and the neutrino flux estimate

In practice it is fairly hard to determine the Z profile directly as outlined in the previous section, since the dependence of the equation of state on Z is weak. Alternatively, Watanabe & Shibahashi (2002) constructed a series of solar models with various Z profiles by imposing the constraint of the seismically determined sound-speed profile, and searched among them for the model for which the density profile fits best with the seismically determined density profile $\rho_{\text{inv}}(r)$. If $Z(r)$ is given, X is represented in terms of $(P, T, c_{\text{inv}}, Z)$ by equation (16.14), and in turn the density, the opacity and the nuclear reaction rates are given as well in terms of $(P, T, c_{\text{inv}}, Z)$ by equations (16.10) – (16.12), then equations (16.6) – (16.9) can be solved. The problem is changed to finding, among various seismic solar models, the model which minimizes $F \equiv \displaystyle\int_0^{r_{\text{conv}}} [\rho_{\text{inv}}(r) - \rho_{\text{model}}(r)]^2/\sigma_\rho^2 dr$, where σ_ρ is the standard error of the helioseismically determined density. By carrying out numerical experiments reproducing the solar model of Bahcall et al. (2001) with a constraint of the model's true sound-speed profile, Watanabe & Shibahashi (2002) demonstrated that the model's Z profile could be determined no better than to within 0.001. In applying the recipe for the realistic seismic data they restricted themselves to trying stepwise Z profiles, varying the height with a step of 0.001; the step width $\Delta r/R_\odot$ was 0.1 except for the innermost step which was taken to be 0.2.

Figure 16.2 shows the X, Y, and Z profiles determined in this way. The thick solid lines correspond to the best fit model, and the error bars in the X and Y profiles are the 1-σ uncertainty level due to all the uncertainties in microphysics and the errors in the inverted profiles of the sound speed and of the density. The 1-σ uncertainty level in the Z profile is shown by dotted lines. For comparison, the evolutionary model by Bahcall et al. (2001) is also shown with the dashed lines in each panel. As seen in the panel (c) of Figure 16.2, the uncertainty in the Z profile is quite large. But the very important point is that even the Z profile can also be deduced from helioseismology. It should be noted here that the main source of the error is not the error in the helioseismically determined profile of the sound speed nor that of the density, but the nuclear cross-section of the pp-reaction (S_{11}-factor).

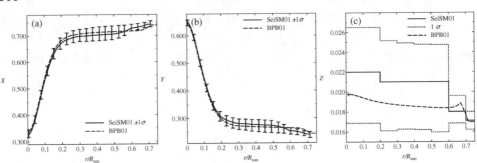

Fig. 16.2. Chemical composition profiles in the radiative core ($r \leq 0.713R_\odot$) of the seismic solar model by Watanabe & Shibahashi (2002). The outer convective envelope is assumed to be chemically homogeneous. The best-fit model corresponds to the thick lines and the 1-σ error levels are shown by tick bars in the panels (a) and (b) and by dotted lines in panel (c). For comparison, the evolutionary model by Bahcall et al. (2001) is shown by the dashed line.

We can say therefore that, with the help of knowledge of microphysics used in stellar evolution, helioseismology works well to construct a very accurate model of the present-day sun.

Once a good seismic solar model is constructed, one can estimate the neutrino fluxes based on it. Figure 16.3 shows the theoretically expected neutrino fluxes based on Watanabe & Shibahashi's (2002) seismic solar model for the gallium experiment (GALLEX, GNO and SAGE), the chlorine experiment (Homestake), and the ^{8}B-neutrino (SK and SNO). In each group, the theoretical value based on the seismic solar model is shown as the second left bar and the numerical value is given at its top with the units of SNU for the gallium and the chlorine experiments and $10^6 \mathrm{cm}^{-2}\mathrm{s}^{-1}$ for the ^{8}B-neutrino. (A SNU – solar neutrino unit – is defined to be 10^{-36} interactions per second per target atom.) The amount of contribution from each neutrino source is shown with different colour tones. The error bars correspond to 1-σ level. For reference, the theoretical estimate based on the evolutionary solar model by Bahcall et al. (2001) is shown as the left-most bar in each group. The values themselves are quite close to each other. However, it should be stressed again here that concepts of these two models are different — the seismic solar model is more directly constructed by using the experimentally well-measured quantities about the present-day sun. The error bars are apparently longer in the seismic solar model than the evolutionary model. As described previously, the main source of the error is the uncertainty in the S_{11}-factor. This is also true in the case of the evolutionary model; here, in order to keep the luminosity, the slight change in the S_{11}-factor can be

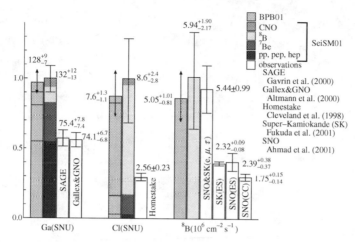

Fig. 16.3. Comparison of theoretically expected neutrino fluxes with the observations. Each of the neutrino fluxes is scaled with respect to the value of the seismic solar model by Watanabe & Shibahashi (2002).

compensated by changes in temperature and in density. On the other hand, in the case of the seismic solar model, since the density is constrained, the change in the nuclear reaction rate can be compensated only by changes in temperature. Hence, the uncertainty in the central temperature due to the uncertainty in the nuclear reaction rate is apparently larger in the case of the seismic solar model than the evolutionary solar model, and this leads to the apparently large error bars concerning the neutrino fluxes.

In Figure 16.3, the observed fluxes are also shown. As concerns the ^{8}B-neutrino flux measurements, the flux measured at SNO through the charged current reaction, $\phi_{\mathrm{SNO}}^{\mathrm{CC}}$, and the fluxes measured at SK and SNO through elastic scattering of electrons are also shown. Furthermore, based on the MSW effect hypothesis, the original ^{8}B-neutrino flux estimated from the combination of the SNO CC data and the SK data (see equation (16.5)) is also shown. As seen in Figure 16.3, the combined data of the Super-Kamiokande and SNO is consistent with the seismic solar model, which justifies the assumption about the thermal equilibrium of the model.

16.8 Future prospects

Together with evidence for neutrino oscillation obtained from atmospheric neutrinos that are described in Section 16.3, the consistency of the SNO-SK combined data with the neutrino flux estimate based on the seismic solar model strongly implies that the solar neutrino problem should be interpreted in terms of neutrino oscillation. The next thing to do is deduction of the neutrino parameters which determine the probability that neutrinos oscillate from one type to another, 'mixing angles' and masses, by comparing the theoretically expected fluxes and the observed ones. The seismic solar model provided by helioseismology should also be used in this process. Remembering that the solar oscillation itself had not been imagined before 1960, I cannot help feeling that it is amazing that looking carefully at the solar surface is eventually linked tightly to elementary particle physics.

Acknowledgements I thank M. Takata and S. Watanabe for help with producing the diagrams.

References

Ahmad, Q. R. et al., 2001, *PRL*, **87**, 071301-1.

Altmann, M. et al., 2000, *Phys. Lett. B*, **490**, 16.

Bahcall, J. N., 1989, *Neutrino Astrophysics*, Cambridge University Press.

Bahcall, J. N. & Davis, R. Jr., 1982, in Barnes, C. A., Clayton, D. D. & Schramm, D. N., eds, *Essays in Nuclear Astrophysics*, Cambridge University Press, p. 243.

Bahcall, J. N., Pinsonneault, M. H. & Basu, S., 2001, *ApJ*, **555**, 990.

Basu, S. & Antia, H. M., 1997, *MNRAS*, **287**, 189.

Berthomieu, G., Provost, J., Rocca, A., Cooper, A. J., Gough, D. O. & Osaki, Y., 1980, in Hill, H. A. & Dziembowski, W. A., eds, *Nonlinear and Nonradial Stellar Pulsation, Lecture Notes in Physics*, **125**, Springer-Verlag, p. 307.

Christensen-Dalsgaard, J., Gough, D. O. & Thompson, M. J., 1991, *ApJ*, **378**, 413.

Christensen-Dalsgaard, J. et al., 1996, *Science*, **272**, 1286.

Cleveland, B. T. et al., 1998, *ApJ*, **496**, 505.

Deubner, F.-L. & Gough, D. O., 1984, *ARAA*, **22**, 593.

Dilke, F. W. & Gough, D. O., 1972, *Nature*, **240**, 262.

Fukuda, S. et al., 2001, *PRL*, **86**, 5651.

Fukuda, Y. et al., 1998, *PRL*, **81**, 1562.

Gavrin, V. N., 2001, *Nuclear Phys. B Proc. Suppl.*, **91**, 36.

Gough, D. O. & Thompson, M. J., 1991, in Cox, A. N., Livingston, W. C. & Matthews, M., eds, *Solar Interior and Atmosphere*, Space Science Series, University of Arizona Press, p. 519.

Gough, D. & Toomre, J., 1991, *ARAA*, **29**, 627.

Hill, F. et al., 1996, *Science*, **272**, 1292.

Kosovichev, A. G. et al., 1997, *Solar Phys.*, **170**, 43.

Leighton, R. B., Noyes, R. W. & Simon, G. W., 1962, *ApJ*, **135**, 474.

Mikheyev, S. & Smirnov, A. Yu., 1986, *Nuovo Cimento*, **9C**, 17.

Shibahashi, H., 1993, in Suzuki, Y. & Nakamura, K., eds, *Frontiers of Neutrino Astrophysics*, Universal Academy Press, p. 93,

Takata, M. & Shibahashi, H., 1998, *ApJ*, **504**, 1035.

Watanabe, S. & Shibahashi, H., 2002, *PASJ*, submitted.

Wolfenstein, L., 1978, *Phys. Rev.*, **17D**, 2369.

17

Helioseismic data analysis

JESPER SCHOU

W. W. Hansen Experimental Physics Laboratory, Stanford University,
HEPL Annex A201, Stanford, CA 94305-4085, USA

The last decade has seen an impressive improvement in the quality and quantity of helioseismic data. While much of the progress has come from a new generation of instruments, such as GONG and MDI, data analysis has also played a major role. In this review I will start with a brief discussion of how the basic analysis of helioseismic data is done. I will then discuss some of the data analysis problems, their influence on our inferences about the Sun and speculate on what improvements may be expected in the near future. Finally I will show a selection of recent results.

17.1 Introduction

Until recently most research in helioseismology has used modes in the low ($l \leq 3$) and medium ($3 < l \leq 200$) degree (l) ranges. Here I will concentrate on the methods and problems in the study of medium-degree modes as well as show selected results. Most studies of modes of high degree ($l > 200$) have used entirely different analysis methods, such as time-distance analysis, which is discussed elsewhere in this volume (Kosovichev 2003). However, I will touch on some of the issues regarding the analysis of the high-degree modes by methods similar to those used for the medium-degree modes. The reader is also referred to Haber et al. (2002) for results from a technique known as ring diagrams which also uses high-degree modes.

I will start by providing some background material on solar oscillations in Section 17.2. In Section 17.3 I will describe some of the instruments used for the observation of solar oscillations, followed in Section 17.4 by a description of some of the procedures used for determining mode parameters and why this is harder than one might have thought. I will also discuss some of the problems encountered in this analysis because of mode properties, instrumental problems and limitations in the analysis procedures. While not directly related to helioseismology, I will discuss briefly some intriguing results obtained from the study of supergranulation in Section 17.5. Finally I will provide some concluding remarks and speculate on some of the improvements we may expect in the near future from improved instruments, data analysis methods and understanding of the underlying physics.

17.2 Background

The Sun oscillates in a large number of normal modes known as solar oscillations. The angular variation of the radial displacement of one of these modes at the solar surface is given by

$$Re\{A(t)Y_l^m(\theta,\phi)\},$$

where

$$Y_l^m(\theta,\phi) = P_l^m(\cos\theta)\exp(im\phi)$$

is a spherical harmonic, l is the degree of the mode, m the azimuthal order, A the mode amplitude, t time, θ the co-latitude and ϕ the longitude. In the absence of damping and excitation $A(t) = A_0\exp(i(\omega_{nlm}t - \psi_0))$, where $\omega_{nlm} = 2\pi\nu_{nlm}$ is the mode frequency and n the radial order. The radial variation of the eigenfunctions is more complicated and depends on l, n and the details of the solar model.

In the spherically symmetric case the mode frequency ω_{nlm} is independent of m, $\omega_{nlm} = \omega_{nl}$. The main perturbation lifting this degeneracy is the solar rotation. Specifically a slow rotation Ω causes a frequency splitting

$$\delta\omega_{nlm} = \omega_{nlm} - \omega_{nl} = \int_0^R \int_0^\pi \Omega(r,\theta)K_{nlm}(r,\theta)\mathrm{d}\theta\mathrm{d}r,$$

where r is radius and the kernels K can be determined from a solar model.

In reality the modes are damped and stochastically excited by the near-surface convection. In the Fourier domain the damping leads to a broadening of the spectral lines, changing them from delta functions to Lorentzians. To complicate things further the excitation process also leads to line asymmetries (e.g., Nigam et al. 1998).

For a more thorough review of the theoretical aspects of helioseismology see, e.g., the review by Gough & Toomre (1991).

17.3 Instruments

Most of the results obtained over the last several years have come from the analysis of data from only 2 projects: The Global Oscillation Network Group

(GONG) and the Solar Oscillations Investigation (SOI) using the Michelson Doppler Imager (MDI) instrument. In the following I will describe these projects and briefly mention some of the others.

17.3.1 *GONG*

The GONG project (Harvey et al. 1996) uses 6 instruments around the Earth to obtain nearly continuous (85% – 90% coverage) observations. When regular observations started in October 1995 the cameras provided 256^2 images (with rectangular pixels). However, in 2001 the cameras were upgraded to 1024^2 and the diameter of the solar images is now ≈ 800 pixels.

GONG produces Doppler velocity, intensity and modulation images on a one minute cadence at each site. Time series of spherical-harmonic transforms of the velocity images (Dopplergrams) are produced for each site (see Section 17.4), corrected for instrumental and site variations and merged to provide a single time series for each mode. For a variety of reasons the production of these time series has been limited to $l \leq 200$.

17.3.2 *MDI*

The MDI instrument (Scherrer et al. 1995) onboard the Solar and Heliospheric Observatory (SOHO) spacecraft was launched in December 1995 and started routine observations in May 1996. MDI produces 1024^2 images mostly with a cadence of one per minute. The MDI instrument can operate in one of several modes generating various combinations of Doppler images, intensity images and magnetograms in full disk and higher resolution. In the full-disk mode the whole visible disk of the Sun is observed while a (fixed) area is imaged with a magnification of 3.2 in the high-resolution mode.

While being located above the disturbing influence of the atmosphere at the Earth-Sun L1 Lagrange point means that the Sun can be observed continuously and with uniform quality, the telemetry is limited and only a subset of the images can be transmitted. For two to three months per year and roughly 8 hours per day a high telemetry rate is available and, depending on the observing program, Dopplergrams are available each minute.

In order to provide continuous data for the whole year a number of lower-resolution data products are produced, stored on board the spacecraft and downlinked several times per day. Most important for helioseismology is the so-called Medium-*l* program, in which the velocity images are convolved with a Gaussian, cropped at 90% of the solar radius and sampled at every fifth pixel to provided roughly 200^2 pixel images. The temporal coverage ranges from 95% to well over 99% on time scales of months. The time series from the Medium-l program are produced up to $l = 300$.

17.3.3 Other projects

A number of other projects have also been observing solar oscillations over the years and several are still operating, including ECHO, TON, BISON and IRIS all of which are ground-based networks and VIRGO and GOLF on-board the SOHO spacecraft. The observations from several of these projects have been important for the analysis of, in particular, the solar core.

17.4 Normal mode analysis

The analysis of the basic image data is generally done in a standard series of steps. First the images are calibrated, a spherical-harmonic transform (SHT) is performed, the resulting time series are Fourier transformed in time (and sometimes turned into power spectra) and the spectra are analyzed to estimate the various mode parameters. Each of these steps is described below followed by a discussion of some of the problems encountered in the analysis. For more details see Schou (1998) and Schou et al. (2002).

After this basic analysis an inverse procedure is typically applied to infer the properties of the solar interior, see Sekii (2003).

17.4.1 Time series generation

The first step in the analysis is to calibrate the basic images, typically Dopplergrams. Also basic image parameters such as image radius and center need to be determined.

The next step is to apodize and remap the images to a uniform grid in longitude and $\sin\theta$. The apodization, which is most often a cosine taper in fractional radius, is done to avoid sharp edges and the resulting ringing in the subsequent transform. The remapping takes into account the image center and apparent solar radius, the P- and B-angles, the finite observer-Sun distance and the non-uniform angular speed of the observer as seen from the Sun. (The P-angle is the position angle of the solar pole and the B-angle is the latitude of the sub-observer point.) The non-uniform angular speed of the observer is important since assuming a constant speed would lead to an apparent annual variation in the observed rotation rate.

The SHT is generally done by calculating inner products of the individual remapped images and the spherical harmonics of the target modes. In practice this is performed (Brown 1985) by using a Fourier transform in longitude and multiplying by the P_l^m's in latitude. This multiplication is quite time consuming and while faster algorithms (Mohlenkamp 1999) do exist, they are not used, mainly due to their complexity and the fact that the subsequent peakbagging, rather than the SHT, is the limiting factor in the overall data analysis.

The Fourier transform of the time series and the construction of power spectra is in principle straightforward. However, the time series typically have to be detrended and gap-filled in order to remove long-term drifts and gaps. In the MDI case the detrending is done by subtracting low-order polynomials, while GONG uses first differencing. GONG has also investigated using multi-tapers to construct the power spectra (Komm et al. 1999).

17.4.2 Peakbagging

The final and by far the most complex part of the data analysis is what has become known as peakbagging and ridge fitting where the mode parameters such as frequencies, linewidths and amplitudes are determined from the Fourier transforms or power spectra. This complexity is caused by a combination of properties of the oscillations and the way they are observed.

First of all the fact that we can only observe less than half of the solar surface means that, although the spherical harmonics form an orthonormal basis on a sphere, the SHT is not able to separate completely the individual modes. This leads to what is known as leaks in the time series and power spectra where, in addition to the target mode, several modes with nearby l and m appear. To be specific let x be the time series of the oscillations on the Sun and y the observed time series, then

$$y_{lm} = \sum_{l',m'} c_{lml'm'} x_{l',m'},$$

where $c_{lml'm'}$ is the so-called leakage matrix (see Schou & Brown 1994).

Furthermore, the power spectrum of a single damped and stochastically excited oscillator (such as a single mode) is given by the product of a Lorentzian and a random function with an exponential distribution. The width of the Lorentzian is given by the inverse damping time while the integral under the profile is given by the average power in the mode. For typical modes in the middle of the p-mode band the linewidths are of the order $1\,\mu$Hz.

The combination of the above two issues and the fact that the frequency spacings between adjacent modes are often comparable to the linewidths

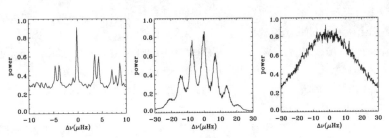

Fig. 17.1. Examples of m-averaged spectra for $l = 150$ using 72 days of MDI data. From left to right is $n = 0$, $n = 4$ and $n = 8$. The individual spectra were shifted by the measured splittings before being averaged. Notice the different scales on the frequency axis. On the left panel the peak around $\Delta\nu = 0$ is the target mode, the pairs around $\pm 4\,\mu$Hz are for $l = \pm 1$ (the splitting is caused by $m' - m = \pm 1$) and the one at $8\,\mu$Hz is from $l = 2$ (the split components are $m' - m = -2, 0$ and 2.

means that the fitting methods need to include assumptions about the frequency variations, simultaneous fits of different modes, and/or elaborate modeling of the effects of the atmosphere, instrument and the SHT. These problems are illustrated in Figure 17.1, which shows power spectra in the three main regimes: low frequency where the m leaks (those from the same l but different m) can be separated, moderate l and ν where the l leaks (those from different l) can be separated but the m leaks blend together and the ridge fitting regime where both l and m leaks blend together.

In the low-frequency regime simple algorithms should work well were it not for the fact that the signal-to-noise ratio (S/N) is quite low. Also the linewidth is sometimes less than the frequency resolution (i.e., the lifetime is longer than the observation time) and thus any given mode may or may not be excited.

Most of the normal analysis has so far been in the regime where the m leaks blend together but where the l leaks can still be separated. The methods developed here have also been used in the low-frequency regime. At present two main algorithms are being used: the MDI algorithm (Schou 1992) and the GONG algorithm (Anderson et al. 1990; Hill et al. 1996). These algorithms and some of their associated problems have been described by Schou et al. (2002).

17.4.2.1 The MDI algorithm

In the MDI algorithm all m values at a given (l, n) are fitted simultaneously and a leakage matrix calculated from our knowledge of the Sun, the instrument and the data analysis is used together with a maximum likelihood minimization which takes into account the correlation between the modes. In order to take the leaks and the associated correlations into account the Fourier transforms rather than the power spectra are fitted. The frequency interval used is generally quite small (typically 5 linewidths) in order to speed up the algorithm. The l leaks are calculated from the leakage matrix and the parameters fitted from the leaks. Since the parameters fitted for a mode depend on that of other modes, the whole procedure is repeated until (essentially) all the desired modes have converged. For the first iteration and for any modes outside the desired range a model is used the determine the mode parameters.

While not inherently a part of the MDI algorithm, individual mode frequencies are generally not fitted directly. Rather, the individual mode frequencies are expanded using so-called a coefficients

$$\nu_{nlm} = \nu_{nl} + \sum_{j=1}^{j_{\max}} a_j(n,l) P_j^{(l)}(m),$$

where $P_j^{(l)}$ are suitably chosen polynomials of order j (Schou et al. 1994), ν_{nl} is the mean multiplet frequency and the a_j are the a coefficients. Also, rather than fitting individual linewidths and amplitudes the inherent values are assumed to be independent of m. The correlation in the background noise between different m's is determined from the data far from the modes. Fitting the m's simultaneously and limiting the number of free parameters greatly stabilizes this method and allows it to operate at low S/N.

17.4.2.2 The GONG algorithm

The GONG algorithm is comparatively simple. Each mode (n, l, m) is fitted separately using a rather large frequency interval around the initial guess frequency. As part of the fit a number of adjacent modes (mostly l leaks) are fitted together with the target mode. The existence of m leaks is ignored, and the resulting peak is fitted as single mode, effectively averaging the underlying frequencies. No consistency is imposed between e.g. the frequency of a mode when it is the target mode and the frequency when it appears as a leak in another fit. Due to the mix of different m's in a given peak this would have required the use of a leakage matrix.

17.4.2.3 Ridge fitting

Finally in the region where both l and m leaks blend together so-called ridge fitting algorithms have been used. These have ranged from fairly simple cross-correlation algorithms (e.g., Korzennik 1990) to more sophisticated algorithms making use of leakage matrices (Rhodes et al. 2001). As discussed in the next section the results from these methods have, due to systematic errors, not been extensively used. However, recent progress (Rhodes et al. 2001) indicated that this may be possible in the near future, leading, one may hope, to significantly improved knowledge of the near-surface layers.

It should be mentioned that the technique known as ring diagrams is in many ways quite similar to this. In that technique a patch of the Sun is tracked and passed through a three-dimensional Fourier transform. The ridges in the resulting power spectra are then fitted to obtain frequency shifts, etc. For some results from this technique, see Haber et al. (2002).

17.4.3 Analysis problems

All the stages of the analysis can contribute to systematic errors in the final results. While it may be argued that all the problems are in the data analysis (the Sun does not make mistakes and many instrumental problems can be corrected in the data analysis), it is nonetheless useful to divide these problems into various categories including unknown physics or parameters, instrumental problems and approximations in the analysis. In addition, obvious problems of unknown origin seem ever present. Below I will briefly address some of these problems and give more details for a few selected ones.

17.4.3.1 Bad physics and parameters

One such problem is that the assumed direction of the solar rotation axis, which is needed for proper remapping, appears to be off by around 0.1° (Giles 1999). This is the most likely cause for the annual variation in the f-mode frequencies shown in Figure 17.2, which has been the cause of some confusion.

Other problems include the unknown ratio of the horizontal to vertical displacement near the solar surface, the exact place in the solar atmosphere where the oscillations are observed and the degree to which magnetic fields affect the observations.

17.4.3.2 Instrumental problems

While much has been done to measure or estimate the instrumental properties from first principles, there are many problems left. Among these are the cubic distortion introduced by the optics, distortion caused by the CCD not being perpendicular to the optical axis, non-square CCD pixels, errors in the absolute P-angle from the mounting of the instrument, the point-spread function (the PSF which describes how a point source is imaged) and its spatial/temporal variations, Doppler sensitivity variations and so forth.

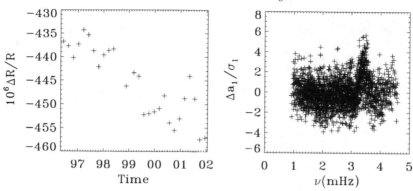

Fig. 17.2. Examples of systematic errors. Left panel shows the difference between the effective radius estimated from f modes and that of a solar model. The results were derived from 72-day time series of MDI Medium-l data covering the period 1996 May 1 to 2002 January 17. The right-hand panel shows the normalized residuals in a_1 from an inversion of 000 days of MDI data covering the period 1996 May 1 through 1997 April 25.

As an example consider the combination of a cubic distortion, a radius error and PSF uncertainties that has caused difficulties for MDI.

Consider a ray of light entering the optical system at an angle θ away from the optical axis. In an ideal optical system the distance y from the optical axis at which the light intersects the image plane is given by $y = ax$, where $x = \tan\theta$ and a is related to the focal length. One of the lowest order aberrations is the cubic distortion where $y = ax + bx^3$ and b/a describes the strength of the cubic distortion. Depending on the sign of b/a this is known as either pincushion or barrel distortion, named after the appearance of a square photographed through such an optical system.

To understand the effect of the cubic distortion note that it corresponds to a radially varying image scale and that the overall image scale is determined by measuring the apparent solar radius. However, since the data processing weights different parts of the image differently the effective image scale of the data actually used is different from the overall average and the main effect of the distortion is the same as that of an incorrect radius.

In the case of the ridge fitting what one effectively does is to determine the center of gravity of the ridge. The knowledge of the leakage matrix and assumptions regarding the smoothness of the frequency as a function of l then allows one to determine the frequency of the target mode, by correcting for the offset of the center of gravity of the leaks relative to the target mode. The results of numerical calculations of this shift for a number of cases are shown in Figure 17.3. As can be seen the shift is close to linear in l. To

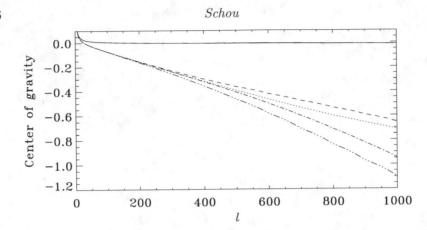

Fig. 17.3. The center of gravity of the ridges (in units of spherical-harmonic degree) as estimated from leakage matrices with various assumptions. The solid lines (on top of each other) are for the ideal case without and with a gaussian PSF. The dotted and dashed lines are the corresponding cases for a radius error of 0.057%. The dash-dotted and dash-triple dotted lines are for a cubic distortion similar to that in MDI.

see why this is the case consider that since the individual modes cannot be observed the only way to determine l is to used the apparent wavelength of the oscillations. With a fractional image scale error of ϵ one might thus expect l to be shifted by an amount ϵl. This is close to what is seen.

Also illustrated in Figure 17.3 is the effect of a cubic distortion. As can be seen it is indistinguishable from a radius error at $l \leq 200$ where the leaks can be measured. At higher degrees the results diverge and even show the opposite effect when a PSF is introduced. On the other hand the solid lines show that the PSF has little influence when the geometry is correct. It is clear that the underlying problem needs to be understood in order to estimate the corrections at higher degrees. The amount of cubic distortion has recently been determined from a ray trace of the optics. Assuming that there are no other similar problems, the necessary parameters may therefore be determined from the medium-degree fits and ridge fitting may see more extensive use in the future.

17.4.3.3 Algorithm problems

These are generally problems where a known effect is ignored or incorrectly modeled or where approximations have to be made in order to make the computational burden reasonable.

Among the effects that are commonly ignored, in addition to the instrumental problems mentioned above, are the horizontal displacement in the modes (see earlier), the distortion of the modes by the differential rotation and meridional flows (Woodard 1989, 2000) and line asymmetry. While some of these can be dealt with in the leakage matrix this is often not done and in the case of the GONG algorithm the leakage matrix is ignored. A difficulty in the MDI algorithm is the treatment of the background noise. Since the algorithm models the covariance between different spectra an estimate of the noise covariance matrix is needed. This is done by analyzing the noise far from the peaks of interest. Unfortunately, the covariance is significantly frequency-dependent and it is hard to find areas of the spectra without leaks except at very low and high frequencies.

The line asymmetry has been implemented in several of the codes. A problem for the MDI method is that it needs to calculate the profiles of the leaks far from the peak where the approximate equations often proposed break down At the same time it is not computationally feasible to fit the modes that far from the leaks and the distant behavior has to be extrapolated from the behavior near the peaks. While an algorithm has been implemented, some testing is still needed. On the positive side it appears that the effect on anything but the mode frequencies is negligible.

Other unpleasant problems include the gapfilling and the fact that the algorithms become unstable when the peaks start to turn into ridges.

17.4.3.4 Problems of unknown source

This is definitely the most insidious form of errors. The perhaps most infamous example is the bump seen in some of the a coefficients near 3.5 mHz shown in Figure 17.2. There are at least three reasons to believe that this bump is unphysical. One is that it is only seen when the MDI algorithm is used to analyze MDI data. The GONG algorithm or the MDI analysis of the GONG data do not give a bump. The second is that the exact shape of the bump depends on the width over which the modes are fitted. The third is that features in the Sun generally depend on ν/L ($L^2 = l(l+1)$), not on ν alone. To reproduce this bump would require extreme variations of the rotation rate with radius; however, while significant efforts have been spent on explaining this apparent systematic error, no definite cause has been determined yet (see also Schou et al. 2002).

17.4.4 Results

Having been comforted by the preceding I will show a few results from the normal mode analysis. Results regarding the interior rotation are shown elsewhere in this volume (Sekii 2003), so I will concentrate on a few results

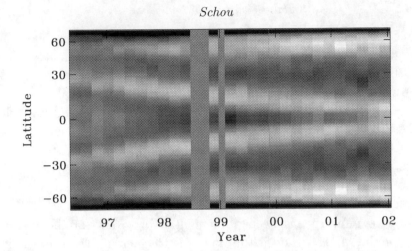

Fig. 17.4. The near-surface rotation rate as determined from the splittings of the f modes used to make the left panel of Figure 17.2. In order to show the weak time variations a smooth fit to the time-averaged rotation has been subtracted. Black corresponds to $-9\,\mathrm{m\,s^{-1}}$ while white corresponds to $+9\,\mathrm{m\,s^{-1}}$. It should be noted that due to the properties of the modes only the north-south symmetric component of the rotation can be determined. The vertical bands are due to missing data.

regarding the near-surface rotation. These results were derived using f-mode frequency splittings from a number of 72 day MDI time series. The f mode is easy to use because it has low linewidths (because of the low frequency) and therefore low random and presumably systematic errors (l leaks are largely unimportant). Also the mode kernels for the f mode have a single peak located very near the surface allowing them to be averaged together. Also a simple one-dimensional inversion in latitude can be used to determine the rotation rate. For the results shown here the f-mode splittings were averaged over the range $160 \leq l \leq 250$, and represent an average over the outer $\approx$1.5% of the Sun by radius. Details were given by Schou (1999).

Shown in Figure 17.4 is the near-surface rotation rate, as determined above, as a function of time and latitude over 6 years. As can be seen weak bands, known as torsional oscillations, are migrating towards the equator. It has been observed that the latitudes at which sunspots emerge appear to follow these bands and it is thus likely that they will provide hints as to the nature of the solar cycle.

Another result, shown in Figure 17.5 is that the near polar rotation rate is quite slow and appears to vary significantly with the solar cycle. An examination of the rotation rate with latitude (see Schou 1999) shows that the rotation rate starts to drop rapidly at around 70° compared to that derived from lower latitudes. As can be see from Figure 17.5 this drop is

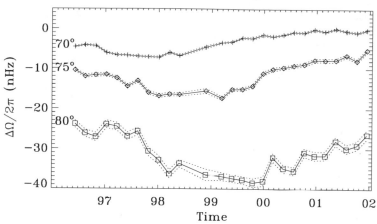

Fig. 17.5. The rotation rate from Figure 17.4 near the pole. The dotted lines show the $\pm 1\sigma$ random errors.

significantly time-dependent and it appears that the variations at higher latitudes lag behind those at lower latitudes, as if though a disturbance is traveling polewards. At 70° the rotation rate has a minimum in the middle of 1997 while the minimum at 80° does not occur until 1999. Also the rate at 70° appears to be reaching a maximum around 2001. Again it is probably reasonable to assume that this behavior near the pole is closely related to the operation of the solar cycle.

A final result from the f modes is shown in Figure 17.2. Apart from the systematic error mentioned in Section 17.4.3 there is a significant offset and a slow time-dependent variation, probably related to the solar cycle. This variation has been the subject of some study, and the reader is referred to Antia et al. (2000) and Dziembowski, Goode & Schou (2001) for interpretations of these results.

17.5 Supergranulation studies

This section actually has little to do with the title of this paper. That is, it is not helioseismology. Rather it is an intriguing result derived by filtering out the oscillations.

Supergranulation has for a long time been a poorly understood phenomenon. It is the dominant signal seen on the Sun when Dopplergrams are averaged together and appears as a convection like pattern with a scale of tens of Mm, a lifetime of the order 24 hours and a predominantly horizontal velocity at the surface. The traditional way of studying the supergranula-

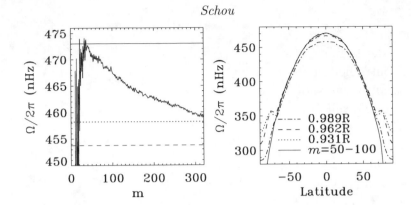

Fig. 17.6. The rotation rate as determined from supergranulation tracking and averaged over the 1996, 1997 and 1998 MDI dynamics runs (lengths of 60, 90 and 90 days, respectively). The left panel shows the near equatorial rotation rate as a function of m. The horizontal lines are from Snodgrass & Ulrich (1990). The dashed line is the spectroscopic rate, the dotted line the magnetic tracers and the solid line their supergranulation rate. The right-hand panel shows various rotation rates as a function of latitude. The inversion results (indicated by their radii) are from 360 days of MDI data covering the period 1996 May 1 through 1997 April 25. The solid line is the supergranulation rotation rate averaged over $50 \leq m \leq 100$.

tion is to average the Dopplergrams over several minutes and tracking the pattern to determine the rotation rate etc.

With the availability of continuous high-quality Dopplergrams from MDI, Beck & Schou (2000) and Schou & Beck (2001) decided to do a large-scale analysis to see if previous reports that the supergranulation is rotating faster than the photosphere could be confirmed. Briefly the analysis goes as follows. First a Gaussian filter with a FWHM of 8 minutes in time is applied to the images in order to suppress the oscillations and the resulting images are sampled every 15 minutes. These images are then remapped to a uniform grid in longitude and $\sin\theta$. At a fixed latitude a pattern such as the supergranulation will be moving diagonally in the longitude-time plane with a slope given by the rotation rate. To find this rotation rate each latitude slice is Fourier transformed in longitude and time and turned into a m-ν power spectrum. A cut in this spectrum at a fixed m will now show a peak at a frequency of $m\Omega$, thus allowing a measurement of the rotation rate as a function of m. Some results are shown in Figure 17.6.

Two main results are obvious from Figure 17.6. One is that the rotation rate is strongly m-dependent. Clearly supergranulation is not simply a pattern advected by the rotation at some particular depth. The other is that, at least at some m values, the rotation rate is higher than any of the surface rotation rates. As the right-hand panel of Figure 17.6 shows one has to go to the maximum in the rotation rate a depth of 7% of the solar radius in order to find a matching rotation rate. Such a deep anchoring is in apparent contradiction with both theoretical considerations suggesting that their vertical extent should be less than their horizontal extent and measurements by Duvall (1998) which gave a depth extent of only ≈1% of the radius. Regarding these results it should, however, be noted that Hathaway (2002) found similar measurements subject to significant systematic errors.

Other results derived include that the torsional oscillation is clearly visible in the supergranulation rotation rate (Schou & Beck 2001). This allows us to obtain much higher latitude resolution than does the normal-mode analysis and allows for a separate determination of the rotation rate in the two hemispheres. Another, more tentative, result from tracking supergranulation is that a meridional flow appears to be visible.

17.6 Conclusion and future prospects

As can be seen from the foregoing and from the other contributions to this volume, we have made very significant progress in our ability to infer solar properties using helioseismology and other techniques.

However, it has also become clear that there is still much to be learned and it is probably still true that we are more limited by our analysis techniques than by the availability of data. This is not to say that more data are not needed, especially in certain areas, but rather that much can still be learned from improvements in data analysis techniques.

If we are to continue to make significant progress we will need to both develop new analysis methods and to work hard on the known problems. In particular we will need better calibration of the instruments, an understanding of the remaining instrumental problems, such as distortions and the PSF and to implement this knowledge, as well as better physical models, into the analysis algorithms.

Having said that, I believe that the next few years will yield very significant improvements in our understanding of the Sun. Understanding the distortion, radius errors and the PSF should make the ridge fitting more reliable and allow us to probe the near-surface layers in more detail. Eliminating systematic errors such as the annual variations and the bump in the a coefficients around 3.5 mHz should improve our inferences about the interior. On the instrumental side we will soon have continuous observations covering a whole solar cycle, hopefully allowing us to understand the solar cycle better. But most importantly improved instruments and algorithms will allow us to study the Sun in ways we have not yet thought of.

Acknowledgements Much of the work described here was done by others or build on their work. Some are mentioned in the reference list, while others have undoubtedly been left out. I wish to take this opportunity to apologize for any inadvertent omissions and offer my thanks for their contributions. This work was supported by NASA grant NAG5-10483 to the SOI project at Stanford University. SOHO is a project of international cooperation between ESA and NASA.

References

Anderson, E. R. et al., 1990, *ApJ*, **364**, 699.

Antia, H. M., Basu, S., Pintar, J. & Pohl, B., 2000, *Solar Phys.*, **192**, 459.

Beck, J. G. & Schou, J., 2000, *Solar Phys.*, **193**, 333.

Brown, T. M., 1985, *Nature*, **317**, 591.

Duvall, T. L., Jr, 1998, in Korzennik, S. G. & Wilson, A., eds, ESA SP-418, *Structure and Dynamics of the Interior of the Sun and Sun-like Stars: Proc. SOHO 6 / GONG 98 Workshop.* ESA Publications Division, Noordwijk, the Netherlands, 581.

Dziembowski, W. A., Goode, P. R. & Schou, J., 2001, *ApJ*, **553**, 897.

Giles, P. M., 1999, PhD Dissertation, Stanford University.

Gough, D. O. & Toomre, J., 1991, *ARAA*, **29**, 627.

Haber, D. A. et al., 2002, *ApJ*, **570**, 855.

Harvey, J. W. et al., 1996, *Science*, **272**, 1284.

Hathaway, D. H., 2002, private communications.

Hill, F. et al., 1996, *Science*, **272**, 1292.

Komm, R. W. et al., 1999, *ApJ*, **519**, 407.

Korzennik, S.G., 1990, PhD Dissertation, Univ. Calif. Los Angeles.

Kosovichev, A. G., 2003, in this volume.

Mohlenkamp, M. J., 1999, *J. Fourier Anal. Appl.*, **5(2/3)**, 159.

Nigam, R. et al., 1998, *ApJ*, **495**, L115.

Rhodes, E.J., Jr. et al., 2001, *ApJ*, **561**, 1127.

Scherrer, P. H. et al., 1995, *Solar Phys.*, **162**, 129.

Schou, J., 1992, PhD Dissertation, Aarhus University, Denmark.

Schou, J., 1998, in Korzennik, S. G. & Wilson, A., eds, ESA SP-418, *Structure and Dynamics of the Interior of the Sun and Sun-like Stars: Proc. SOHO 6 / GONG 98 Workshop.* ESA Publications Division, Noordwijk, the Netherlands, 47.

Schou, J., 1999, *ApJ*, **523**, L181.

Schou, J. & Beck, J. G., 2001, in Wilson, A. ed, ESA SP-464, *Helio- and Asteroseismology at the Dawn of the Millennium: Proc. SOHO 10 / GONG 2000 Workshop.* ESA Publications Division, Noordwijk, the Netherlands, 677.

Schou, J. & Brown, T. M., 1994, *A&AS*, **107**, 541.

Schou, J., Christensen-Dalsgaard, J. & Thompson, M. J., 1994, *ApJ*, **433**, 389.

Schou, J. et al., 2002, *ApJ*, **567**, 1234.

Sekii, T., 2003, in this volume.

Snodgrass, H. B. & Ulrich, R. K., 1990, *ApJ*, **351**, 309.

Woodard, M. F., 1989, *ApJ*, **347**, 1176.

Woodard, M. F., 2000, *Solar Phys.*, **197**, 11.

18

Seismology of solar rotation

TAKASHI SEKII

Solar Physics Division, National Astronomical Observatory of Japan,
Mitaka, Tokyo 181-8588, Japan

Helioseismology provides us with means to investigate the otherwise invisible solar interior. The seismic approach is indispensable for the study of internal structure and evolution of the sun. It is even more so, however, for the study of dynamical aspects of the sun, because of the lack of other reliable means. The current status of seismology of solar rotation is reviewed and outstanding problems are discussed.

18.1 Introduction

In 1984, Douglas Gough started his paper, entitled 'On the rotation of the Sun', by pointing out our lack of understanding of the dynamical history of the sun (Gough 1984). The question of how the sun has evolved dynamically, since its arrival on the main sequence, still stands as one of the big questions in astronomy. With an increased level of interest attracted by the issue of how our solar system (and other 'solar' systems) formed and evolved, it may be a problem of even greater importance today.

Another big problem regarding the solar rotation is what is behind the solar cycle, and if a dynamo mechanism is responsible, as is generally believed, how it works. Here, too, the problem seems to be recognized in a wider community because of the great interest currently shown towards the solar-terrestrial study.

In tackling both problems, an important key is the dynamical structure of the sun *today*, and in particular how it rotates. Observational clues are not many. We observe the surface differential rotation using tracers such as sunspots and other magnetic features, or by measuring Doppler velocity. There have been reports that coronal holes seem to rotate more or less rigidly (Weber et al. 1999), which presumably is related to how the base of open magnetic field lines rotates in the solar interior. These are important observational constraints, but all of them are either limited to the near-surface layers, or are difficult to interpret.

On the theoretical side, the picture is not too bright either. Direct simulations of solar and stellar dynamical evolution have been attempted (Deupree & Guzik 2000; see also Maeder & Meynet 2000). MHD modelling of the convection zone of the present sun has been progressing (Toomre 2003). One fundamental difficulty in any such attempt is that, according to the Kolmogorov argument (see e.g. Frisch 1995), turbulent convection in three dimension, with Reynolds number $R_e \sim 10^{12}$, implies the degree of freedom of $\sim (L/\eta)^3 \sim R_e^{9/4} \sim 10^{27}$, where L is the length scale of the system and η is the Kolmogorov dissipation scale. This means that, in a straightforward gridding method, one requires 10^{27} grid points, which is a huge number even for modern computers. One cannot take a head-on approach.

Douglas Gough's 1984 paper was about changing this situation by helioseismology. Helioseismology is a revolutionary tool to investigate the solar interior, and it has revealed statical and dynamical properties of the solar interior, and has constrained input physics (see, e.g., Gough et al. 1996). From the discussion above, however, it seems that the study of solar dynamics benefits from helioseismology much more than do stellar evolution studies.

Seismology of solar rotation is a two-step process. First, we need rotational splitting measurement. Second, we have to carry out inversion to obtain a spatially resolved view of solar rotation. On accomplishing these two steps, one can then proceed to investigate the physics of solar rotation. Below I shall summarize the background and current status of, and discuss problems in and around, seismology of solar rotation.

18.2 Helioseismic measurement of solar internal rotation

Helioseismology, as it is practiced today, is mainly about using eigenmodes of acoustic oscillations as a probe into the solar interior. Wave propagation in the sun is affected by the solar rotation, and then the quantization condition for forming standing waves changes, thereby shifting the eigenfrequencies. This is done principally by advection in the case of high-frequency p modes.

Since the sun is close enough to a sphere, we can identify the normal modes of the solar oscillation by radial order n, spherical-harmonic degree l and azimuthal order m ($= 0, \pm 1, \pm 2, \ldots, \pm l$). Without rotation or any other symmetry-breaking agent, the eigenfrequency depends only on n and l. We may write this (hypothetical) frequency as ω_{nl}. In the presence of rotation, however, this degeneracy with respect to m is lifted and the frequency depends on m as well. Let us denote this frequency by ω_{nlm}. If we forget magnetic field and other structural asphericity for the moment, we can say that the difference between ω_{nlm} and ω_{nl} is caused by rotation. Let us write down, in spherical polar coordinates (r, θ, ϕ), the displacement vector associated with mode (nlm) as $\boldsymbol{\xi}_{nlm}(r, \theta, \phi) \exp(i\omega_{nlm}t)$, where t is time. Then the pulsation equation can be written in the form

$$\mathcal{L}\boldsymbol{\xi}_{nlm} = \rho\omega_{nlm}^2\boldsymbol{\xi}_{nlm} ,$$

where ρ is the density and $\mathcal{L}$ is a linear operator. For the expression of $\mathcal{L}$ see, e.g., Unno et al. (1989).

The frequency shift due to rotation $\boldsymbol{\Omega}_{\mathrm{rot}}(r, \theta) = \Omega(r, \theta)\boldsymbol{e}_z$ ($\boldsymbol{e}_z$ is a unit vector in the z-direction) is calculated by taking account, by linear perturbation theory, of the perturbation $\Delta\mathcal{L}(\boldsymbol{\xi}) = 2i\omega_{nl}\rho(\boldsymbol{v}_{\mathrm{rot}} \cdot \boldsymbol{\nabla})\boldsymbol{\xi}_{nlm}$ to the linear operator in the pulsation equation caused by the rotational velocity $\boldsymbol{v}_{\mathrm{rot}}$. This results in

$$\Delta\omega_{nlm} \equiv \omega_{nlm} - \omega_{nl} = \frac{m}{I_{nl}} \int |\boldsymbol{\xi}_{nlm}|^2\Omega\rho\mathrm{d}V - \frac{i}{I_{nl}} \int \boldsymbol{\xi}_{nlm}^* \cdot (\boldsymbol{\Omega}_{\mathrm{rot}} \times \boldsymbol{\xi}_{nlm})\rho\mathrm{d}V ,$$

where $\mathrm{d}V = r^2 \sin\theta\mathrm{d}r\mathrm{d}\theta\mathrm{d}\phi$ and

$$I_{nl} = \int |\boldsymbol{\xi}_{nlm}|^2\rho\mathrm{d}V ,$$

which does not depend on m (integrations are over the whole volume of the sun). This is often rewritten in the following form:

$$\Delta\omega_{nlm} = m \int K_{nlm}(r, \theta)\Omega(r, \theta)\mathrm{d}r\mathrm{d}\theta . \tag{18.1}$$

The rotational shift of the frequency $\Delta\omega_{nlm}$ is essentially a weighted average of $\Omega(r, \theta)$, and the weighting function $K_{nlm}(r, \theta)$ is the splitting kernel, which can be derived from an equilibrium model and its eigenfunctions after solving the eigenvalue problem under appropriate boundary conditions. In the solar case $\Delta\omega_{nlm} = m\langle\Omega\rangle \sim m \times 400\,\mathrm{nHz}$, which already gives a measure of the solar rotation rate without any further analysis (here $\langle\cdots\rangle$ denotes an average). However, to disentangle contributions from different parts of the sun, so that we have any spatial resolution, we need an inversion procedure.

18.3 Inversion for internal rotation

Since in equation (18.1) $\Delta\omega$ is expressed as a linear functional of Ω, the inverse problem for Ω is linear. There are various inversion methods that have been proposed or used for the rotation inversion, such as i) asymptotic methods, which rely on asymptotic approximations to yield semi-analytically invertible expressions, ii) methods based on optimally localized averaging (OLA), and iii) methods based on least-squares fitting, such as regularized least-squares fitting (RLSF). Normally, model fitting is distinguished from inversion but the boundary can be fuzzy, as can be seen from the inclusion of the RLSF in the list above. There are certain important properties of linear inverse problems, some very general, some depending on which linear inverse problem one attempts to solve, and some depending on which method one uses. For any linear inverse problem,

- there exist functions orthogonal to all the kernels (the space spanned by such functions is called the annihilator),
- there is a trade-off between resolution and error magnification,
- the region where no kernel has amplitude is inaccessible, and
- data redundancy can be caused by kernels having similar structure, even locally

which all limits the amount of information we can extract. Properties specific to the inversion of rotational splitting are derived from the properties of the p-mode splitting kernels K_{nlm}, which are

- K_{nlm} has no amplitude close to the centre or to the pole,
- K_{nlm} has large amplitude near the surface,
- K_{nlm} contains only those terms that are even in m ,
- K_{nlm} is sensitive only to the north-south symmetric component.

It is interesting to compare the first two of the list immediate above, and the last two of the previous list. From the third item in the above, we see that any even component in the frequency shift is due to magnetic field or asphericity, including the second-order effect of rotation (cf. Dziembowski & Goode 1992). Observationally, this even component is small but detectable. In order to remove any effect of the even component from rotation inversion, we can use $\Delta\omega_{nlm} \equiv (\omega_{nlm} - \omega_{nl,-m})/2$ as the new definition of $\Delta\omega_{nlm}$.

For the convenience of a more formal discussion, let us rewrite the set of linear constraints on $\Omega(r,\theta)$ in a more symbolic fashion. Introducing discretization at this point is not quite necessary, but it will save the need of switching between discussion of functions and discussion of vectors of finite dimension, in what follows. Suppose we discretize equation (18.1) using N grid points and obtain

$$\sum_{j=1}^{N} K_{ij}\Omega_j = d_i \ ,$$

where the mode index i, which corresponds to a certain set of (nlm), runs from 1 to M (the number of the modes). Here the right-hand side represents $\Delta\omega_{nlm}/m$ for the mode i. Then we can introduce an $M \times N$ matrix K, an N-vector $\boldsymbol{\Omega}$ and an M-vector $\boldsymbol{d}$ to write

$$K\boldsymbol{\Omega} = \boldsymbol{d} \ . \tag{18.2}$$

Since in reality any observation is associated with error, we must replace $\boldsymbol{d}$ by

$$\hat{\boldsymbol{d}} = \boldsymbol{d} + \boldsymbol{e} \ ,$$

where the term $\boldsymbol{e}$ represents observational error.

Solving a linear inverse problem by a linear process is equivalent to finding, somehow, another matrix R to obtain an estimate of $\hat{\boldsymbol{\Omega}}$ as

$$\hat{\boldsymbol{\Omega}} - R\hat{\boldsymbol{d}} \ . \tag{18.3}$$

Obviously, the component of $\hat{\boldsymbol{\Omega}}$ that is due to the term $\boldsymbol{e}$ is $R\boldsymbol{e}$. We would like $\delta\hat{\boldsymbol{\Omega}} \equiv |R\boldsymbol{e}|$ as small as possible. However, this alone does not guarantee that $\hat{\boldsymbol{\Omega}}$ resembles $\boldsymbol{\Omega}$. For this there are two other important aspects we need to check. One is the degree of misfit: $|\hat{\boldsymbol{d}} - K\hat{\boldsymbol{\Omega}}| = |(I_M - KR)\hat{\boldsymbol{d}}|$, where I_M is the identity matrix of size M. The other is resolution. Since $\hat{\boldsymbol{\Omega}} = RK\boldsymbol{\Omega} + \delta\hat{\boldsymbol{\Omega}}$, we can examine RK to see how well features in $\boldsymbol{\Omega}$ are resolved in $\hat{\boldsymbol{\Omega}}$.

Simplistically put, methods based on optimally localized averaging (OLA) aims at rendering RK as diagonal as possible ($RK \to I_N$), while keeping the magnitude of $|R\boldsymbol{e}|$ at a reasonable level.

On the other hand, methods based on least-squares fitting aims to minimize the misfit ($KR \to I_M$), also without increasing the magnitude of $|R\boldsymbol{e}|$ too much. When applied with an additional constraint on smoothness, we have RLSF.

The OLA and the RLSF, including their variations, are currently the two most often used inversion methods. For more details and discussion, see e.g. Sekii (1997).

One of the often overlooked issues is correlation of data errors (Gough 1996, Gough & Sekii 1998) and correlation of the estimates (Howe & Thompson 1996).

Correlation of data errors is mainly caused by mode leakage (see also Schou 2003). Spherical harmonics are orthogonal to each other on a complete sphere, but since we do not observe the entire surface of the sun, or even the half of it in a uniform manner, spherical-harmonic decomposition of the surface wavefield leads to leakage of power in one set of (lm) to other sets.

This is in principle tractable, but even so the sheer size of the covariance matrix, comprising $M^2/2$ independent elements (although not all of them of significant magnitude) renders it challenging in real cases to estimate and then utilize the covariance matrix. However, as Gough (1996) pointed out, a certain type of data correlation can lead to spurious features in inversions. It is necessary to proceed with caution.

From equation (18.3) we have the covariance of $\hat{\boldsymbol{\Omega}}$:

$$\langle \delta\hat{\boldsymbol{\Omega}}\delta\hat{\boldsymbol{\Omega}}^T \rangle = R\langle \boldsymbol{ee}^T \rangle R^T \;,$$

where $\langle \cdots \rangle$ denotes a statistical average. Even if there is no correlation in the data errors, i.e., $\langle \boldsymbol{ee}^T \rangle$ is diagonal, it does not mean that the covariance matrix of $\hat{\boldsymbol{\Omega}}$ is diagonal, irrespective of how RK or KR resemble identity matrices. In general, there is correlation between elements of $\hat{\boldsymbol{\Omega}}$, that is to say, between the estimates of Ω at a certain part of the sun and another. As Howe & Thompson (1996) pointed out, this correlation can also lead to a spurious feature in inversion that looks like a systematic trend but is in fact an artifact.

18.4 Solar internal rotation observed by helioseismology

18.4.1 Observational data

The observational data that are currently used widely for seismology of solar rotation are obtained mainly through Doppler measurements, such as GONG (`http://www.gong.noao.edu/`), a ground-based network observation, and SOI/MDI (`http://soi.stanford.edu/`), on-board the SOHO satellite. It has turned out that measuring individual frequencies $\{\omega_{nlm}\}$ is rather hard. The GONG project determined $\sim$200,000 mode frequencies around 1995 with a mixed result, although they have recently restarted their attempt (Hill, private communication). On the other hand, so far there has not been any such attempt from the SOI/MDI project. Instead of measuring individual frequencies, a polynomial expansion is used:

$$\Delta\omega_{nlm} = \sum_{k=1}^{N_p} a_k(n,l)\mathcal{P}_k(m;l) \;,$$

where $\mathcal{P}_k(m;l)$ is, e.g., a degree-k polynomial of m/l. The coefficients are called *a coefficients*. The fitting for the $a_k(n,l)$ is done in Fourier space (or power) directly, rather than painstakingly producing a poorly measured table of $\{\omega_{nlm}\}$ first. Since the rotationally induced component of $\Delta\omega_{nlm}$ is an odd function of m (see previous section), the effect of rotation appears only in the odd coefficients a_1, a_3, a_5, ..., while the even coefficients a_2, a_4, a_6, ... are measures of asphericity in solar structure.

18.4.2 How to tackle 2-dimensional (2D) inversions

The current datasets from GONG or SOI/MDI comprise a few thousands of (n, l) multiplets of degree l up to ~ 250. The a coefficients are most reliably measured in the frequency range of $\sim 2\,\mathrm{mHz}$ to $\sim 4\,\mathrm{mHz}$. The number of terms in expansion is ~ 20 or more.

In the previous section various inversion methods were briefly mentioned. The classification there was based on what each method does mathematically. When discussing 2D inversions, we can also classify inversion methods in terms of strategy they take in tackling the 2D problem.

The first is the most obvious — direct 2D approach whose only fault is that it is computationally most expensive (Sekii 1990, 1991, Schou 1991, Christensen-Dalsgaard et al. 1995).

In the so-called 1.5D approach, $\Omega(r, \theta)$ is expanded to reduce the 2D problem to a set of one-dimensional (1D) problems (Brown et al. 1989):

$$\Omega(r, \theta) = \sum_k \Omega_k(r) Q_k(\cos \theta) ,$$

which, if Q_k and $\mathcal{P}_k(m; l)$ have been chosen properly, leads to a set of 1D problems

$$a_k(n, l) = \int \tilde{K}_k^{nl}(r) \Omega_k(r) dr \ (k = 1, 3, 5, \ldots) .$$

As Sekii (1995) pointed out, with this approach one does not have control over the angular resolution.

Then there is the 1D×1D approach. By exploiting the fact that, for p modes,

$$K_{nlm}(r, \theta) \simeq K_{nl}(r)[P_{lm}(\cos \theta)]^2 ,$$

the 2D inverse problem is separated into radial and latitudinal 1D problems (Sekii 1993, Pijpers & Thompson 1996). It is then possible to control both the angular and radial resolution. Note that the 1D×1D approach is not necessarily limited by the accuracy of the approximate formula above.

There are many combinations of choices of which basic method one uses and which strategy one takes for a 2D case. A good news is that when applied sensibly, they all agree reasonably well (cf. Schou et al. 1998). A bad news is that none of these combinations can perform magic; irrespective

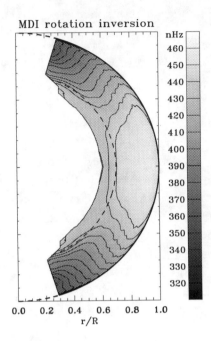

MDI rotation inversion

Fig. 18.1. Differential rotation in the sun inferred from inversion of MDI data. Adapted from Schou et al. (1998).

of the way it is solved, the central region and the high-latitude region are difficult to access because few splitting kernels have any amplitude there.

18.4.3 *What we have learned*

Figure 18.1 shows an example of rotation inversion of MDI data (adapted from Schou et al. 1998). Let us summarize what we have learned from such inversions of the current helioseismic data.

(i)   The surface differential rotation pattern holds through most of the the solar convection zone. In particular, there is no Taylor-Proudman-type cylindrical profile in the convection zone, which compels numerical modelling of a dynamical dynamo to be reconsidered.

(ii) Differential rotation is weak, if present, in the radiative interior.

(iii) In low- to mid-latitude, the rotation rate is maximum around $r/R_\odot \simeq 0.95$, with a sub-photospheric shear layer immediately above. It has been suggested that this layer supports a small-scale dynamo.

(iv) There is another shear layer at the base of the convection zone, named the *tachocline* (Spiegel & Zahn 1992, Gough & McIntyre 1998).

The sound-speed anomaly found also at the base of the solar convection zone has been interpreted as evidence for extra mixing associated with the tachocline (Gough et al. 1996). The classical α-ω dynamo has difficulty in explaining equatorward migration of a dynamo wave, both due to the magnitude and the sign of the radial gradient of angular velocity in the bulk of the solar convection zone. An interface dynamo mechanism at the base of the convection zone has been suggested (Choudhuri 1990, Parker 1993).

There are some other features such as high-latitude jet (Schou et al. 1998), rotation slowing down near the pole (Schou et al. 1998) and zonal flow (Kosovichev & Schou 1997), which experts agree to be as real as the data, in spite of their potentially controversial nature. The physical implications of these features are yet to be understood. The variation over the solar cycle has also been investigated (Howe et al. 2000). It seems that banded zonal flows are moving towards the equator as a cycle progresses, in the region down to $\sim 0.1 R_\odot$ depth.

18.5 Rotation in the the solar convection zone

Initially, there was some doubt expressed as to the genuineness of the non-cylindrical rotation profile in the solar convection zone. Some discussed that it might be an artifact of the seismic analysis due to lack of spatial resolution (Gough et al. 1993, Sekii, Gough & Kosovichev 1995). With improved data and more spatial resolution, however, it now seems an unlikely explanation.

So why does the rotation rate in the solar convection zone not follow the Taylor-Proudman behaviour? Let us start from the momentum equation for the fluid velocity v in a rotating frame:

$$\frac{\partial v}{\partial t} + (v \cdot \nabla)v = -\frac{\nabla P}{\rho} - \nabla \Phi - 2\Omega_0 \times v \ ,$$

where P and Φ are pressure and gravitational potential, respectively, and Ω_0 denotes the angular velocity of the rotating frame, which is to be identified with the local angular velocity of the point under consideration. We drop the LHS for a stationary flow and with a small inertia term, and take the curl of the RHS which has to vanish. From its ϕ component we have

$$2\Omega_0 \frac{\partial v_\phi}{\partial z} = \frac{(\nabla P \times \nabla \rho)_\phi}{\rho^2} = \frac{(\nabla T \times \nabla P)_\phi}{\rho T} \tag{18.4}$$

(v_ϕ is the ϕ component of v), where a simple equation of state $P \propto \rho T$ is

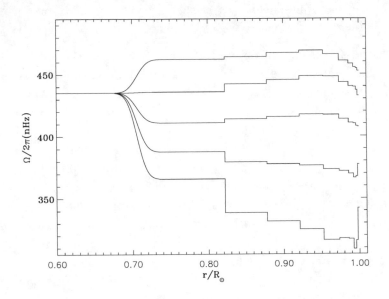

Fig. 18.2. Rotation rate obtained by a non-linear model fitting to MDI data. From top to bottom, angular velocities at the latitudes of $0°$, $30°$, $45°$, $60°$ and $90°$.

assumed. If the fluid is barotropic, ∇T and ∇P are parallel and $\partial v_\phi / \partial z$ must vanish. In fact, if we take a rather naive view of a perfectly spherical sun, ∇T and ∇P are parallel irrespective of the barotropic nature of the fluid. However, the statement that $\partial v_\phi / \partial z$ vanishes, because of the adiabatic thermal structure brought about by convection, is more robust and has to be valid irrespective of how fast the sun might be rotating.

In deriving equation (18.4) we dropped a few terms. Also, there are terms missing in the first place, such as those due to magnetic field or Reynolds stress, which are certainly important in some parts of the sun, and may be important in many other places in the current context of considering a fine balance between quantities. Even so, let us suppose that we take the equation seriously. Throughout the convection zone, $\partial v_\phi / \partial z$ is negative, and so therefore is $(\nabla T \times \nabla P)_\phi$ (in the northern hemisphere). This means that the temperature gradient is inclined to the pressure gradient in such a way that if we move up towards the pole along an isopressure surface, temperature increases; the pole is warmer than the equator, which might have been caused by anisotropy of the convective flux. Coming back to the missing terms in the equation, the baroclinicity may of course be due to a magnetic field. Or, it may be the effect of Reynolds stress and this may be a way to start observational study of turbulent convection at high R_e (see,

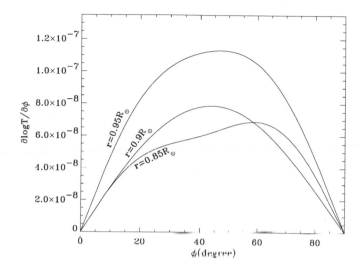

Fig. 18.3. Nominal latitudinal temperature gradient along isopressure surface at various radii, calculated from rotation rate presented in Figure 18.2.

however, Elliott 2003). As for the terms we dropped, note that v_ϕ, the main component of the flow on a global scale, cannot produce, on its own, any additional term in equation (18.4) through the time-dependent term or the inertia term. The r- and θ-components of a small-scale velocity field can contribute via the inertia term.

Let us go further to see how big is this baroclinicity. We maintain axisymmetry but there is an angle between the temperature gradient and the pressure gradient; we call it χ. We assume that the asphericity is so small that for everything other than the angle χ and what are derived from this angle, we can use a value of a spherical model of the sun. Then, the latitudinal gradient of temperature, along local isopressure surface, is

$$\frac{1}{r}\frac{\partial \log T}{\partial \phi} \approx -\frac{\partial \log T}{\partial r}\sin\chi = \frac{\sin\chi}{H_T} \ ,$$

where H_T denotes temperature scale height. On the other hand, from equation (18.4) we have (the subscript '0' has been dropped from Ω)

$$2\Omega\frac{\partial v_\phi}{\partial z} = \frac{(\nabla T \times \nabla P)_\phi}{\rho T} \approx \frac{1}{\rho T}\frac{\partial T}{\partial r}\frac{\partial P}{\partial r}\sin\chi = \frac{P\sin\chi}{\rho H_P H_T}$$

(H_P is density scale height). From these two equations it follows that

$$\frac{\partial \log T}{\partial \phi} \approx 2\frac{\rho r H_P}{P}\Omega\frac{\partial v_\phi}{\partial z} \ .$$

From

$$H_P = -\frac{r^2 P}{\rho G M_r} \ ,$$

where G is the gravitational constant and M_r the mass contained in a sphere of radius r, and using $v_\phi = r\Omega\cos\phi$, we finally have

$$\frac{\partial \log T}{\partial \phi} \approx \frac{2R_\odot^3}{GM}\left(\frac{M_r}{M}\right)^{-1}x^4\Omega\frac{\partial\Omega}{\partial\zeta}\cos\phi \qquad (18.5)$$

($x = r/R_\odot$ and $\zeta = z/R_\odot$). In the middle of the convection zone around the mid-latitude,

$$x \sim 0.8 \ , \ M_r/M \sim 1 \ , \ \Omega \sim 400\,\text{nHz} \ , \ \frac{\partial\Omega}{\partial\zeta} \sim 100\,\text{nHz} \ ,$$

and by dropping (but not completely forgetting) $\cos\phi$, we have a rough estimate of the temperature gradient:

$$\frac{\partial \log T}{\partial \phi} \sim 10^{-7} \ .$$

If we measure the latitudinal gradient along an isobar, the only change is an additional multiplying factor that is the ratio of the density scale height to the pressure scale height.

Figure 18.2 presents a result of a non-linear model-fitting to MDI rotational splitting data. A model with a rigidly rotating core, differentially rotating (of the form $\Omega_0(r) + \Omega_1(r)\cos^2\theta + \Omega_2(r)\cos^4\theta$) outer zone and a smooth transition between the two, has been fitted to the MDI data. Figure 18.3 shows latitudinal variation of $\partial \log T/\partial\phi$ at three radii in the convection zone calculated using equation (18.5) from the rotation rate presented in Figure 18.2.

Taken at their face value, the results above imply a relative difference in temperature, between the pole and the equator, of the order of 10^{-7}. This does not change much in the outer layers, as can be guessed from equation (18.5) and from inversions. This is very small compared to the relative difference of the order of 10^{-4}, which was obtained from a surface temperature observation (Kuhn 1988). The number is so small because, essentially, the sun rotates so slowly.

18.6 Line-blending problem

Among the global p modes, low-degree and high-frequency modes are the ones that penetrate most deeply into the interior of the sun, and therefore the most useful probes into the deep interior. However, for average splitting measurements at low degree there was a disagreement between full-disk experiments: some experiments came up with an average splitting of around 450 nHz (Toutain & Kosovichev 1994, Elsworth et al. 1995, Appourchaux et al. 1995), which is basically the same as the rotation rate in the outer part of the sun, but some came up with numbers close to or exceeding 500 nHz (Toutain & Fröhlich 1992, Loudagh et al. 1993, Jiménez et al. 1994). The former implies a more-or-less flat rotation profile in the core, while the latter requires a rapidly spinning core.

So which numbers are correct? Indeed, are any of them correct? There was some concern because the discrepancies arose in the high-frequency range where the mode linewidth became comparable with the rotational splitting, which they were trying to measure. The analysis technique might have been inadequate, or an accurate measurement might be simply impossible.

In an attempt to answer these questions, Chang, Gough & Sekii (1998) carried out a Monte-Carlo simulation. A simple model of a damped oscillator with stochastic excitation was adopted:

$$\ddot{x}(t) - 2\kappa \dot{x}(t) + \omega_0^2 x(t) = s(t) ,$$

from which it follows that, approximately,

$$X(\omega) \propto \frac{S(\omega)}{\omega - \omega_0 - i\kappa} .$$

where $X(\omega)$ and $S(\omega)$ are Fourier transforms of $x(t)$ and $s(t)$, respectively. The source term $S(\omega)$ was assumed to be Gaussian to obtain a single profile for a single value of the central frequency ω_0. By adding two profiles with the central frequencies differing by $\Delta\omega$, a double-peaked artificial profile with an input splitting $\Delta\omega$ is generated, to which uniform Gaussian noise is added. Chang et al. (1998) generated sets of such profiles for various ranges of parameters, each set containing 10000 noisy profiles, and carried out maximum-likelihood fitting with double Lorentzian. What they found was that, with $\kappa/\Delta\omega$ approaching and then exceeding unity,

(i) the splitting is overestimated when double Lorentzians with independent heights are fitted, and

(ii) when double Lorentzians with equal heights are fitted, the splitting is *still* overestimated although to a lesser degree.

The amount of the overestimate decreases with lower noise level and longer observation period. One might hope that we can use the results of such Monte-Carlo simulations to correct the measurements. However, as Appourchaux et al. (2000a) argued, it is difficult to go beyond a simple and rough correction.

This is a technical but unsolved issue that is important for seismology of rotation in the solar core. For the sun, in fact, the simplest way to go around the problem is to use resolved measurement. However, full-disk measurements of low-degree modes still have an edge over resolved measurements. In any case, resolved measurements are not available for stars, so this will pose a greater problem for asteroseismology for slow rotators.

18.7 Summary

Of the two steps in seismology of solar rotation that I mentioned at the end of the introduction, on the issue of measurement please refer to Schou (2003).

Inversion techniques are already fairly mature, although the 2D OLA, the most favoured method, is still too CPU-intensive for everyone to use everyday. Helioseismic inversions have revealed how the sun rotates in the bulk of the solar convection zone, and below the tachocline.

From a purely seismological point of view, the outstanding issues are how the deep interior and the polar region rotate, and what is going on in the solar tachocline. From a more general point of view, why the solar convection zone rotates like it does now, has to be explained.

So, can we rule out a fast-spinning core? It seems less and less likely (Chaplin et al. 2001), is all we can say at the moment. To investigate this further, a better low-degree splitting measurement that overcomes the line-blending problem certainly helps. However, there may not be enough room for improvement. The search for g modes continues, but currently there is not a lot of optimism. The upper limit of $1\,\mathrm{cm\,s^{-1}}$ at $200\,\mu\mathrm{Hz}$ has been reported (Appourchaux et al. 2000b).

And what is taking place in the polar region? There seems to be a polar slow-down and a jet-like structure. Like in the above, more precise measurement of low-m modes (with higher amplitudes near the poles) in principle helps but with a limited room for improvement. Local(-area) helioseismology at high latitudes, such as the one proposed for the Solar Orbiter mission (Gizon et al. 2001), seems to be the most promising path. The polar rotation can become the last missing piece in the puzzle of the dynamics of the solar convection zone.

Then there is the meridional flow. Hathaway et al. (1996) reported, from GONG data, a poleward, $\sim 20\,\mathrm{m\,s}^{-1}$ flow up to around the latitude of 60°. Giles et al. (1997) also found a similar flow in the outer 4 % of the sun, by a time-distance technique applied to MDI data. Will we ever find the return flow?

Finally, what about other stars? In case of white dwarfs, we may be just beginning to go beyond measuring rotational splitting (Kawaler, Sekii & Gough 1999). However, it will be a long time before we will know as much about the white-dwarf rotation as we currently know about the sun.

Acknowledgments I am infinitely grateful to Douglas Gough for his continuous encouragement and numerous enlightening discussions throughout my nine and a half years in Cambridge, during which we carried out together many works on the subject of the current article.

References

Appourchaux, T., Chang, H.-Y., Gough, D. O. & Sekii, T., 2000a, *MNRAS*, **319**, 365.

Appourchaux, T., Fröhlich, C., Andersen, B., Berthomieu, G., Chaplin, W. J., Elsworth, Y., Finsterle, W., Gough, D. O., Isaak, G. R., Kosovichev, A. G., Provost, J., Scherrer, P. H., Sekii, T. & Toutain, T., 2000b, *ApJ*, **538**, 401.

Appourchaux, T., Toutain, T., Telljohann, U., Jiménez, A., Rabello-Soares, M. C., Andersen, B. N. & Jones, A. R., 1995, *A&A*, **294**, L13.

Brown, T. M., Christensen-Dalsgaard, J. Dziembowski, W. A., Goode, P., Gough, D. O. & Morrow, C. A., 1989, *ApJ*, **343**, 526.

Chang, H.-Y., Gough, D. O. & Sekii, T., 1998, in Provost, J. & Schmider, F.-X., eds, *IAU Symp. 181: Sounding Solar and Steller Interiors (Poster Volume)*, Observatoire de la Côte d'Azur, p. 13.

Chaplin, W. J., Elsworth, Y. P., Isaak, G. R., Marchenkov, K. I., Miller, B. A. & New, R., 2001, *MNRAS*, **327**, 1127.

Choudhuri, A. R., 1990, *ApJ*, **355**, 733.

Christensen-Dalsgaard, J., Larsen, R. M., Schou, J. & Thompson, M. J., 1995, in Ulrich, R. K., ed., *GONG 1994: Helio- and Astero-Seismology from Earth and Space*, the Astronomical Society of the Pacific, San Francisco, p. 70.

Deupree, R. G. & Guzik, J. A., 2000, in İbanoğlu, C., ed., *Variable Stars as Essential Astrophysical Tools*, Kluwer Academic, Dordrecht, p. 253.

Dziembowski, W. A. & Goode, P. R., 1992, *ApJ*, **394**, 670.

Elliott, J. R., 2003, in this volume.

Elsworth, Y. P., Howe, R., Isaak, G. R., McLeod, C. P., Miller, B. A., Wheeler, S. J., New, R. & Gough, D. O., 1995, in Ulrich, R. K., ed., *GONG 1994: Helio- and Astero-Seismology from Earth and Space*, the Astronomical Society of the Pacific, San Francisco, p. 43.

Frisch, U., 1995, *Turbulence*, Cambridge University Press.

Gizon, L., Birch, A. C., Bush, R. I., Duvall, T. L., Jr., Kosovichev, A. G., Scherrer, P. H. & Zhao, J., 2001, in Battrick, B. & Sawaya-Lacoste, H., eds,

Solar encounter: Proceedings of the First Solar Orbiter Workshop, ESA Publication Division, Noordwijk, p. 227.

Giles, P. M., Duvall, T. L., Jr., Scherrer, P. H. & Bogart, R. S., 1997, *Nature*, **390**, 52.

Gough, D. O., 1984, *Phil. Trans. Roy. Soc. London, Ser. A*, **313**, 27.

Gough, D. O., 1996, in Roca Cortés, T. & Sánchez, F., eds, *The Structure of the Sun*, Cambridge University Press, p. 141.

Gough, D. O. & McIntyre, M. E., 1998, *Nature*, **394**, 755.

Gough. D. O., Kosovichev, A. G., Sekii, T., Libbrecht, K. G. & Woodard, M. F., 1993, in Brown, T. M., ed., *GONG 1992: Seismic Investigation of the Sun and Stars*, the Astronomical Society of the Pacific, San Francisco, p. 213.

Gough, D. O. et al., 1996, *Science*, **272**, 1296.

Gough, D. O. & Sekii, T., 1998, in Korzennik, S. & Wilson, A., eds, *SOHO 6/GONG 98 Workshop: Structure and Dynamics of the Interior of the Sun and Sun-like Stars*, ESA Publication Division, Noordwijk, p. 787.

Hathaway, D. H. et al., 1996, *Science*, **272**, 1306.

Howe, R. & Thompson, M. J., 1996, *MNRAS*, **281**, 1385.

Jiménez, A., Pérez Hernández, F., Claret, A., Pallé, P. L., Régulo, C. & Roca Cortés, T., 1994, *ApJ*, **435**, 874.

Kawaler, S. D., Sekii, T. & Gough, D. O., 1999, *ApJ*, **516**, 349.

Kosovichev, A. G. & Schou, J., 1997, *ApJ*, **482**, L207.

Kuhn, J. R., 1988, *ApJ*, **331**, L131.

Loudagh, S. et al., 1993, *A&A*, **275**, L25.

Maeder, A. & Meynet, G., 2000, *ARAA*, **38**, 143.

Parker, E. N., 1993, *ApJ*, **408**, 707.

Pijpers, F. P. & Thompson, M. J., 1996, *MNRAS*, **279**, 498.

Schou, J., 1991, in Gough, D. O. & Toomre, J., eds, *Challenges to Theories of the Structure of Moderate-Mass Stars*, Springer-Heidelberg, p. 93.

Schou, J. 2003, in this volume.

Schou, J. et al., 1998, *ApJ*, **505**, 390.

Sekii, T., 1990, in Osaki, Y. & Shibahashi, H., eds, *Progress of Seismology of the Sun and Stars*, Springer-Verlag, Berlin, p. 337.

Sekii, T., 1991, *PASJ*, **43**, 381.

Sekii, T., 1993, *MNRAS*, **264**, 1018.

Sekii, T., 1995, in Ulrich, R. K., ed., *GONG 1994: Helio- and Astero-Seismology from Earth and Space*, the Astronomical Society of the Pacific, San Francisco, p. 74.

Sekii, T., 1997, in Provost, J. & Schmider, F.-X., eds, *IAU Symp. 181: Sounding Solar and Steller Interiors*, Kluwer Academic, Dordrecht, p. 189.

Sekii, T., Gough D. O. & Kosovichev A. G., 1995, in Ulrich, R. K., ed., *GONG 1994: Helio- and Astero-Seismology from Earth and Space*, the Astronomical Society of the Pacific, San Francisco, p. 59.

Spiegel, E. A. & Zahn, J.-P., 1992, *A&A*, **265**, 106.

Toomre, J. 2003, in this volume.

Toutain, T. & Fröhlich, C., 1992, *A&A*, **257**, 287.

Toutain, T. & Kosovichev, A. G., 1994, *A&A*, **284**, 265.

Unno, W., Osaki, Y, Ando, H., Saio, H. & Shibahashi, H., 1989, *Nonradial Oscillations of Stars (second edition)*, University of Tokyo Press.

Weber, M. A., Acton, L. W., Alexander, D. & Kubo, S., 1999, *Solar Phys.*, **189**, 271.

19

Telechronohelioseismology

ALEXANDER KOSOVICHEV

W. W. Hansen Experimental Physics Laboratory, Stanford University,
HEPL Annex A201, Stanford, CA 94305-4085, USA

Telechronohelioseismology (or time-distance helioseismology) is a new diagnostic tool for three-dimensional structures and flows in the solar interior. Along with the other methods of local-area helioseismology, the ring diagram analysis, acoustic holography and acoustic imaging, it provides unique data for understanding turbulent dynamics of magnetized solar plasma. The technique is based on measurements of travel time delays or wave-form perturbations of wave packets extracted from the stochastic field of solar oscillations. It is complementary to the standard normal mode approach which is limited to diagnostics of two-dimensional axisymmetrical structures and flows. I discuss theoretical and observational principles of the new method, and present some current results on large-scale flows around active regions, the internal structure of sunspots and the dynamics of emerging magnetic flux.

19.1 Introduction

Telechronohelioseismology (or telechronoseismology) is defined as a subdiscipline of helioseismology by Gough (1996) in his reply to criticism of the term 'asteroseismology' (Trimble 1995). Gough argued that, being derived from all classical Greek words, 'thoroughbred' telechronohelioseismology should be preferred to 'oedipal combinations' of Greek and Latin words. Telechronohelioseismology belongs to a new class of helioseismic measurements, broadly defined as epichorioseismology† (also called local-area helioseismology), which provides three-dimensional diagnostics of the solar interior.

Helioseismology is originally based on interpretation of the frequencies of

---

† The main root comes from χωριον, meaning a particular place. The prefix epi-, from the preposition επι (with smooth breathing), has a host of meanings, but denotes direction (Gough, private communication).

normal modes of solar oscillation. This approach allows us to estimate global axisymmetrical components of the Sun's properties such as the sound speed (e.g. Christensen-Dalsgaard et al. 1985), the density (Kosovichev, 1990), the adiabatic exponent (Elliott & Kosovichev 1998), the element abundances (Gough & Kosovichev 1988), and the rotation rate (Duvall et al. 1984).

The first idea that helioseismology can measure local properties of the solar interior was suggested by Gough & Toomre (1983). They estimated the influence of large-scale convective eddies on the wave patterns of five-minute oscillations of high degree and showed that the distortion to the local $k - \omega$ relation has two constituents: one depends on the horizontal component of the convective velocity and has a sign which depends on the sign of ω/k; the other depends on temperature fluctuations and is independent of the sign of ω/k. They concluded that by studying the distortions it would be possible to reveal some aspects of the large-scale flow in the solar convection zone. This idea is developed now in a sophisticated measurement procedure called ring-diagram analysis (Hill 1988; González Hernández et al. 2000; Haber et al. 2000).

Further developments of epichorioseismology led to the idea to perform the measurements of local wave distortions in the time-distance space instead of the traditional Fourier space (Duvall et al. 1993). In this case the wave distortions can be measured as perturbations of wave travel times. However, because of the stochastic nature of solar waves it is impossible to track individual wave fronts. Instead, it was suggested to use a cross-covariance time-distance function that provides a statistical measure of the wave distortion. Indeed, by calculating a cross-covariance of solar oscillation signals at two points one may hope that the main contribution to this cross-covariance will be from waves propagating between these points. However, the interpretation of these measurements is extremely challenging, and various approximations are used to relate the observed perturbations of travel times to internal properties such as sound-speed perturbations and flow velocities.

In this brief review, I discuss basic principles and procedures of telechronoseismology and some initial results.

19.2 Observational and Theoretical Principles

The basic idea of telechronohelioseismology is to measure the acoustic travel time between different points on the solar surface, and then to use the measurement for inferring perturbations of the sound speed and flow velocities in the interior along the wave paths connecting the surface points (Fig. 19.1a).

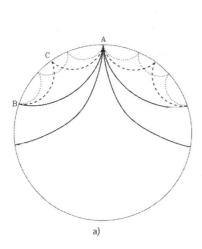

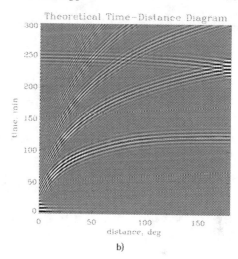

a)

b)

Fig. 19.1. (a) A sample of ray paths of acoustic waves propagating through the Sun's interior from surface point A. (b) The cross-covariance as a function of the travel distance and lag time. The lowest ridge corresponds to wave packets propagating between two points on the solar surface directly, e.g. along ray path AB (solid curve). The second ridge from below corresponds to acoustic waves that have an additional reflection at the surface ('second bounce'), e.g. along ray path ACB. This ridge appears reflected at the distance of 180°, because the propagation distance is measured in the interval from 0° to 180°.

This idea is similar to seismology of the Earth. However, unlike in the Earth, the solar waves are generated stochastically by numerous acoustic sources in the subsurface layer of turbulent convection. Therefore, the wave travel time and other wave propagation properties are determined from the cross-covariance function, $\Psi(\tau, \Delta)$, of the oscillation signal, $f(t, d\mathbf{r})$, between different points on the solar surface (Duvall et al. 1993):

$$\Psi(\tau, \Delta) = \frac{1}{T} \int_0^T f(t, d\mathbf{r}_1) f^*(t + \tau, d\mathbf{r}_2) dt, \qquad (19.1)$$

where Δ is the angular distance between the points with coordinates $d\mathbf{r}_1$ and $d\mathbf{r}_2$, τ is the delay time, and T is the total time of the observations. Because of the stochastic nature of excitation of the oscillations, function Ψ must be averaged over some areas on the solar surface to achieve a signal-to-noise ratio sufficient for measuring travel times τ. The oscillation signal, $f(t, d\mathbf{r})$, is usually the Doppler velocity or intensity. A typical cross-covariance function averaged over the whole disk is shown in Figure 19.1b. It displays a set of ridges. The lowest ridge corresponds to wave propagating directly between two surface points, (the lowest ridge). The second ridge from below is formed by waves which experience one additional reflection at the surface on their way from point A to B, e.g. wavepath ACB in Fig. 19.1a, (so-called 'second bounce' ridge). The upper ridges correspond to waves with multiple reflections from the surface. The cross-covariance function represents a 'helioseismogram'.

For an unperturbed solar model the cross-covariance function can be represented by a superposition of wave packets (Kosovichev & Duvall 1997):

$$\Psi(\tau, \Delta) \propto \sum_{\delta v} \cos\left[\omega_0\left(\tau - \frac{\Delta}{v}\right)\right] \exp\left[-\frac{\delta\omega^2}{4}\left(\tau - \frac{\Delta}{u}\right)^2\right]. \tag{19.2}$$

Here δv is a narrow interval of the phase speed, $v = \omega_{nl}/L$, where $L = \sqrt{l(l+1)}$, $u \equiv (\partial\omega/\partial k_h)$ is the horizontal component of the group velocity, $k_h = 1/L$ is the angular component of the wave vector, and ω_0 is the central frequency of a Gaussian frequency filter applied to the oscillation data, and $\delta\omega$ is the characteristic width of the filter. The phase and group travel times are measured by fitting individual terms of eq. (19.2), represented by a Gabor-type wavelet, to the observed cross-covariance function using a least-squares technique.

Usually the phase travel time is measured more accurately and, thus, is more commonly used for inferring internal properties of the Sun. Typically the measured variations of the travel time do not exceed 5%. These variations can be related to perturbations of the sound speed $\delta c/c$ and flow velocity U via a linear integral equation:

$$\delta\tau = \int_V K_c \frac{\delta c}{c} dV + \int_V (K_u U) dV, \tag{19.3}$$

where $K_c(r)$ and $K_u(r)$ are sensitivity functions for $\delta c/c$ and U; the integration is over the interior volume, V.

The sensitivity kernels are derived by using some approximations, e.g. the ray theory or the first Born approximation. In the Born approximation, the sensitivity kernels can be expressed in terms of the unperturbed eigenfunctions of solar oscillation modes (Birch & Kosovichev 2000). Examples of the Born sensitivity kernels for sound-speed perturbations are shown in Fig. 19.2. Calculations of these kernels involve double summation over a large set of normal modes, and represent a significant computing task. For sufficiently large distances the kernels are relatively simple (Fig. 19.2a) and can be approximated by a semi-analytical formula (Jensen et al. 2000). However, for shorter distances and multiple bounces (Fig. 19.2b) the Born kernels are quite complicated.

One unexpected feature of the first-bounce kernels calculated in the Born

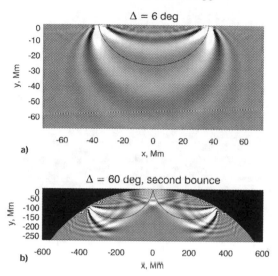

Fig. 19.2. Samples of the travel time sensitivity function in the Born approximation for (a) a direct waves and (b) a second-bounce waves. The solid curves show the corresponding ray paths.

approximation is that these kernels have zero values along the ray path. Such kernels are called 'banana-doughnut kernels' by Marquering et al. (1999) and Dahlen et al. (2000). This curious property is related to a 'wave-front healing' effect (Hung et al. 2001) illustrated in Fig. 19.3 which shows the results of numerical simulations of propagation of a spherical wave from a point source through a localized positive sound-speed perturbation. After scattering on the perturbation, the wave front is split into an accelerated direct wave and a retarted diffracted wave. However, later the structure of the wave front is gradually restored and becomes close to the original spherical shape. The wave healing can be explained by the Huygens principle according to which the sharp edges of the perturbed wave front generate secondary waves that fill in the break in the wave front caused by the perturbation.

The simple numerical model of Hung et al. was used by Birch et al. (2001) to test the accuracy of the Born and ray approximation (Fig. 19.4). The results (Fig. 19.4) show that for typical perturbations in the solar interior the Born approximation is sufficiently accurate, while the ray approximation significantly overestimates the travel times for perturbations smaller than the size of the first Fresnel zone. That means that the inversion results based on the ray theory may significantly underestimate the strength of the small-scale perturbations. Recently, Gizon & Birch (2002) obtained sensitivity kernels for distributed sources.

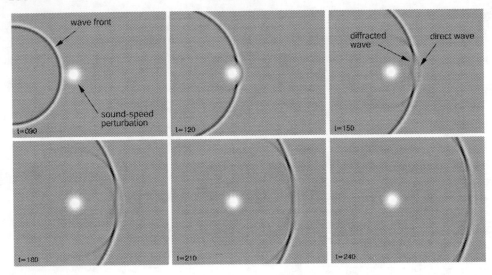

Fig. 19.3. Numerical simulation illustrating the wave-front healing effect. A circular wave initiated by a point source at the left boundary propagates through a strong (50%) localized sound-speed spherical perturbation (the white object in the middle). The perturbed wave front consists of two parts, an accelerated direct wave and a diffracted wave which lags behind. At larger distances the wave becomes close to the original spherical shape again.

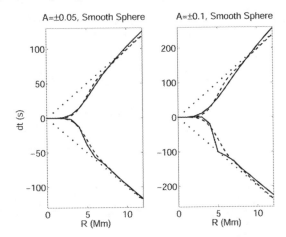

Fig. 19.4. Tests of the ray and Born approximations: travel-time perturbations due to spherically symmetric sound-speed perturbations, as functions of the radius of the perturbing region. The sound-speed perturbation has a cosine-bell profile, with maximum amplitude A and half-width-at-half-maximimum R. The solid lines are the numerical results. The dashed curves are the travel-time perturbations calculated in the Born approximation and the dotted lines are those calculated in the first-order ray approximation. The left panel is for two perturbations of the relative amplitude $A = \pm 0.05$. The right panel is for $A = \pm 0.1$. (Birch et al. 2001).

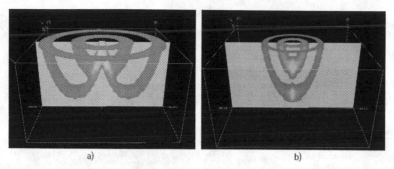

Fig. 19.5. Samples of travel-time sensitivity functions in the ray approximation for two schemes of calculations of the cross-covariance function: (a) "surface-focusing" scheme in which the central-area oscillation signal is the cross-correlated with the signals of surrounding annuli; (b) the "deep-focusing" scheme in which the signals of the opposite sides of an annulus are cross-correlated. The light colour of the sensitivity functions corresponds to high absolute values of the sensitivity functions, and darker colour corresponds to the low values. Each picture shows the sensitivity functions for two distances.

In the ray-theoretical approximation, which is still often used for helioseismic inferences, the first-order perturbations to the phase travel time is given by (Kosovichev & Duvall 1997):

$$\delta\tau = -\int_{\Gamma}\left[\frac{(\boldsymbol{n}\cdot\boldsymbol{U})}{c^2} + \frac{\delta c}{c}S + \left(\frac{\delta\omega_c}{\omega_c}\right)\frac{\omega_c^2}{\omega^2 c^2 S} + \frac{1}{2}\left(\frac{c_A^2}{c^2} - \frac{(\boldsymbol{k}\cdot\boldsymbol{c_A})^2}{k^2 c^2}\right)S\right]\mathrm{d}s,$$
(19.4)

where $\boldsymbol{n}$ is a unit vector tangent to the ray, $S = k/\omega$ is the phase slowness, $\boldsymbol{k}$ is the wave vector and k its magnitude, ω_c is the acoustic cut-off frequency, $\boldsymbol{c_A} = \boldsymbol{B}/\sqrt{4\pi\rho}$ is the vector Alfvén velocity, $\boldsymbol{B}$ is the magnetic field strength, c is the adiabatic sound speed, and ρ is the plasma density. The integration is performed along the unperturbed ray path Γ according to Fermat's principle (e.g. Gough 1993). The effects of flows and structural perturbations are separated by taking the difference and the mean of the reciprocal travel times. Magnetic field causes anisotropy of the mean travel times, which allows us to separate, in principle, the magnetic effects from the variations of the sound speed (or temperature).

The travel time measured at a point on the solar surface is the result of the cumulative effects of the perturbations in each of the traversed rays of the 3D ray systems. This pattern is then translated for different surface points in the observed area, so that overall the travel times are sensitive to all subsurface points in some depth range. In a "surface-focusing" scheme the travel time is measured between a small central area and a set of surrounding

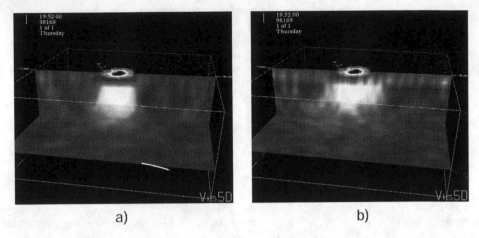

a)   b)

Fig. 19.6. Sound-speed perturbations beneath a sunspot, obtained using (a) the surface-focusing and (b) deep-focusing schemes. The white colour corresponds to positive perturbation; the dark colour regions just beneath the spot are regions of the negative sound-speed perturbation. The upper semi-transparent plane shows a white-light image of the spot, a dark umbra surrounded by a lighter penumbra. The lower horizontal plane shows the sound-speed distribution at approximately 15 Mm beneath the surface. The horizontal size of the box is 158 Mm.

annuli (Fig. 19.5a). These measurements are mostly sensitive to a region just beneath the central area. Therefore, the diagnostics of the deep interior in such a scheme rely entirely on inversion procedures. Another scheme initially suggested by Duvall (1995) is based on cross-correlating signals from the opposite sides of the annuli and does not involve the central area. These measurements are mostly sensitive to a small region localized in the deep interior, where the ray paths intersect (Fig. 19.5b). This "deep-focusing" scheme has many advantages. One of them is that it allows us to study the structure and mass flows directly beneath sunspots without using the oscillation signal inside the spots which may be strongly affected by magnetic fields. The idea of deep focusing is also used in two other methods: acoustic imaging (Chou 2000; Chou & Duvall 2000) and acoustic holography (Lindsey & Braun 2000a; Braun & Lindsey 2000).

Figure 19.6 shows inversion results for the sound-speed variations in an isolated sunspot by using both the surface-focusing (panel a) and the deep-focusing (panel b) schemes in the ray approximation. It is important that the two different measurement schemes reveal the similar vertical structure beneath the spot: the sound-speed is lower than in the surrounding plasma down to 4 Mm below the photosphere, and is higher in deeper layers. This provides confidence in the time-distance measurements. Evidently, the deep-focusing results are noiser close to the surface because this scheme does not cover the immediate subsurface layers well. Perhaps for the best results a combination of these schemes should be used.

19.3 Current Inferences

Telechronohelioseismology has been successfully used to infer local properties of large-scale zonal and meridional flows (Giles et al. 1997; Giles & Duvall 1998; Giles 1999; Chou & Dai 2001), convective flows and structures (Duvall et al. 1997; Kosovichev & Duvall 1997; Gizon et al. 2000), structure and dynamics of active regions (Kosovichev 1996; Kosovichev et al. 2000), flows around sunspots (Duvall et al. 1996; Zhao et al. 2001). Here we present some results of inversion of travel time data for large-scale flows, developing active regions and sunspots. These results are obtained in the ray approximation. We also discuss a new interesting application of detecting active regions on the far side of the Sun, which is potentially important for space weather forecasts.

19.3.1 Large-scale flows and solar activity

One of the first studies by telechronoseismology was to study large-scale flows and structures associated with active regions using observations at South Pole in January 1991 by (Jefferies et al. 1994). The observed area is shown in the Ca^+K-line intensity map in Fig. 19.7a. The travel times in this area have been measured by Duvall et al. (1996) using the surface-focusing scheme (Fig. 19.5) for four sets of rays propagating from each of 72×52 points equally spaced in the observed area of $123° \times 106°$. The four radial distance ranges Δ were 2.5–4.25, 4.5–7, 7.25–10, and 10.25–15 heliographic degrees.

The inversion results for the three components of the flow velocity in the upper layer, inferred from the data, are illustrated in Fig. 19.7b (Kosovichev 1996). The results show strong converging downflows of $\simeq 1\,\mathrm{km\,s}^{-1}$ around the sunspots, and some much weaker isolated upflows at the boundaries of the active latitudes. These results were later confirmed from SOHO/MDI data by Gizon et al. (2001), and Haber et al. (2000). It is important that the azimuthally averaged flows maps reproduce the pattern of 'torsional oscillations' determined by the normal-mode 'global' analysis (Kosovichev and Schou 1997; Schou 1999). This provides an important cross-check between the 'global' and 'local' approaches.

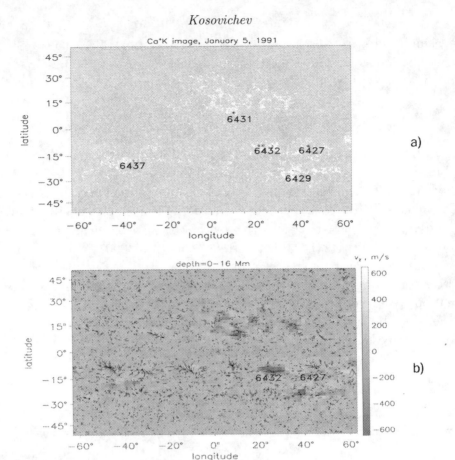

Fig. 19.7. (a) Ca$^+$K image of the Sun on January 5, 1991. (b) The corresponding map of flow velocity averaged over the top 16 Mm; the background gray-scale map shows the vertical component of velocity (darker - downward, and lighter - upward); the horizontal components are shown with arrows following the flow field proportional in length to the velocity; the arrows are composed of a number of small segments that follow the streamlines. The flow map is obtained by Kosovichev (1996) from observations by Duvall et al. (1996) at South Pole.

19.3.2 Developing active regions

One of the important tasks is diagnostics of emerging active regions in the interior. For space weather predictions it would be very important to detect active regions before they emerge. However, this task has proven to be very difficult. Here we present the results for an emerging active region observed in January, 1998. This active region (NOAA 8131) was a high-latitude region of the new solar cycle which began in 1997.

Figure 19.8 shows the sound speed variations in a vertical cross-section in

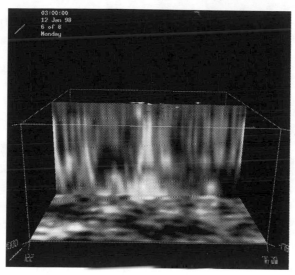

Fig. 19.8. Image of the sound-speed perturbation in an emerging active region in the solar convection zone obtained from the SOHO/MDI data on January 12, 1998, from 02:00 to 04:00 UT. The horizontal size of the box is approximately 560 Mm, and the depth is 18 Mm. The (mostly) transparent panel on the top is an MDI magnetogram showing the surface magnetic field where it is stronger than 200 Gauss. The vertical and bottom panels show perturbations of the sound speed which are approximately in the range from -1.3 to $+1.3$ km s^{-1}. The positive variations are shown in white and light gray, and the negative ones - in dark colour. A large active region formed at this location within a day after these observations.

the region of the emerging flux and in a horizontal plane at a depth of 18 Mm, averaged for a 2-hour interval. The perturbations of the magnetosonic speed are associated with the magnetic field and temperature variations in the emerging magnetic structures. In the subsurface layer the magnetic flux propagates very rapidly. Only reducing the integration time to 2 hours were we able to detect a strong perturbation at the bottom of our observing region (Kosovichev & Duvall 1999). From the investigation of emerging active regions we conclude that it is necessary to probe much deeper layers of the solar convection zone because the emerging flux propagates very rapidly in the top 20 Mm. The estimated emergence speed is approximately 1.3 km s^{-1}. This is somewhat higher than predicted by theories. The high emergence speed makes the problem of detecting emerging active regions in the solar interior very challenging.

The typical amplitude of the sound-speed variation in the perturbation is about 0.5 km s^{-1}. This may correspond to a 500 G magnetic field at the bottom of the box, or a temperature variation of 300 K. After the emergence we observed a gradual increase of the perturbation in the subsurface layers, and the formation of sunspots. The observed development of the active region suggests that the sunspots were formed as result of the concentration of magnetic flux close to the surface.

19.3.3 Structure and dynamics of sunspots

The high-resolution data from SOHO/MDI has allowed structures and flows beneath sunspots to be investigated in some detail (Kosovichev et al. 2000; Zhao et al. 2001). Figure 19.9 shows an example of the internal structure of a large sunspot observed on June 17, 1998. An image of the spot taken in the continuum is shown at the top. The sound-speed perturbations under the sunspot are much stronger than these of the emerging flux, and reach more than $3\,\mathrm{km\,s^{-1}}$. It is interesting that beneath the spot the perturbation is negative in the subsurface layers and becomes positive in the deeper interior. One can suggest that the negative perturbations beneath the spot are probably due to the lower temperature. It follows that magnetic inhibition of convection that makes sunspots cooler is most effective within the top $2-3\,\mathrm{Mm}$ of the convection zone. The strong positive perturbation below suggests that the deep sunspot structure is hotter than the surrounding plasma. However, the effects of temperature and magnetic field have not been separated in these inversions. These data also show at a depth of $4\,\mathrm{Mm}$ connections to the spot of small pores, A and B, which have the same magnetic polarity as the main spot. Pore C of the opposite polarity is not connected. This suggests that sunspots represent a tree-like structure in the upper convection zone.

Figure 19.10 shows flow maps beneath the sunspot. In Fig. 19.10a which shows results for the first layer corresponding to an average of depth of $0-3\,\mathrm{Mm}$, one can clearly identify a ring of strong downflows around the sunspot, with relatively weaker downflows inside the ring. Converging flows at the sunspot centre can also be seen. Figure 19.10b shows the flows in the layer corresponding to a depth of $6-9\,\mathrm{Mm}$. The sunspot region is occupied by a ring of upflows with almost zero velocity at the centre. Outside this region, the results are much noisier, and no clear upflows or downflows can be identified. At the same time, strong outflows from the sunspot centre can be seen. Clearly, the motion to the South-East direction is much stronger.

Figures 19.10c-d show two vertical cuts, one in the East-West direction, the other in the North-South direction, through the centre of the sunspot. in between two layers by use of linear interpolation. Converging and downward flows can be seen in both diagrams right below the sunspot region from

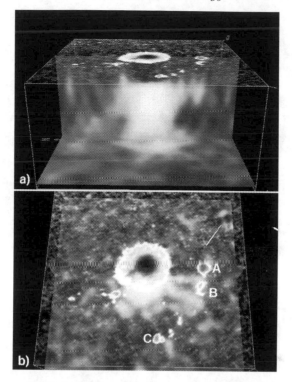

Fig. 19.9. The sound-speed perturbation in a large sunspot observed on June 20, 1998, are shown as vertical and horizontal cuts. The horizontal size of the box is 13 degrees (158 Mm), the depth is 24 Mm. The positive variations of the sound speed are shown in light gray, and the negative variations (just beneath the sunspot) - in dark. The upper semitransparent panel is the surface intensity image (dark colour shows umbra, and light colour shows penumbra). In the lower panel the horizontal sound-speed plane is located at the depth of 4 Mm, and shows long narrow structures ('fingers') connecting the main sunspot structure with surrounding pores A and B of the same magnetic polarity as the spot. Pores of the opposite polarity (e.g. C) are not connected to the spot.

1.5 Mm to about 5 Mm. Below that, the horizontal outflows seem to dominate in this region, though relatively weaker upflows also appear. Below a depth of $\sim$ 10 Mm, the flows seem not to be concentrated in the region vertically below the sunspot. This can be seen more clearly in the East-West cut (Fig. 19.10c). It is intriguing that an upflow toward the East dominates in the region from 10 Mm to 18 Mm. In South-North cut (Fig. 19.10d), this pattern is not so clear but still can be seen, with the upflow toward the South stronger than toward the North.

The inversion result shows converging flows in the top layer of the sunspot region which are opposite to the well-known diverging Evershed flow ob-

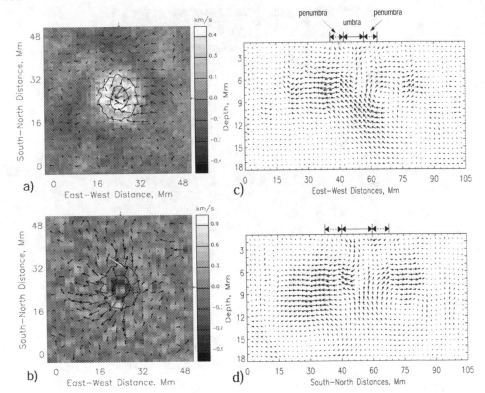

Fig. 19.10. The mass flows in a sunspot area at the depth of $0-3$ Mm (a) and depth of $6-9$ Mm (b). The arrows show the magnitude and directions of the horizontal flows, and the background gray-scale map shows the vertical flows (light colour indicates downflow). The contours at the centre correspond to the umbra and penumbra boundaries. The longest arrow represents $1.0\,\mathrm{km\,s^{-1}}$ in panel (a), and $1.6\,\mathrm{km\,s^{-1}}$ in panel (b). The arrows outside the frame indicate where the vertical cuts shown in panels (c) and (d) are made. The vertical cuts are made through the sunspot centre with the cut direction of East-West (panel (c), with East on the left side) and South-North (panel (d), with South on the left side). The range covered by the line arrows indicate the area of umbra, and the range covered by the dotted arrow indicate the area of penumbra. The longest arrow represents $1.4\,\mathrm{km\,s^{-1}}$ (Zhao et al. 2001).

served spectroscopically on the solar surface. Gizon et al. (2000) have studied the horizontal flows in sunspots using the f-mode diagnostics. The velocity amplitude of the Evershed flow inferred from the f-mode is systematically lower than that at the surface. This suggests that the Evershed flow is a shallow phenomenon. This can explain the apparent disagreement with the results obtained from acoustic modes.

The observational evidence of converging flows beneath the spot supports

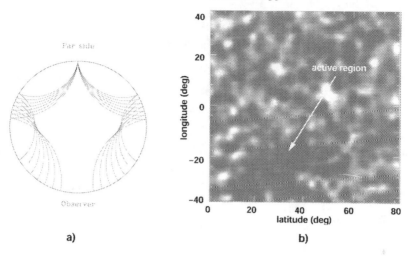

Fig. 19.11. (a) A "deep-focusing" scheme with the acoustic ray focus point on the far side of the Sun. (b) A map of the travel-time variations (white - positive, and black - negative) due to perturbations on the far-side surface. A dark region at -20° longitude and 30° latitude is caused by shorter travel times due to an active region. (Duvall & Kosovichev 2001).

a cluster model of sunspots suggested by Parker (1979). According to this model the field of a sunspot divides into many separate tubes within the first $1 - 2$ Mm below the surface, and a converging downward flow beneath the sunspot holds the separate tubes in a loose cluster. A numerical model of these flows is developed by Hurlburt & Rucklidge (2000).

19.3.4 Far-side imaging

A new approach for detecting new active regions on the far side of the Sun was recently demonstrated by Lindsey & Braun (2000b) and Braun & Lindsey (2001). They used a holographic technique to image the far side of the Sun. This allows early detection of new active regions which a few days later will rotate to the front side and affect the space weather. The holographic method is discussed by Braun & Lindsey (2000). We only mention that the time-distance technique also can provide measurements of the travel-time perturbations on the far side. We illustrate this in Fig. 19.11, the left panel of which shows the measurement scheme. This is essentially a deep-focusing scheme with the focus point chosen on the far-side surface. The right panel of Fig. 19.11 shows a map of the travel time variations on the far side. The travel time becomes shorter by $\simeq 6$ sec when there is an active region in the focus. The far-side signal was recovered from 48-hour time series of medium resolution images. Currently, the daily far-side images are posted on the MDI web site (http://soi.stanford.edu/data/farside/recent.html).

19.4 Conclusion

During its first 8 years, telechronohelioseismology demonstrated unique capabilities to explore the three-dimensional dynamics of the solar interior. It has confirmed the prediction made by Douglas Gough at the NASA Solar Physics Exploration Seminar, April 10, 1991, that future helioseismic observations would change the simplistic view of structures inside the Sun and show that the inside of the Sun is very complicated, as complicated as the outside. Obtaining precise measurements of these complicated structures requires deeper understanding of wave propagation in the magnetized solar plasma and the development of more accurate procedures of data analysis and inversion. This will be achieved through a combination of high-precision observations and numerical simulations.

Developing telechronohelioseismology is one of the most challenging and exciting problems of solar physics. The initial results are very promising. They reveal interesting dynamics of supergranulation, meridional circulation, emerging active regions and the formation of sunspots in the upper convection zone. Further studies will shed more light on the dynamics of the Sun's interior and the mechanisms of solar activity.

References

Birch, A.C. & Kosovichev, A.G., 2000, *Solar Phys.* **192**, 193

Birch, A.C., Kosovichev, A.G., Price, G.H. & Schlottmann, R.B., 2001, *ApJ*, **561**, L229

Braun, D.C. & Lindsey, C., 2000, *Solar Phys.*, **192**, 307

Braun, D.C. & Lindsey, C., 2001, *ApJ*, **560**, L189

Chou, D.-Y., 2000, *Solar Phys.*, **192**, 241

Chou, D.-Y. & Dai, D.-C., 2001, *ApJ*, **559**, L175

Chou, D.-Y. & Duvall, T.L., Jr., 2000, *ApJ*, **533**, 568

Christensen-Dalsgaard, J., Duvall, T L., Jr., Gough, D.O., Harvey, J.W. & Rhodes, E.J., Jr., 1985, *Nature*, **315**, 378

Dahlen, F.A., Hung, S.-H. & Nolet, G., 2000, *Geophys. J. Int.*, **141**, 157

Duvall, T.L., Jr., 1995, in Ulrich, R.K., Rhodes, E.J., Jr. & Däppen, W., eds, *GONG '94: Helio- and Astero-Seismology from the Earth and Space*, ASP Conf. Series, **76**, (Astronomical Society of the Pacific, San Francisco), p. 93

Duvall, T.L., Jr., Dziembowski, W.A., Goode, P.R., Gough, D.O., Harvey, J.W. & Leibacher, J.W., 1984, *Nature*, **310**, 22

Duvall, T.L., Jr. & Kosovichev, A.G., 2001, in Brekke, P., Fleck, B. & Gurman, J.B., eds, Proc. IAU Symp. 203 *Recent Insights into the Physics of the Sun* and Heliosphere: Highlights from SOHO and Other Space Missions* (Astronomical Society of the Pacific), p. 159

Duvall, T.L., Jr., Jefferies, S.M., Harvey, J.W. & Pomerantz, M.A., 1993, *Nature*, **362**, 430

Duvall, T.L., Jr., D'Silva, S., Jefferies, S.M., Harvey, J.W. & Schou, J., 1996, *Nature*, **379**, 235

Duvall,T.L., Jr., Kosovichev, A.G., Scherrer, P.H., Bogart, R.S., Bush, R.I., De Forest, C., Hoeksema, J.T., Schou, J., Saba, J.L.R., Tarbell, T.D., Title, A.M., Wolfson, C.J. & Milford, P.N., 1997, *Solar Phys.*, **170**, 63

Elliott, J.R. & Kosovichev, A.G., 1998, *ApJ*, **500**, L199

Giles, P.M., 1999, D. Phil. thesis, Stanford University

Giles, P.M., & Duvall, T.L., Jr., 1998, in Deubner, F.-L., Christensen-Dalsgaard, J. & Kurtz, D., eds, Proc. IAU Symposium 185 *New Eyes to See Inside the Sun and Stars. Pushing the Limits of Helio- and Asteroseismology with New Observations from the Ground and from Space* (Kluwer Academic Publishers, Netherlands), p. 149

Giles, P.M., Duvall, T.L., Jr, Scherrer, P.H. & Bogart, R.S., 1997, *Nature*, **390**, 52

Gizon, L. & Birch, A.C., 2002, *ApJ*, **571**, 966.

Gizon, L., Duvall, T.L., Jr. & Larsen, R.M., 2000, *J. Astrophys. & Astron.*, **21**, 339

Gizon L., Duvall T.L., Jr. & Larsen R.M., 2001, in Brekke, P., Fleck, B. & Gurman, J.B., eds, Proc. IAU Symp. 203 *Recent Insights into the Physics of the Sun and Heliosphere: Highlights from SOHO and Other Space Missions* (Astronomical Society of the Pacific), p. 189

González Hernández, I., Patrón, J., Roca Cortés, T., Bogart, R.S., Hill, F. & Rhodes, E.J., Jr., 2000, *ApJ*, **535**, 454

Gough, D.O, 1993, in Zahn, J.-P. & and Zinn-Justin, J., eds, *Astrophysical Fluid Dynamics* (Elsevier Science Publ.), p. 339

Gough, D.O., 1996, *The Observatory*, **116**, 313

Gough, D.O. & Kosovichev, A.G., 1988, in Berthomieu, G. & Cribier, M., eds, Proc. IAU Colloq. 121 *Inside the Sun* (Kluwer), p. 327.

Gough, D.O. & Toomre, J., 1983, *Solar Phys.*, **82**, 401

Haber, D.A., Hindman, B.W., Toomre, J., Bogart, R.S., Thompson, M.J. & Hill, F., 2000, *Solar Phys.*, **192**, 335

Hill, F., 1988, *ApJ*, **333**, 996

Hung, S.-H., Dahlen, F.A. & Nolet, G., 2001, *Geophys. J. Int.*, **146**, 289

Hurlburt, N.E. & Rucklidge, A.M., 2000, *MNRAS*, **314**, 793

Jefferies, S.M., Osaki, Y., Shibahashi, H., Duvall, T.L., Jr., Harvey, J.W. & Pomerantz, M.A., 1994, *ApJ*, **434**, 795

Jensen, J.M., Jacobsen, B.H. & Christensen-Dalsgaard, J., 2000, *Solar Phys.*, **192**, 231

Kosovichev, A.G., 1990, in Osaki, Y. & Shibahashi, H., eds, *Progress of Seismology of the Sun and Stars* (Springer), LNP vol. **367**, 319

Kosovichev, A.G., 1996, *ApJ*, **461**, L55

Kosovichev, A.G. & Duvall, T.L., Jr., 1997, in Pijpers, F.P., Christensen-Dalsgaard, J. & Rosenthal, C.S., eds, *SCORe'96 : Solar Convection and Oscillations and their Relationship* (Kluwer) p. 241

Kosovichev, A.G., & Duvall, T.L., Jr., 1999, *Current Science*, **77**, 1467

Kosovichev, A.G., & Schou, J., 1997, *ApJ*, **482**, L207

Kosovichev, A.G., Duvall, T.L. & Scherrer, P.H., 2000, *Solar Phys.*, **192**, 159

Lindsey, C. & Braun, D.C., 2000a, *Solar Phys.* **192**, 261
Lindsey, C. & Braun, D.C., 2000b, *Science*, **287**, 1799
Marquering, H., Dahlen, F.A., & Nolet, G., 1999, *Geophys. J. Int.*, **137**, 805
Parker, E. N., 1979, *ApJ*, **230**, 905
Schou, J., 1999, *ApJ*, **523**, L181
Trimble, V., 1995, *Publ. Astr. Soc. Pac.*, **107**, 1012
Zhao, J., Kosovichev, A.G. & Duvall, T.L., Jr., 2001, *ApJ*, **557**, 384

V Large-scale numerical experiments

20

Bridges between helioseismology and models of convection zone dynamics

JURI TOOMRE

JILA and Department of Astrophysical and Planetary Sciences,
University of Colorado, Boulder, CO 80309-0440, USA

The sun is a magnetic star whose variable activity has a profound effect on our technological society. The high speed solar wind and its energetic particles, mass ejections and flares that affect the solar-terrestrial interaction all stem from the variability of the underlying solar magnetic fields. We are in an era of fundamental discovery about the overall dynamics of the solar interior and its ability to generate magnetic fields through dynamo action. This has come about partly through guidance and challenges to theory from helioseismology as we now observationally probe the interior of this star. It also rests on our increasing ability to conduct simulations of the crucial solar turbulent processes using the latest generation of supercomputers.

20.1 Introduction

The intensely turbulent convection zone of the sun, occupying the outer 30% by radius or 200 Mm in depth, exhibits some remarkable dynamical properties that have largely defied theoretical explanation. The most central issues concern the differential rotation with radius and latitude that is established by the convection redistributing angular momentum, and the manner in which the sun achieves its 22-year cycles of magnetic activity. These dynamical issues are closely linked: the global dynamo action is most likely very sensitive to the angular velocity Ω profiles realized within the sun. It is striking that the underlying solar turbulence can be both highly intermittent and chaotic on the smaller spatial and temporal scales, and yet achieve a large-scale order that is robust in character (e.g. Brummell, Cattaneo & Toomre 1995).

The interaction of highly turbulent flows and magnetism within the solar convection zone yields a complex dynamical system. Deciphering the operation of solar magnetism is made most challenging by the lack of detailed models that describe the origin and evolution of these magnetic fields. Yet several elements are coming together to help characterize the flows that contribute to their generation and evolution. Firstly, helioseismology has provided tools to probe the differential rotation and large-scale flows that occur within the convection zone. This has led to the discovery of both a tachocline of rotational shear at the base of the convection zone which is the likely seat of global dynamo action, and a near-surface shear layer that contains intricate large-scale flows called Solar Subsurface Weather (SSW). Such flows exhibit complex behaviour, including evolving meridional cells with reversing circulations, propagating banded zonal flows, and meandering flows that may be associated with the largest scales of deep convection. Neither shear layer had been anticipated. Their existence is forcing reconsideration of how magnetic fields are generated in the solar interior, emerge at the solar surface, and evolve within the solar atmosphere. Secondly, rapid advances in supercomputing are enabling 3–D simulations of turbulent convection coupled to rotation and magnetism with sufficient spatial resolution to begin studying in detail aspects of the solar dynamo and the redistribution of magnetic fields by the large-scale flows associated with SSW. Thus we foresee an emerging era in which helioseismic and magnetic observations will interact closely with major 3–D spherical-shell simulations of the solar convection zone that are now becoming tractable with massively-parallel computer architectures.

20.2 Differential rotation: tachocline and near-surface shear

The solar acoustic p modes involve a wide range of horizontal scales (e.g. Gough & Toomre 1991). Modes possessing large horizontal wavelengths (low harmonic degree l) travel around the sun many times within their measured lifetimes. Thus accurate determination of their temporal frequencies ν provides globally averaged measures of sound speed and rotation in the layers of the sun where their amplitudes are appreciable. Through the measurement and analysis of such frequencies, global helioseismology is providing a detailed picture of differential rotation throughout most of the solar interior that is unlike any anticipated by prior convection theory. The angular velocity Ω obtained by inversion of frequency splittings of the p modes (e.g. Libbrecht 1989; Tomczyk, Schou & Thompson 1995; Thompson et al. 1996; Schou et al. 1998; Howe et al. 2000a) has the striking behaviour shown in Fig. 20.1. The latitudinal variation of Ω observed near the surface, where the rotation is considerably faster at the equator than near the poles, extends through much of the convection zone with relatively little radial dependence.

A region of strong shear, now known as the *tachocline*, exists at the base of the convection zone where Ω adjusts to apparent solid body rotation in the deeper radiative interior. The tachocline also exhibits temporal varia-

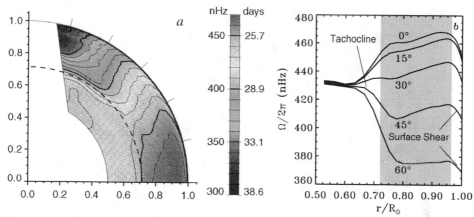

Fig. 20.1. (*a*) Angular velocity profile Ω as deduced from helioseismology using SOI–MDI data, with equatorial regions exhibiting fast rotation and the high latitudes the slowest rotation [Schou et al. 1998]. (*b*) Time-averaged rotation rates from five years of helioseismic data from the Global Oscillations Network Group (GONG), plotted against radius at different latitudes. The near-surface shear layer is clear, as is the tachocline near the base of the convection zone [Howe et al. 2000a].

tions: we have recently discovered changes in Ω just above and below the tachocline with a period of about 1.3 years (Howe et al. 2000a). This is of great interest since the detected variations are occurring near the site favoured for the operation of the global magnetic dynamo. Figure 20.1 also shows that there exists a thin but important *near-surface shear layer* (about 35 Mm in depth, or 5% in radius) in which Ω increases rapidly with depth at intermediate and low latitudes. The intense smaller scales of convection involving granulation, mesogranulation and supergranulation are embedded within it, and may contribute both to its existence and to local magnetic dynamo action there. It is within the upper portions of this rotational shear layer that ring-diagram analyses have revealed the presence of SSW flows.

20.3 Solar dynamo: ordered and chaotic emergence of flux

Observations and simplified modelling suggest a strong connection between solar flows and solar magnetism. Such links range from the generation of the field itself by dynamo processes within the solar interior to the shearing of the magnetic field in the solar corona through footpoint motions. Active regions result from organized toroidal magnetic fields that emerge at the photosphere (e.g. Fisher et al. 2000). Such toroidal field is thought to be generated from existing poloidal field in the strong rotational shear occurring within the tachocline; the weaker poloidal field is then regenerated either

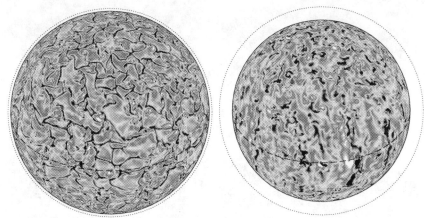

Fig. 20.2. (*a*)–(*b*) Snapshots of radial velocity on spherical surfaces near the top and middle of a solar convection zone simulation carried out with ASH in a deep spherical shell. Darker shades denote downflows, lighter shades upflows; dashed curve indicates the equator. The flow fields are dominated by intermittent plumes of upflow and stronger downflow extending over much of the shell depth, some possessing distinctive cyclonic swirl [Brun & Toomre 2002].

throughout the convection zone by the action of cyclonic turbulence (Parker 1993) or possibly near the surface by the break-up of active regions (Babcock 1961, Leighton 1969). Recent mean-field dynamo models (Tobias 1996, Charbonneau & MacGregor 1997, Beer et al. 1998) have shown that an *interface dynamo* (in which the regions of generation of toroidal and poloidal field are separated) can circumvent the problem of strong α-quenching by mean magnetic fields (Cattaneo & Hughes 1996) and so produce field strengths similar to those inferred from observations. They can also give rise to modulation of the basic cycle.

Coexisting with the large-scale ordered magnetic fields are other compact and intense flux structures that emerge randomly over much of the solar surface, and are often swept into networks of field at the periphery of supergranules (e.g. Schrijver et al. 1997, Berger et al. 1998). This diverse range of activity is most likely generated by two conceptually distinct magnetic dynamos: a *small-scale dynamo*, functioning within the intense turbulence of the convection zone, that builds the chaotic components, and a *global dynamo*, seated within the tachocline, that builds the more ordered fields. The prominent rotational shear within the tachocline, combined with a stable density stratification, makes it likely that strong toroidal magnetic fields can be stretched into existence and can reside there for some time before magnetic buoyancy and other instabilities disrupt these flux concentrations and drag them upward, ultimately to emerge as loops at the surface. The

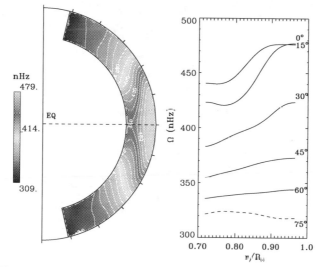

Fig. 20.3. Differential rotation achieved in a 3-D simulation of rotating turbulent convection. The resulting time-averaged angular velocity Ω is displayed both as contours and latitudinal cuts. These profiles are now making good contact in the bulk of the convection zone with the helioseismic deductions [Brun & Toomre 2002]. The depth range studied coincides with the shaded region in Fig. 20.1*b*.

magnetic fields that pierce the photosphere and reach into the solar corona continue to be affected by solar flows. The shearing and converging of the footpoints of magnetic field lines by subphotospheric motions can twist magnetic loops or arcades, adding energy to the magnetic field which is eventually released explosively (e.g. Mikic & Linker 1994, Antiochos et al. 1999, Amari et al. 2000, Tokman & Bellan 2002). There is also evidence that the destabilization of magnetic structures within the corona is accomplished by the gradual accumulation of helicity through the continual emergence through the photosphere of magnetic flux that is already twisted (van Ballegooijen & Martens 1990, Low 1996). In such a case it is likely that the twist or helicity is imparted to the field by flows during its buoyant rise through the convection zone. The complex and shearing subsurface flow patterns seen as SSW may thus play a crucial role in the distribution and evolution of the magnetic fields that lace the solar atmosphere.

20.4 Tachocline: boundary layer of strong shear

The tachocline has been one of the more surprising discoveries of helioseismology, and current theoretical approaches to explain its presence are still only innovative sketches. Substantial theoretical work is required to understand both how this shear layer is maintained, and the likely operation of the global magnetic dynamo within this region. The first models of the tachocline invoked strong horizontal viscosity due to anisotropic turbulence in the stable layer, accompanied by a weak meridional circulation, to circumvent the diffusive spreading of the differential rotation radially inward (Spiegel & Zahn 1992; Elliott 1997). Gravity waves driven by penetrative plumes provide another route to achieve angular momentum redistribution near the base of the convection zone (Zahn, Talon & Matais 1997; Kumar & Quataert 1997). More recently, Gough & McIntyre (1998) argued that even a weak remnant of a primordial magnetic field in the radiative interior would force it to rotate uniformly, as was also appreciated by Mestel & Weiss (1987). They worked out properties of the magnetic skin layer that would be formed at the base of the tachocline of shear, with the primordial field excluded from diffusing upward by a weak meridional circulation.

In contrast, Gilman (2000) has argued that such *slow dynamical processes*, involving meridional circulations with overturning times of order one million years, may have trouble competing with *fast dynamical processes*, such as internal gravity waves triggered by combined shear and magnetic instabilities in the vicinity (e.g. Charbonneau, Dikpati & Gilman 1999). Further, intermittent penetrative plumes of downflow near the base of the convection zone (e.g. Brummell, Clune & Toomre 2002), some serving to transport magnetic flux downward (Tobias et al. 2001), may disrupt the slow processes. We should also note that it is not yet clear from the inversions of helioseismic data whether the tachocline of shear in Ω overlaps with the base of the convection zone or not. Helioseismic estimates place the midpoint of the tachocline at radius $0.692R$, with a thickness estimated to be of order 0.02 to $0.05R$ (e.g. Kosovichev 1996, Basu 1997, Elliott & Gough 1999), whereas from SOI and GONG data the base of the convection zone is at a radius of $0.713R$. These issues emphasize that the tachocline is likely to be the site of complex and possibly competing physical processes, with helioseismic observations having a great potential for guiding the theory.

20.5 Contact with 3-D simulations of turbulent convection

The largest current 3–D simulations of turbulent convection coupled to rotation, studying localized planar domains, have revealed that a surprising degree of *coherent structure* can be embedded in otherwise chaotic flow fields: long-lived networks of downflow, persistent plumes and vortices, and overturning circulations and zonal jets (e.g. Brummell, Hurlburt & Toomre 1996, 1998). These large-scale structures arise from *self-organizing processes*

Fig. 20.4. Evolution of radial velocity sampled at five-day intervals in a 3–D spherical shell simulation of solar turbulent convection. Concentrated downflows are shown in dark tones, revealing that the large-scale patterns are modified noticeably even over such a time span. Both the induced differential rotation and fairly rapid changes in convective structures lead to these changes [Brun & Toomre 2002].

(inverse cascades) that can operate in highly turbulent systems, and appear to have major implications for the transport of energy, angular momentum and magnetic fields within the convection zone. Using our anelastic spherical harmonic (ASH) code (Clune et al. 1999), we have recently studied full spherical shells of rotating turbulent convection (e.g. Elliott et al. 2000; Miesch et al. 2000; Brun & Toomre 2001, 2002). Our current 3–D simulations have attained a spatial resolution more than twenty-fold greater in each dimension than those of the progenitor studies (e.g. Glatzmaier 1987), using spherical harmomics Y_l^m (for horizontal expansion) up to degree $l \sim 680$. The models are a highly simplified description of the solar convection zone: solar values are taken for the heat flux, rotation rate, mass and radius, and a perfect gas is assumed since the upper boundary of the shell lies well below the hydrogren and helium ionization zones. The radial extent of most shells studied range from $0.72R$ to $0.96R$, though some models have also considered penetration into a lower stable zone.

The resulting convection is highly time dependent and the flows are intricate. As evident in Fig. 20.2, the convection is dominated by intermittent plumes of upflow and stronger downflow, some possessing a distinctive cyclonic swirl; the role of coherent plumes, first revealed in planar geometry, is now becoming apparent. The convective patterns, as shown in the sequence of radial velocities near the top of the simulation domain in Fig. 20.4, also evolve over fairly short time scales compared to the solar rotation period, and are advected and sheared by the strong differential rotation that they drive. This suggests that the largest global scales of convection are unlikely to be readily recognizable after one full rotation. These ASH simulations are the first to begin to make serious contact with helioseismic findings about differential rotation within the deep interior of the sun. The time-averaged angular velocity Ω in one of our simulations (Fig. 20.3) is nearly constant on radial lines throughout much of the convection zone at mid-latitudes, and there is a systematic decrease of rotation rate with latitude from the equator to the poles (Brun & Toomre 2001, 2002).

The strong latitudinal contrast in Ω achieved in the bulk of the convection zone (Fig. 20.3) results primarily from Reynolds stresses serving to redistribute angular momentum. Such transport is achieved by correlations in velocity components arising from convective structures, particularly the fast downflow plumes, that are tilted toward the rotation axis and depart from the local radial direction and away from the meridional plane. The spatial resolution of the simulations is adequate to begin to realize such coherent structures, which are crucial in sustaining strong Reynolds stresses at higher turbulence levels. The simulations have not yet tackled in detail issues concerning the role of penetrative convection in the formation of a tachocline near the base of the convection zone.

20.6 Near-surface shear layer and solar subsurface weather

The near-surface shear layer will be the subject of intense scrutiny by several upcoming helioseismology experiments. The discovery of the layer and of SSW within it was the result of steadily accumulating spacecraft and from GONG. Recently, the GONG array has been upgraded with new high-resolution cameras. This enhanced array, called GONG+, will permit continuous probing of SSW in the near-surface shear layer as the solar cycle progresses. Later in the decade these observations should be augmented by the planned Solar Dynamics Observatory (SDO) mission. We anticipate that in the next few years it will be possible to extend 3–D global modelling of deep shells of convection into spatial regimes which deal explicitly with supergranulation and large-scale flows within the upper reaches of the convection zone. These should help to provide the theoretical framework to interpret the rich dynamics that are part of the near-surface shear layer, addressing such issues as how the strong shearing layer is established, why the layer spans a depth of about 5% in solar radius, and how the SSW flows arise and evolve.

The local helioseismic techniques of *ring-diagram* and *time-distance analysis* used to study SSW are complementary. Time-distance tomography (e.g. Giles et al. 1997; Giles 1999; Duvall & Gizon 2000; Korzennik 2001; Chou & Dai 2001) yields the finer spatial resolution of flows and structures, though at considerable computational expense. The ring-diagram procedure (e.g.

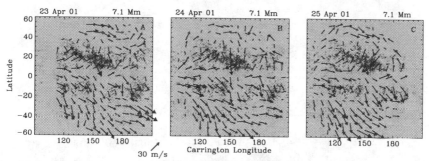

Fig. 20.5. Local helioseismic probing of subsurface flows with ring-diagram methods. (a)–(c) Dense-pack maps for three successive days in April 2001 showing the horizontal velocities determined through inversion at a depth of 7.1 Mm below the photosphere. All velocities are shown relative to the surface differential rotation rate. Estimated velocity errors are about 5% at this depth. Underlying these flow maps are surface magnetic field patterns. The flows are richly evolving, yet possess spatial patterns which can be recognized from day to day [Haber et al. 2002].

Hill 1988; Basu, Antia & Tripathy 1999; Haber et al. 1998, 2000) provides the means to probe and map subsurface flows rapidly, measuring properties of the high-degree acoustic wave field over small localized domains or tiles. Inversion of the displacements of the rings in power arising from frequency splittings can be used to infer the horizontal flows with depth below that domain. By repeating the analysis over many such tiles, a 3–D map of the velocity field within the near-surface layers may be generated. We have used such *dense-pack* ring-diagram analyses to sample an overlapping mosaic of regions, 15° in diameter, which are equally spaced 7.5° apart, roughly filling the solar disk with 189 circular tiles out to 60° from disk center; the evolution of the flows in time can be studied by repeating the dense-pack analyses on a daily interval. Solar rotation brings new sectors into view, with given sites remaining visible for about seven days. We have performed such analyses on the uninterrupted full-disk Doppler velocity data from the SOI–MDI Dynamics Programs, which are typically limited to two to three months each year starting in 1996 (Haber et al. 2000, 2001, 2002). The flows of SSW exhibit a range of features and time scales that constitute a major challenge to theory.

Rapid Day-to-Day Variability in SSW Figure 20.5 presents a dense-pack mosaic of flow fields for three successive days. The flows near active region complexes can be especially intricate, and it appears that magnetic activity and the ordering of SSW flows in their vicinity are closely linked. Some of the flow structures are ephemeral and are seen to emerge, develop and propagate over several days, and then disappear. Other longer-lived

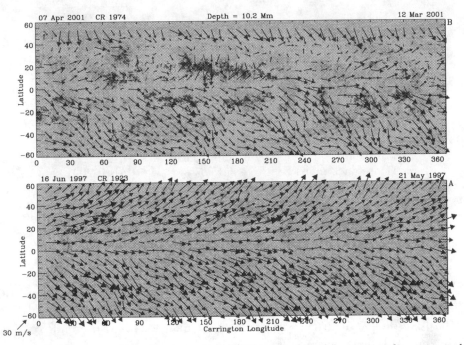

Fig. 20.6. Synoptic maps of horizontal SSW flows deduced from ring-diagram probing at a depth of 10.2 Mm for Carrington rotations (*a*) 1923 (from the year 1997) and (*b*) 1974 (year 2001). The large-scale patterns of meandering flows possess strikingly different character as the magnetic activity has intensified in 2001, as compared to flows in the relatively quiet 1997, with distinctive flow deflections in the vicinity of active complexes. The northern hemisphere in 2001 shows the reversed circulation of a meridional cell directed equatorward at high latitudes [Haber et al. 2002].

structures arise and vanish on a time scale between a week and a month; still others persist longer than a month and may evolve as the solar cycle progresses (Haber et al. 2000).

Monthly Synoptic Global Mappings of SSW Large-scale flow properties are revealed by averaging velocities from the dense-pack mosaic at fixed longitude and latitude over the seven days that a given location remains within the observing matrix, yielding synoptic maps for a range of depths for each Carrington rotation. Figure 20.6*a* is such a synoptic map for one rotation in 1997 when the magnetic activity was relatively weak, showing meandering jets and wavy flow structures. There exists a poleward meridional circulation in both hemispheres, and a net prograde zonal flow (here to the right) due to the increasing rotation rate with depth that occurs within the surface shear layer. Coexisting with these flows is a low-order undula-

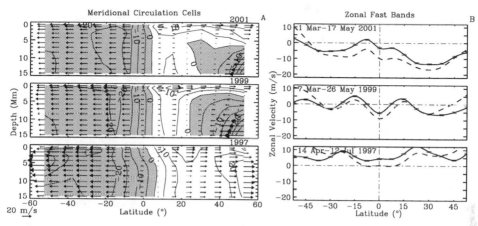

Fig. 20.7. (*a*) Meridional flows with latitude and depth for the three years 1997, 1999 and 2001, obtained by longitudinal and temporal averaging of ring-diagram mappings. Gray regions denote southerly flow and white regions northerly. The meridional flow is consistently poleward (southerly) in the southern hemisphere (negative latitudes). It is likewise poleward near the surface in the northern hemisphere, but at greater depths starting in 1999 the circulations are reversed within a submerged mid-latitude cell. (*b*) Variation of the zonal flows with latitude at depths of 0.9 Mm (dashed) and 7.1 Mm (solid) as sampled during the three years, revealing migrating bands of fast rotation [Haber et al. 2002].

tion in longitude, somewhat reminiscent of the jet stream flow present in the earth's atmosphere; possibly it is an inertial wave response (e.g. Ulrich 2001). This undulatory pattern changes from one rotation to the next. Figure 20.6*b* presents a synoptic map for another rotation in 2001 when the sun was magnetically active. Many of the large-scale flow patterns are clearly correlated with the presence of magnetic active regions, which appear as zones of convergent flow and probably subduction.

Year-to-Year Evolving Global Circulations   There are also pronounced changes with solar cycle in the mean zonal and meridional flows (obtained by averaging over all longitudes). Figure 20.7*b* shows the mean zonal flow with latitude at two depths, confirming the presence of fast bands of zonal flow that propagate toward the equator as the cycle advances (Haber et al. 2001, 2002). These bands correspond to the 'torsional oscillations' known from Doppler measurements of the surface (e.g. Howard & LaBonte 1980, Ulrich 1998) and from global f-mode studies (Kosovichev & Schou 1997, Schou 1999): these orderly but weak flows can be sensed down to radius $0.90R$ using inversions of global f- and p-mode data (Howe et al. 2000b). Turning to the mean meridional flows (Fig. 20.7*a*), in 1997 the circulation is remarkably constant with depth and primarily poleward in both hemispheres. With advancing magnetic cycle, in 1999 and 2001 there is a striking change of behaviour in the northern hemisphere: at midlatitudes and at greater depths, a flow cell with reversed meridional circulation has appeared. The flow just below the surface remains poleward. Finding such a submerged cell with reversed circulation emphasizes that the flow responses in the near-surface shear layer can possess prominent differences (or symmetry breaking) in the two hemispheres.

20.7 Origin of near-surface shear layer

Supergranulation most likely plays a crucial role in the formation of the near-surface shear zone for two reasons. Firstly, of the granular, mesogranular and supergranular scales of convection clearly visible at the surface, one expects that the largest cells extend deepest below the surface: some effects of supergranulation may be experienced at depths of $15 - 40\,\mathrm{Mm}$. Thus supergranulation may be able to influence the angular velocity Ω over the shear-zone depth (about $0.05R$ from helioseismology). However, the nature of downflow networks is intricate, involving sheets which become plumes at the interstices of the cell boundaries, and the flow patterns associated with supergranulation may be likewise quite complicated: thus the depth of influence must be determined by detailed simulations. Secondly, that zone of negative radial gradient in Ω may result from fluid parcels, such as within supergranulation, partially conserving their angular momentum as they move toward or away from the rotation axis (Gilman & Foukal 1979).

Turbulent Convection in Thin Shells To examine some of the convective dynamics that may yield the near-surface rotational shear layer, we have carried out preliminary studies with ASH within thin spherical shells, such as positioned between $0.90R$ and $0.98R$ (DeRosa & Toomre 2001; DeRosa, Gilman & Toomre 2002). Figure 20.8 shows the level of complexity in the turbulent convection of many scales realized in those simulations: the smallest resolvable cellular flows are close to supergranulation in size. The largest scales of convection visible near the top are associated with a connected network of downflow lanes having a spatial scale of about $200\,\mathrm{Mm}$. The large areas enclosed by the downflow lanes each contain many smaller-scale upflows measuring about $15 - 30\,\mathrm{Mm}$ across (comparable to supergranular scales). These features tend to be advected laterally by the horizontal outflow motions associated with the broader cells. At greater depths the broad downflow networks have fragmented into more isolated and compact plume-like structures while retaining vestiges of the network patterns seen

Case D2: Radial Velocity

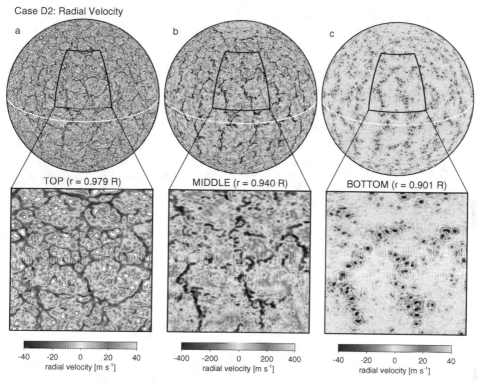

Fig. 20.8. Preliminary simulations with ASH of multi-scale turbulent convection in a thin shell. Shown are instantaneous convective patterns in radial velocities near the top, middle and bottom of that layer which extends in radius from $0.90R$ to $0.98R$. The varying nature of the fast downflow networks (darker tones) with depth are most evident in the close-up views [DeRosa, Gilman & Toomre 2002].

near the top. There are some large-scale streaming motions that come and go, and the meridional circulations involve several cells in each hemisphere. These solutions exhibit partial conservation of angular momentum by scales of convection such as supergranulation which only sense the Coriolis forces weakly, thereby yielding spindown as fluid is carried radially outward and away from the rotation axis at low latitudes. We find that the predominance of faster downflows does serve to build gradients $d\Omega/dr$ at lower latitudes with the same sense, but larger amplitudes, as those deduced from helioseismic observations. These shells are the first to exhibit a decrease in Ω with radius at low- and mid-latitudes that are in the spirit of the helioseismic findings, strongly encouraging the deep shell simulations that we have recently initiated, capable of both resolving supergranulation and of dealing with angular momentum redistribution within the full shell.

Deep Shells with Resolved Supergranulation The deep shell modelling with ASH will need to explore new treatments for the sub-grid scales (SGS) of turbulence, such as granulation and mesogranulation, that cannot be resolved explicitly in such global approaches. We will calibrate our SGS approaches using local planar domain 3–D modelling highly turbulent convection (e.g. Brummell et al. 1998, Tobias et al. 2001). We will also turn to local domain studies of granulation (e.g. Rast et al. 1993, Stein & Nordlund 2001) incorporating realistic equations of state, ionization and radiative transfer to guide how low-entropy downflowing material may be introduced as a stochastic upper boundary condition or as forcing within the deep-shell simulations seeking to capture the near-surface shear layer.

20.8 Reflections

Helioseismology is providing a remarkable window for sampling the rich dynamical processes occurring within the solar convection zone. The surprises have been many, ranging from the presence of a tachocline to the near-surface shear layer exhibiting the intricate flows of SSW, the gradually propagating bands of faster zonal flow, and the evolving multi-celled meridional circulations. The magnetic fields observed at the surface are clearly embedded in the SSW. Seeking to understand how these magnetic fields evolve into configurations that may become unstable and lead to eruptions requires knowing the dynamical properties of the near-surface zone in which the field footpoints reside. The helioseismic observations with GONG+ and SDO will devote much attention to probing and mapping the subsurface flows and surface magnetic fields in order to unravel the complicated interactions in this near-surface zone. It is crucial too that there exist a theoretical framework which can explain the nature and variability of the shear layer and the flows within it. The continued development of 3–D global simulations of solar convection, utilizing for instance the spherical-shell ASH code with advanced SGS representations for unresolved scales, should provide such a framework to help interpret the intrinsic variability and complexity of dynamical processes occurring within the sun and its atmosphere.

Acknowledgements Deborah Haber and Bradley Hindman have been primary contributors here at JILA to the helioseismic studies of SSW, joined by the sage advice provided by Douglas Gough and Michael Thompson in sustained collaborations. Sacha Brun, Tom Clune, Marc DeRosa, Julian Elliott, Gary Glatzmaier, Peter Gilman and Mark Miesch have all been principal players in the studies of turbulent convection in spherical shells using ASH, and Nicholas Brummell and Steve Tobias in the studies of turbulent magneto-convection and dynamo processes. This research was partly supported by NASA through grants NAG 5-7996 and NAG 5-8133, by NSF through grant ATM-9731676, and the simulations carried out with NRAC allocations on supercomputers at SDSC and NCSA.

References

Amari, T., Luciani, J.F., Mikic, Z. & Linker, J., 2000, *ApJ*, **529**, L49
Antiochos, S.K., DeVore, C.R. & Klimchuk, J.A., 1999, *ApJ*, **510**, 485
Babcock, H.W., 1961, *ApJ*, **130**, 364
Basu, S., 1997, *MNRAS*, **288**, 572
Basu, S., Antia, H. M., & Tripathy, S. C., 1999, *ApJ*, **512**, 458
Beer, J., Tobias, S.M., & Weiss, N.O., 1998, *Solar Phys.*, **181**, 237
Berger, T.E., Löfdahl, M.G., Shine, R.A., & Title, A.M., 1998, *ApJ*, **506**, 439
Brummell, N.H., Cattaneo, F., & Toomre, J., 1995, *Science*, **209**, 1370
Brummell, N.H., Clune, T. & Toomre, J., 2002, *ApJ*, **570**, 825
Brummell, N.H., Hurlburt, N.E., & Toomre, J., 1996, *ApJ*, **473**, 494
Brummell, N.H., Hurlburt, N.E., & Toomre, J., 1998, *ApJ*, **493**, 955
Brun, A.S. & Toomre, J., 2001, in Eff-Darwich, A. & Wilson, A., eds, 'Helio- and Asteroseismology at the Dawn of the Millenium', *ESA SP–464*, 619
Brun, A.S. & Toomre, J., 2002, *ApJ*, **570**, 865
Cattaneo, F. & Hughes, D.W., 1996, *Phys. Rev. E*, **54**, 4532
Charbonneau, P., Dikpati, M., & Gilman, P.A., 1999, *ApJ*, **526**, 513
Charbonneau, P. & MacGregor, K.B., 1997, *ApJ*, **486**, 502
Chou, D.-Y. & Dai, D.-C., 2001, *ApJ*, **559**, L175
Clune, T.L., Elliott, J.R., Miesch, M.S., Toomre, J., & Glatzmaier, G.A., 1999, *Parallel Computing*, **25**, 361
DeRosa, M.L., Gilman, P.A., & Toomre, J., 2002, *ApJ*, in the press.
DeRosa, M.L. & Toomre, J., 2001, in Eff-Darwich, A. & Wilson, A., eds, 'Helio- and Asteroseismology at the Dawn of the Millenium', *ESA SP–464*, 595
Duvall, T.L. Jr. & Gizon, L., 2000, *Solar Phys.*, **192**, 177
Elliott, J.R., 1997, *A&A*, **327**, 1222
Elliott, J.R. & Gough, D.O., 1999, *ApJ*, **516**, 475
Elliott, J.R., Miesch, M.S., & Toomre, J., 2000, *ApJ*, **533**, 546
Fisher, G.H., Fan, Y., Longcope, D.W., Linton, M.G. & Pevtsov, A.A., 2000, *Solar Phys.*, **192**, 119
Giles, P.M., 1999, Ph.D. Thesis, Stanford University
Giles, P.M., Duvall, T.L. Jr., Scherrer, P.H., & Bogart R.S., 1997, *Nature*, **390**, 52
Gilman, P.A., 2000, *Solar Phys.*, **192**, 27
Gilman, P.A. & Foukal, P.V., 1979, *ApJ*, **229**, 1179
Glatzmaier, G.A., 1987, in Durney, B.R. & Sofia, S., eds, *The Internal Solar Angular Velocity*, Reidel, 263
Gough, D.O. & Toomre, J., 1991, *ARAA*, **29**, 627
Gough, D.O. & McIntyre, M.E., 1998, *Nature*, **394**, 755
Haber, D.A., Hindman, B.W., Toomre, J., Bogart, R.S., Schou, J., & Hill, F.,

1998, in Korzennik, S. & Wilson, A., eds, 'Structure and Dynamics of the Sun and Sun-like Stars', *ESA SP–418*, 791

Haber, D.A., Hindman, B.W., Toomre, J., Bogart, R.S., Thompson, M.J., & Hill, F., 2000, *Solar Phys.*, **192**, 335

Haber, D.A., Hindman, B.W., Toomre, J., Bogart, R.S., & Hill, F., 2001, in Eff-Darwich, A. & Wilson, A., eds, 'Helio- and Asteroseismology at the Dawn of the Millenium', *ESA SP–464*, 213

Haber, D.A., Hindman, B.W., Toomre, J., Bogart, R.S., Larsen, R.M., & Hill, F., 2002, *ApJ*, **570**, 855

Hill, F., 1988, *ApJ*, **333**, 996

Howard, R. & LaBonte, B.J., 1980, *ApJ*, **239**, L33

Howe, R., Christensen-Dalsgaard, J., Hill, F., Komm, R.W., Larsen, R.M., Schou, J., Thompson, M.J., & Toomre, J., 2000a. *Science*, **287**, 2456

Howe, R., Christensen-Dalsgaard, J., Hill, F., Komm, R.W., Larsen, R.M., Schou, J., Thompson, M.J., & Toomre, J., 2000b. *ApJ*, **533**, L163

Korzennik, S.G., 2001, in Eff-Darwich, A. & Wilson, A., eds, 'Helio- and Asteroseismology at the Dawn of the Millenium', *ESA SP–464*, 149

Kosovichev, A.G., 1996, *ApJ*, **469**, L61-L64

Kosovichev, A.G. & Schou, J., 1997, *ApJ*, **482**, L207

Kumar, P. & Quataert, E.J., 1997, *ApJ*, **475**, L143

Leighton, R.B., 1969, *ApJ*, **156**, 1

Libbrecht, K.G., 1989, *ApJ*, **336**, 1092

Low, B.C., 1996, *Solar Phys.*, **167**, 217

Mestel, L. & Weiss, N.O., 1987, *MNRAS*, **226**, 123

Miesch, M.S., Elliott, J.R., Toomre, J., Clune, T.L., Glatzmaier, G.A., & Gilman, P.A., 2000, *ApJ*, **532**, 593

Mikic, Z. & Linker, J.A., 1994, *ApJ*, **430**, 898

Parker, E.N., 1993, *ApJ*, **408**, 707

Rast, M.P., Nordlund, A., Stein, R.F., & Toomre, J., 1993, *ApJ*, **408**, L53

Schou, J., et al., 1998, *ApJ*, **505**, 390

Schou, J., 1999, *ApJ*, **523**, L181

Schrijver, C.J., Title, A.M., van Ballegooijen, A.A., Hagenaar, H.J., & Shine, R.A., 1997, *ApJ*, **487**, 424

Spiegel, E.A. & Zahn, J.-P., 1992, *A&A*, **265**, 106

Stein, R.F. & Nordlund, A., 2001, *Solar Phys.*, **192**, 91

Thompson, M.J., et al., 1996, *Science*, **272**, 1300

Tobias, S.M., 1996, *ApJ*, **467**, 870

Tobias, S.M., Brummell, N.H., Clune, T., & Toomre, J., 2001, *ApJ*, **549**, 118

Tokman, M. & Bellan, P.M., 2002, *ApJ*, **567**, 1202

Tomczyk, S., Schou, J., & Thompson, M.J., 1995, *ApJ*, **448**, L57

Ulrich, R.K., 1998, in Korzennik, S. & Wilson, A., eds, 'Structure and Dynamics of the Sun and Sun-like Stars', *ESA SP–418*, 851

Ulrich, R.K., 2001, *AGU Spring Meet.*, Abs. SP31A-01, S386

van Ballegooijen, A.A. & Martens, P.C.H., 1990, *ApJ*, **361**, 283

Zahn, J.-P., Talon, S., & Matais, J., 1997, *A&A*, **322**, 320

21

Numerical simulations of the solar convection zone

JULIAN R. ELLIOTT

Eurobios UK, Sir John Lyon House, 5 High Timber Street,
London EC4V 3NX, UK

Deep convection occurs in the outer one-third of the solar interior and trans-
ports energy generated by nuclear reactions to the surface. It leads to a
characteristic pattern of time-averaged differential rotation, with the poles
rotating approximately 20% slower than the equator. A particularly notable
feature of the solar differential rotation is that it departs significantly from
the Taylor-Proudman state of rotation constant on cylinders aligned with
the rotation axis. Although this observation contrasts with results from early
numerical simulations, such simulations provide the best hope of understand-
ing the observations. Many studies have adopted the DNS (Direct Numer-
ical Simulation) approach and justified the artificially large viscosities and
thermal diffusivities used as modelling transport by unresolved eddies. LES
(Large Eddy Simulation) techniques, which use a suitable turbulence closure
model, offer a superior alternative but face the problem of choosing an ap-
propriate turbulence closure; this can be difficult in the face of complicating
factors such as stratification and rotation. An alternative approach is to shift
responsibility for truncating the turbulent cascade to the numerical scheme
itself. Since this approach abandons the rigorous notions of the LES ap-
proach, we refer to it as a VLES (Very Large Eddy Simulation). This paper
compares results of DNS simulations carried out with a spherical harmonic
code, and preliminary results obtained using a VLES-type code. Both make
the anelastic approximation.

21.1 Introduction

In the outer third of the solar interior, radiative energy transport becomes
inefficient (due to the increasing opacity) and convection takes over as the
primary mechanism of energy transport. This convection is driven by strong
radiative cooling near the surface and compensating warming occurring over

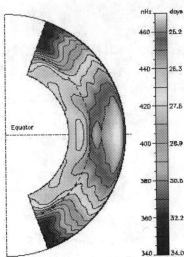

Fig. 21.1. Solar angular velocity in a meridional cut as determined by helioseismology. (Reprinted with permission from Thompson et al., Science, **296**, 1300. Copyright 1996 American Association for the Advancement of Science.)

a broad region in radius near the base of the convection zone. The time scale of the convection is about 30 days, with characteristic velocities of a few hundred $m\,s^{-1}$ and characteristic temperature fluctuations of a few Kelvin. Since the convection and rotation time scales are very similar (leading to a convective Rossby number of order unity), the convection is significantly influenced by rotation—this leads to the observed differential rotation, with the poles rotating significantly more slowly than the equator (see Fig. 21.1). In addition, since the density in the convection zone varies by over ten orders of magnitude, the convection is strongly influenced by stratification effects. Since the solar plasma is a very good electrical conductor (leading to small magnetic diffusivities and high magnetic Reynolds numbers), magnetic fields are yet another factor which influences the dynamics of the solar interior; the interaction of fluid flow and magnetic fields leads to the 22-year cycle of magnetic fields and sunspot activity, although the exact mechanism is as yet unclear.

The combined effects of rotation, stratification and magnetic fields in a high Reynolds number (and therefore highly turbulent) flow leads to a system with a rich spectrum of behaviour that we can only hope to understand through detailed numerical simulations. Since they are computationally intensive, such simulations have only become possible in the last few decades with advances in high-performance computing. In spite of these advances, certain approximations are necessary in carrying out simulations of solar convection—the most important of these is to filter out fast sound waves that play a small part in the overall dynamics of the convection zone but could unreasonably limit the computational time step. In both the codes described here, this is done using the *anelastic* approximation (Gough 1969). Another fundamental simplification is to neglect magnetic fields—the validity of this approximation depends on magnetic fields not being too strong. A final simplification made in the simulations described here is to limit the range of densities modelled, effectively cutting off the simulations near the solar surface, where the density scale height becomes small; this removes the need to use multiple grids to model the different scales of convection and simplifies the problem.

The choice of numerical integration scheme in modelling solar convection is paramount. Interestingly, thermodynamic considerations provide some insight. Radiative heating near the base of the convection zone and compensating cooling near the solar surface act as a source of negative entropy, which in equilibrium must be balanced by the action of irreversible processes. Two irreversible processes dominate—firstly, the transfer of heat from one fluid element to another by radiative diffusion, and secondly, the dissipation of kinetic energy by the action of viscosity. The relative importance of the two is unclear, although since the Prandtl number is low (the radiative diffusivity being much higher than the kinematic viscosity) the first is probably dominant. Unfortunately, both operate at scales far too small to be resolved computationally. For this reason, irreversible entropy production must be accounted for by artificial dissipation, either in the form of enhanced values of the radiative diffusivity and viscosity, or a dissipative numerical scheme. Studies employing non-dissipative numerical schemes (such as pseudo-spectral and CTS—Centred in Time and Space—methods) have often justified the artificial dissipation added as in some way modelling the effect of small eddies on the large-scale flow (e.g. Gilman 1977; Glatzmaier 1984). Since the diffusion and viscosity took the form of Laplacian operators with constant coefficients, these studies may be classified as DNS simulations. The next section summarizes some of the results from Elliott at al. (2000) and Miesch et al. (2000), which fall into this category.

The problem with DNS techniques is that large thermal diffusivities must be introduced everywhere, even where the flow is laminar. A better approach is to adopt the LES technique, where the flow field is split into resolved and unresolved scales with the latter acting back on the former through the divergence of the SGS (Sub-Grid Scale) stress tensor. In general, the components of this stress tensor are expressed in terms of the resolved flow field—this is known as the SGS model. The most well known SGS model is that postulated by Smagorinsky (1963), in which the SGS stress tensor is proportional to the local rate of strain of the resolved flow.

Historically, SGS models have tended to be judged more on their ability to suppress false computational oscillations than on their representation of what is actually happening on sub-grid scales. Once one gives up the notion that SGS models accurately represent the behaviour of unresolved eddies, there are simpler, more effective means of achieving the suppression of false computational oscillations. In particular, there is a class of finite difference methods – Non-oscillatory Forward in Time (NFT) – that have the remarkable property of representing LES/VLES without recourse to any explicit SGS model. Such methods are inherently dissipative and provide an irreversible entropy source, as they must given the constraints mentioned above. The use of this class of numerical methods in geophysical fluid dynamics has been justified by simulations of the Earth's planetary boundary layer (Margolin et al. 1999) that compare well with observations. This paper presents first results for solar convection obtained from a code based on NFT methods, in particular the MPDATA (Multidimensional Positive-Definite Advection and Transport Algorithm) scheme (Smolarkiewicz & Margolin 1998).

This is not the first time that comparisons between different methods of simulating solar convection have been made. For example, Cattaneo et al. (1991) compared three-dimensional Cartesian convection simulated with a pseudo-spectral code and with a code implementing the PPM method (Piecewise Parabolic Method, Collela & Woodward 1984); the latter is another method from the NFT class that can be run stably with no explicit diffusion or viscosity. Although detailed comparisons of results were not presented in this study, a description of the philosophy of NFT methods was included. Much confusion has been introduced into the subject by the inappropriate use of the word "inviscid" to describe NFT methods.

This paper is divided into three further sections. The next covers results from a pseudo-spectral DNS model that uses constant diffusivities and viscosities. Next, results from a VLES simulation of solar convection are described. The final section highlights the conclusions from this work.

21.2 DNS results

This section describes numerical simulations carried out within the DNS framework using a code (see Clune et al. 1999) that expands the prognostic variables (entropy, pressure and velocity) in spherical harmonics in the horizontal, and in Chebyshev polynomials in the vertical. Spherical harmonic expansions have been widely used in meteorological applications and have the advantage of providing uniform resolution over the sphere (at least when triangular truncation is used) — this avoids the so-called pole problem. Non-linear terms are calculated in physical space (the transform method), using a grid with sufficient resolution to avoid quadratic aliasing. Apart from the uniform resolution, spherical harmonic basis functions also offer the advantage of yielding a very easy solution of the elliptic pressure equation, since the linear terms decouple.

We now describe the equation set used. The prognostic variables are velocity and entropy. The continuity equation in the anelastic approximation (which applies to both models described in this paper) is

$$\nabla \cdot (\rho_0 \boldsymbol{v}) = 0 \,, \tag{21.1}$$

where ρ_0 is the density in the reference state (T_0 and θ_0 are the corresponding temperature and potential temperature respectively) and $\boldsymbol{v}$ is the fluid velocity. The momentum equation is

$$\frac{\partial \boldsymbol{v}}{\partial t} + (\boldsymbol{v} \cdot \nabla)\, \boldsymbol{v} = -\frac{1}{\rho_0} \nabla p' + \frac{\rho'}{\rho_0} \boldsymbol{g} + 2\boldsymbol{v} \times \boldsymbol{\Omega} - \frac{1}{\rho_0} \nabla \cdot \boldsymbol{D} \,, \tag{21.2}$$

where p' and ρ' are the departures of the pressure and density from the reference state values, $\boldsymbol{g}$ is the acceleration due to gravity, $\boldsymbol{\Omega}$ is the rotation vector, and $\boldsymbol{D}$ is the viscous stress tensor, given in terms of the viscosity ν and rate-of-strain tensor $\boldsymbol{e}$ by

$$\boldsymbol{D}_{ij} = -2\rho_0 \nu \left[e_{ij} - \frac{1}{3} (\nabla \cdot \boldsymbol{v}) \delta_{ij} \right] . \tag{21.3}$$

The entropy equation is given by

$$\rho_0 T_0 \left[\frac{\partial S'}{\partial t} + (\boldsymbol{v} \cdot \nabla)\, S' \right] = \varepsilon + \nabla \cdot \left(\kappa \rho_0 T_0 \nabla S' \right) + 2\nu \rho_0 \left[e_{ij} e_{ij} - \frac{1}{3} (\nabla \cdot \boldsymbol{v})^2 \right] , \tag{21.4}$$

where S' is the fluctuation of entropy away from the reference state value, κ is the thermal diffusivity, and ε is the divergence of the radiative flux. The last term on the right-hand side of this equation is the viscous heating term.

The domain used for the simulations is a spherical shell with an inner radius of 5×10^5 km and an outer radius of 6.9×10^5 km. The density contrast from the bottom of the domain to the top is about 30, which, although substantially less than the actual contrast across the convection zone, allows for significant stratification effects to be felt. Thermal forcing is modelled by assuming that the mean stratification in the bulk of the

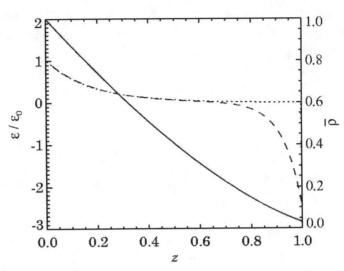

Fig. 21.2. The divergence ε of the radiative flux (dotted line: actual, dashed line: that assumed in the model), relative to its value ε_0 at the bottom of the simulation domain, and the density (solid line); z is the fractional height from the bottom.

convection zone is always nearly adiabatic, allowing the divergence of the radiative flux, ε, to be assumed to be fixed — in the simulations described in this paper, the following form is used for ε:

$$\varepsilon = \alpha \left(\frac{\zeta_1 e^{-\zeta_1 z}}{1 - e^{-\zeta_1}} - \frac{\zeta_2 e^{-\zeta_2(1-z)}}{1 - e^{-\zeta_2}} \right) , \qquad (21.5)$$

where α is an appropriate constant of proportionality, z is the non-dimensional height within the domain, and ζ_1 and ζ_2 are non-dimensional constants, which take the values 5.5 and 13.0 respectively. Figure 21.2 compares the actual radiative flux divergence in the convection zone with the formula used, and also shows the density profile assumed. Near the surface, the assumption of a near-adiabatic stratification breaks down, and there is an extremely narrow region of strong radiative cooling — this is artificially broadened (by making the scale of the region comparable to the local density scale height) to enable it to be resolved on the computational grid used.

Consistent with the DNS framework, the Navier-Stokes equations are solved with entropy diffusion and viscous terms having coefficients that depend only on radius. Tests have been carried out with several different dependencies of these coefficients on radius—the effects on the results are not too substantial. Sufficiently large values of entropy diffusion and vis-

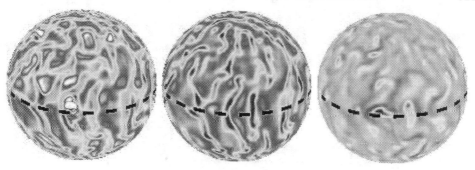

Fig. 21.3. The vertical velocity on three horizontal surfaces (from left to right) near the top, middle, and bottom of the shell, in a DNS simulation of solar convection. Dark shades denote downflows and light shades upflows. The dashed line marks the equator.

cosity are required to produce stable, well-behaved solutions; typical values used lead to Reynolds numbers in the region of 50.

In a linear analysis, the most unstable modes of the system are convective rolls aligned with the rotation axis. These are also seen in the non-linear solutions, albeit with considerable distortion from non-linear effects. Figure 21.3 shows typical snapshots of the vertical velocity on three horizontal levels of a solution from the DNS code. The convective rolls, or "banana cells", can clearly be seen in the left and middle panels of this figure (which correspond to levels near the surface, and near the centre of the convection zone, respectively), with rotationally aligned lanes of downflowing material interspersed by upflowing material.

This simulation used 64 radial grid points and spherical harmonics up to degree 85. The Reynolds number for the largest eddies was about 30. Higher resolution would enable higher Reynolds numbers to be attained; however, the computational effort to increase the resolution is very high, since the number of floating point operations scales with the fifth power of resolution (since the spherical harmonic transforms scale with the cube of the horizontal resolution, and the number of time steps also scales linearly with the resolution).

Turning to other aspects of the results, the time-averaged differential rotation for the same simulation is shown in Fig. 21.4. Comparing with Fig. 21.1, which shows the solar differential rotation as deduced by helioseismology, some similarities are clear. First, the simulation gives a fast equator and slow poles. Secondly, the contrast in angular velocity is similar to that seen in the sun. Unfortunately, there are significant differences in

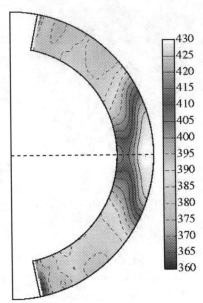

Fig. 21.4. The time-averaged angular velocity from the same simulation as shown in Fig. 21.3. The scale on the right is in nHz.

other respects. Firstly, most of the angular velocity contrast in the simulation is seen near the equator, whereas in the sun there is a continuous change in angular velocity from the equator to the poles. Also, the sun shows considerable tilting of the lines of constant angular velocity relative to the rotation axis, whereas the simulation shows very little. Other simulations (Gilman 1977; Glatzmaier 1984) have also shown little of such tilting, with angular velocity contours nearly parallel to the rotation axis. The reasons for these discrepancies are currently unclear, although what does seem to be the case is that Reynolds stresses alone (in this case correlations of the radial and latitudinal velocity components) are not strong enough, and typically of the wrong sign, to produce the observed tilting—buoyancy forces almost certainly play a part, implying that the poles are warm with respect to the equator.

21.3 VLES results

The simulations described so far rely on large, fixed viscosities and thermal diffusivities to obtain a stable solution. An alternative option is to use an SGS model within an LES framework. This avoids the use of a large amount of dissipation where it is unnecessary, the only drawback being the difficulty of finding an appropriate SGS model (especially in the presence of rotation and stratification). Another way of avoiding unnecessary dissipation, without this drawback, is to use an appropriate numerical advection scheme that is itself responsible for the dissipation. Such advection schemes belong to the general class of NFT methods for which ensuring linear stability (in particular the familiar Courant-Friedrichs-Lewy condition) also assures non-linear stability. We refer to numerical simulations of fluid flow employing such advection schemes as VLES, to distinguish them from LES.

This section describes VLES simulations of solar convection carried out using a code described in Smolarkiewicz et al. (2001). The code is an anelastic, grid-based code, with options for carrying out advection either by means of a semi-Lagrangian scheme or an Eulerian scheme (MPDATA, Smolarkiewicz & Margolin 1998). Because NFT methods are inherently two time-level (i.e. the current and the next time step), accurate (e.g. second order) time integration couples the non-linear terms into the implicit solve, necessitating the solution of a large non-symmetric linear system representing the complex non self-adjoint elliptic pressure equation. For deep fluids, such as the sun's convection zone, such a problem can be solved using standard Krylov subspace methods (although note that in the case of the Earth's atmosphere, the resulting elliptic equations are extremely stiff, and preconditioning is necessary) — that approach is adopted in this code.

In terms of the equation set, potential temperature is used instead of entropy, leading to the following momentum and potential temperature equations:

$$\frac{\partial \boldsymbol{v}}{\partial t} + (\boldsymbol{v} \cdot \nabla)\,\boldsymbol{v} = -\nabla\left(\frac{p'}{\rho_0}\right) + \boldsymbol{g}\frac{\theta'}{\theta_0} + 2\boldsymbol{v} \times \boldsymbol{\Omega} - \frac{1}{\rho_0}\nabla \cdot (\nu\rho_0 \nabla \boldsymbol{v})\ , \quad (21.6)$$

$$\frac{\partial \theta'}{\partial t} + (\boldsymbol{v} \cdot \nabla)\,\theta' = \frac{\theta_0}{\rho_0 K T_0}\epsilon - \frac{1}{\rho_0}\nabla \cdot \left(k\rho_0 \nabla \theta'\right) + 2\nu\rho_0\left[e_{ij}e_{ij} - \frac{1}{3}\left(\nabla \cdot \boldsymbol{v}\right)^2\right]\ , \tag{21.7}$$

where θ' is the fluctuation of potential temperature away from the reference state, K is the thermal capacity of the fluid, and k is the thermal diffusivity. Note the somewhat different form of the viscous stress in the momentum equation. This term is calculated by means of scalar Laplacians on each component, rather than taking the full vector Laplacian — this means that certain parts of the stress (in particular those arising from the curvature of the coordinate system) are neglected.

Coordinate singularities, such as that present at the poles in the spherical polar coordinate system, represent a potential difficulty for NFT methods.

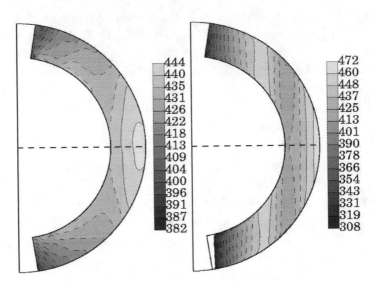

Fig. 21.5. The time-averaged angular velocity in two simulations carried out using the Smolarkiewicz code. Left: a DNS simulation for comparison with Fig. 21.4; right: a genuine VLES simulation.

This is generally not particularly significant when simulating the Earth's atmosphere, since there is not much convective activity near the poles, but could present a problem in the solar case, since convection occurs at all latitudes with nearly equal strength.

Figure 21.5 shows the time-averaged differential rotation in two such simulations, the left-hand of which is a DNS-like simulation for comparison with the results of section 21.2, while the right-hand of which is a genuine VLES simulation, with no added thermal diffusion or viscosity. The DNS-like simulation was carried out using similar values of viscosity and thermal diffusivity to those used in the simulation that yielded Fig. 21.4. Comparing the two, the overall contrast in angular velocity between the equator and poles is similar, amounting to about 15 % of the equatorial value—this is somewhat less than the corresponding contrast seen in the sun (approximately 25 %). Looking beyond this simple measure, the new simulation is more solar-like in two principal ways—first, a larger part of the angular velocity contrast occurs near the poles, as opposed to being almost entirely concentrated at latitudes below 30 degrees. Secondly, the angular velocity contours are considerably tilted with respect to the rotation axis, showing significant departure from the Taylor–Proudman regime.

The right-hand panel of Fig. 21.5 shows the time-averaged angular veloc-

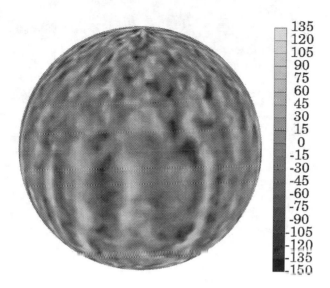

	135
	120
	105
	90
	75
	60
	45
	30
	15
	0
	-15
	-30
	-45
	-60
	-75
	-90
	-105
	-120
	-135
	-150

Fig. 21.6. The vertical velocity on a horizontal surface near the middle of the domain for the VLES simulation. The scale on the right is in ms^{-1}.

ity from the genuine VLES simulation, carried out using the Eulerian advection option. Peaks are seen at the equator and at mid-latitudes, with weak minima at around 40 degrees and strong minima near the poles. The overall contrast in angular velocity between the equator and poles is very large, larger indeed than in both the DNS simulation described in section 21.2, and in the real sun. The angular velocity contours are very nearly parallel to the rotation axis, showing little departure from the Taylor–Proudman state. This simulation clearly shows less similarity with the sun's differential rotation than either of the DNS simulations described—why should this be the case?

At this point it is worth considering a few of the possible reasons why the VLES simulation should give a less realistic differential rotation profile than either of the DNS simulations. The limitations of the VLES simulations can be broadly grouped into two categories—problems with the physics used, and insufficient resolution to capture the smallest eddies in the flow. Problems with the physics include insufficient density contrast between the bottom and top of the domain, incorrect boundary conditions, and the absence of magnetic fields. Insufficient resolution would prevent the simulation from modelling sufficiently small eddies, and would be manifested as a lack

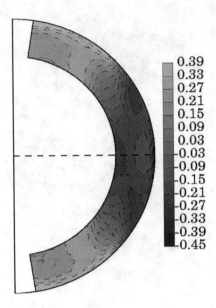

Fig. 21.7. Longitudinally averaged potential temperature departures (in K) from the spherical mean in the VLES simulation.

of good convergence with increasing resolution. The DNS simulations share the same limitations, but suffer the additional limitation of having excessive dissipation acting where it is not strictly necessary. It thus seems that although the DNS simulations are obtaining a better result, it is probably for the wrong reasons.

Considering other aspects of the solutions obtained with the Smolarkiewicz code, Fig. 21.6 shows the vertical velocity from the VLES simulation on a horizontal surface at the midpoint of the domain. Close similarities can be seen with the vertical velocity from the spherical-harmonic code DNS simulation (Fig. 21.3), with elongated "banana cells" near the equator, and less rotationally influenced convection near the poles. The maximum downflow velocity is somewhat higher than that of the upflows , while the absolute magnitudes are of the order of 100 - 150 ms^{-1}. The simulation shows evidence of coordinate problems near the poles, with the typical scale of flow features diminishing in those regions: the danger of this occurring was noted earlier in this section.

Figure 21.7 shows the longitudinally averaged departures of the potential temperature from the spherical mean value in the VLES simulation. This figure demonstrates how small the departures are and how close the convection zone is to spherical symmetry—the relative temperature fluctuations are of the order of 10^{-6}. Attempts to measure the variation of the solar surface temperature and luminosity are thus faced with a very challenging problem. The small latitudinal temperature variation that the figure *does* show produces a thermal wind circulation tending to tilt the angular velocity contours away from the rotation axis; importantly, since the poles are warm with respect to the equator, this tilting is in the right direction (see the comment at the end of section 21.2). Note that little evidence of such tilting is actually seen in the angular velocity plots of Fig. 21.5, presumably because the pole-equator temperature difference is insufficiently large.

21.4 Conclusion

This paper has highlighted the challenges of understanding the complex dynamics of the solar convection zone, and has shown some results from two investigations using large-scale numerical simulations. Many previous investigations of solar convection have adopted the DNS approach, facing the problem of having to introduce large artificial thermal diffusivities and viscosities. The first investigation presented here is of this type, and employs spherical harmonics to expand the prognostic variables. The second investigation presented is of a very different type, employing NFT advection techniques to create an implicitly stable numerical integration scheme that avoids the use of large amounts of artificial thermal diffusivity and viscosity. Simulations carried out under this philosophy are known as VLES—these results are presented along with results obtained using the same code but adding artificial viscosity and diffusion; this enables comparison with the already-presented DNS results.

All three sets of results (the DNS with spherical harmonic code, the DNS with the NFT code, and the VLES with the NFT code) show similar patterns of vertical velocity in the convection zone, with "banana-cell" convective rolls and convection velocities of the order of a few hundred $\mathrm{m\,s^{-1}}$. The three differ, however, in the differential rotation produced: the DNS with the spherical harmonic code gives a somewhat similar pattern to the DNS with the NFT code, but the VLES with the NFT code is rather different. Since the first two are carried out with almost identical equation sets and parameter values, one would expect them to give similar results; the fact that there are differences at all indicates the sensitivity of the solutions to the details of the viscous stresses. On the other hand, the fact that the VLES simulation gives such different results is interesting and illustrates well the strong effect of the artificial viscosity and diffusivity used in DNS simulations. Each of the simulations captures some elements of the observed results that the others do not, so no one could clearly be identified as the most accurate; comparing the three provides some hints as to the sensitivities of solar convection.

Future work in these investigations will concentrate on better understanding the origin and consequences of the discrepancies between the DNS simulation carried out with the spherical-harmonic code (as described in section 21.2), and that carried out with the NFT code (as described in section 21.3). In addition, further studies will be carried out to investigate the degree of convergence of the VLES results; were the results from this simulation to change significantly with resolution, it would indicate a poor degree of numerical convergence. If, even in the converged case, the results fail to show good agreement with the observed solar differential rotation, the suggestion would be that some other aspect of the physics of the problem has been incorrectly modelled, whether it is the lack of magnetic fields, insufficient density contrast across the domain, or other factors. There is clearly much scope for further investigation of this fascinating problem.

Acknowledgements The author is grateful to P. K. Smolarkiewicz who carried out simulations with his VLES code.

References

Cattaneo, F., Brummell, N. H., Toomre, J., Malagoli, A. & Hurlburt, N., 1991. *ApJ*, **370**, 282

Clune, T., Elliott, J. R., Glatzmaier, G. A., Miesch, M. S. & Toomre, J., 1999. *Parallel Computing*, **25(4)**, 361

Colella, P. & Woodward, P. R., 1984. *J. Comput. Phys.*, **54**, 174

Elliott, J. R., Miesch, M. S. & Toomre, J., 2000. *ApJ*, **533**, 546

Gilman, P. A., 1977. *GAFD*, **8**, 93

Glatzmaier, G. A., 1984. *J. Comput. Phys.*, **55**, 461

Gough, D. O., 1969. *J. Atmos. Sci.*, **26**, 448

Margolin, L. G., Smolarkiewicz, P. K. & Sorbjan, Z., 1999. *Physica D*, **133**, 390

Miesch, M. S., Elliott, J. R., Toomre, J., Clune, T. & Glatzmaier, G. A., 2000. *ApJ*, **532**, 593

Smagorinsky, J., 1963. *Monthly Weather Review*, **91**, 99

Smolarkiewicz, P. K. & Margolin, L. G, 1998. *J. Comput. Phys.*, **140**, 459

Smolarkiewicz, P. K., Margolin, L. G. & Wyszogrodzki, A. A., 2001. *J. Atmos. Sci.*, **58**, 349

Thompson, M. J., et al., 1996. *Science*, **272**, 1300

22

Modelling solar and stellar magnetoconvection

NIGEL WEISS

Department of Applied Mathematics & Theoretical Physics,
University of Cambridge, Cambridge CB3 9EW, UK

Numerical experiments on three-dimensional convection in the presence of
an externally imposed magnetic field reveal a range of behaviour that can be
compared with that observed at the surface of the Sun (and therefore expected
to be present in other similar stars). In a strongly stratified compressible
layer small-scale convection gives way to a regime with flux separation as the
field strength is reduced; with a weak mean field magnetic flux is concentrated
into narrow lanes enclosing vigorously convecting plumes. Small-scale dy-
namos, generating disordered magnetic fields, have been found in Boussinesq
calculations with very high magnetic Reynolds numbers; there is a gradual
transition from dynamo action to magnetoconvection as the strength of the
imposed field is increased.

22.1 Introduction

Thirty-seven years ago, when I was a postdoc at Culham, Roger Tayler told
me that he was sending a very bright young research student to spend the
summer there – and so I first met Douglas. When I moved to Cambridge
a year later he was finishing his Ph.D. and then he and Rosanne went off
to the States for a few years. We've been in close contact ever since they
returned to Cambridge and it has been a great pleasure having Douglas as
a colleague and a friend – always stimulating and often argumentative, but
never causing any serious disagreement. So I am very glad to have a chance
of saying 'Thank you' here.

As we have already been reminded, Douglas's third paper (Gough &
Tayler 1966) was on *magnetoconvection*. In those days we were all concerned with linear theory and they used an energy principle to extend the
Schwarzschild criterion for convective stability in a stratified atmosphere
and to establish a sufficient criterion for stability in the presence of a mag-

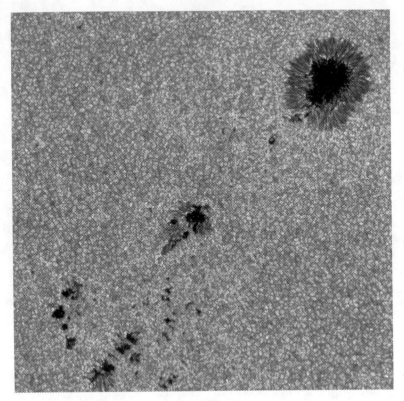

Fig. 22.1. Magnetoconvection at the solar photosphere. This G-band image, obtained on the Swedish Vacuum Solar Telescope at La Palma by T. Berger & G. Scharmer, shows a region about 120 000 km square. Granular convection is interrupted by sunspots and pores, with strong magnetic fields, and locally intense fields show up as small bright points in intergranular lanes. (Courtesy of T. Berger.)

netic field. Nowadays we are all involved with nonlinear problems. What is exciting is that advances in observational techniques and high performance computing have at last made it possible to compare fine structures revealed by high-resolution observations with corresponding features in numerical experiments. So I shall present some recent computational results on nonlinear magnetoconvection and relate them to small-scale fields at the surface of the Sun.

Let me begin with the observational motivation. Figure 22.1 shows a remarkable image of the solar photosphere, obtained in the CH G-band at 4305 Å. The granulation, corresponding to convective plumes with diameters of about 1 000 km, is clearly visible, as are various magnetic features embedded in it. The latter range from a large sunspot, where convection is partially suppressed by a strong, predominantly vertical magnetic field, through small, dark pores to bright points that denote the sites of locally intense magnetic fields. These concentrated fields are embedded in the dark lanes surrounding the bright granules, which form a network of colder sinking gas, and especially at junctions in this network. On this scale, the Sun's rotation has no significant effect.

Observations from the MDI instrument on SOHO have demonstrated that magnetic flux is constantly emerging through the solar surface to form ephemeral active regions, largely uncorrelated with the ordered flux associated with the 11-year activity cycle. This flux is shredded by the granulation to form small flux elements which are transported by the supergranular flow until they end up in the network that encloses supergranules, where oppositely directed fields eventually reconnect and disappear (Simon, Title & Weiss 2001). This whole process raises the question of how and where all this magnetic flux is generated, to which I shall return.

In what follows, I shall first describe some general features of compressible magnetoconvection. Then I shall focus on the issue of flux separation, which we have recently explored at Cambridge. After that, I shall discuss the related problem of turbulent dynamo action and the maintenance of disordered small-scale fields, which has been studied in collaboration with colleagues in Chicago. Finally, I shall attempt to summarize where we stand and what we need to do next.

22.2 Compressible magnetoconvection

Nonlinear convection in a perfect gas, in the presence of an imposed vertical magnetic field, is governed by the equation of motion:

$$\rho \frac{D\mathbf{u}}{Dt} = -\nabla p + \rho \mathbf{g} + \mathbf{j} \times \mathbf{B} + \mathbf{F}_{\text{visc}} ,$$

together with the continuity equation

$$\frac{\partial \rho}{\partial t} = -\text{div}(\rho \mathbf{u}) ,$$

the equation of state $p = \mathcal{R}T\rho$, the entropy equation

$$\rho T \frac{DS}{Dt} = -\text{div}(k\nabla T) + \mathcal{S}_{\text{diss}}$$

and the induction equation

$$\frac{\partial \mathbf{B}}{\partial t} = \text{curl}\,(\mathbf{u} \times \mathbf{B}) + \eta\nabla^2\mathbf{B} ,$$

with div $\mathbf{B} = 0$. Here $\mathbf{u}$, $\mathbf{B}$ and $\mathbf{j}$ are the velocity, magnetic field and electric current, while ρ, p, S and T are the density, pressure, entropy density and temperature, respectively; $\mathbf{g}$ is the gravitational acceleration, $\mathcal{R}$ is the gas constant, k is the thermal conductivity, η is the magnetic diffusivity, $\mathbf{F}_{\mathrm{visc}}$ is the viscous force and S_{diss} includes both viscous and ohmic dissipation. These equations are solved in a cuboidal region $\{0 \leq x \leq \lambda d,\ 0 \leq y \leq \lambda d,\ z_0 \leq z \leq z_0 + d\}$, subject to periodic lateral boundary conditions in the horizontal x- and y-directions. At the upper and lower boundaries the normal velocity and the tangential components of the viscous stress both vanish, while the magnetic field is constrained to be vertical. The temperature is fixed at the lower boundary $(z = z_0 + d)$; at the upper boundary $(z = z_0)$, there are two possibilities: either the temperature is fixed or else a 'radiative' boundary condition is imposed, with the conductive heat flux matched to a flux proportional to T^4.

In the absence of convection there is a polytropic reference atmosphere with $T \propto z$ and $\rho \propto z^m$. For a gas with $\gamma = 5/3$ the layer is superadiabatically stratified if $m < 3/2$: in the calculations described here $m = 1$ and $\theta = d/z_0 = 10$, so that the density increases by a factor of $(\theta + 1) = 11$ across the layer. When the equations are rendered dimensionless the system is defined by five dimensionless parameters. These are the Rayleigh number R, which measures the superadiabatic gradient; the Chandrasekhar number Q, which is proportional to the square of the imposed magnetic field, the Prandtl number σ; the ratio ζ of the magnetic to the thermal diffusivity; and the aspect ratio λ. Since both R and ζ are functions of depth it is convenient to use the values $\hat{R}$ and $\hat{\zeta}$ defined at the middle of the layer $(z = z_0 + d/2)$. All the compressible calculations presented here adopt the parameter values $\sigma = 1$ and $\hat{\zeta} = 1.2$. Thus the actual diffusivity ratio, which is proportional to the density, lies in the range $0.2 \leq \zeta \leq 2.2$; this mimics the effects of ionization near the solar surface, where $\zeta > 1$ at depths between $2\,000$ and $20\,000$ km.

The governing equations are solved numerically for three-dimensional motion in a cuboidal box, using a mixed finite-difference pseudospectral code that has been optimized for parallel processing, with up to $256 \times 256 \times 100$ mesh points. The results described here were obtained using a 64-processor partition of the Hitachi SR2201 computer at the University of Cambridge High Performance Computing Facility.

The onset of instability, followed by an ordered array of weakly nonlinear plumes and eventually by turbulent convection, can be studied either by increasing $\hat{R}$ for fixed Q or by decreasing Q for fixed $\hat{R}$. Rucklidge et al. (2000) set $Q = 1\,000$, and fixed the temperature at the upper boundary. In a small box, with $\lambda = 2$, the initial pattern of steady plumes on a slightly distorted hexagonal lattice, gave way to intermittent behaviour and eventually to broad chaotic plumes as $\hat{R}$ was increased. The results are, however, sensitive to the aspect ratio of the computational box. In a sufficiently wide box, with $\lambda = 8$, a new effect appears: there are distinct regions where the field is strong and convection is suppressed, separated from regions from which magnetic flux has been expelled by vigorous convecting plumes (Tao et al. 1998).

22.3 Flux separation

The regime in which flux separation occurs has been systematically explored by setting $\hat{R} = 100\,000$ and varying Q in a box with $\lambda = 8$ and a 'radiative' thermal boundary condition at $z = z_0$ (Weiss, Proctor & Brownjohn 2002). For these parameter values, linear theory shows that the onset of convection occurs as a stationary bifurcation at $\hat{R} \approx 4\,200$. This is followed by a magnetically dominated regime, for $Q \geq 2200$, with a pattern of steady small-scale convection in narrow hexagonal cells. Figure 22.2 shows the array of steady plumes for $Q = 3\,000$. Owing to the stratification, these plumes expand as they rise, and they are surrounded by a network of cooler sinking fluid. The rising plumes sweep magnetic flux aside and concentrate it into a network at the upper surface. The sinking fluid is focused into slender falling plumes that impinge upon the lower boundary, where magnetic flux is concentrated at the centres of the rising plumes. Small-scale convection remains stable down to $Q = 1\,600$; the pattern illustrated in Figure 22.3 is, however, time-dependent and alternate plumes wax and wane aperiodically in vigour. Such spatially modulated oscillations were first identified in two-dimensional calculations.

When $Q = 1\,400$ this pattern is unstable. Rising plumes amalgamate, expelling magnetic flux to form a vigorously convecting and almost field-free cluster. This is surrounded by a region where the magnetic field is sufficiently strong that only small-scale convection can occur. Thus magnetic fields are segregated from the motion. This process of flux separation is, moreover, associated with hysteresis: Figure 22.4 shows a flux-separated solution for the same parameter values as Figure 22.3. Both types of solution are stable in the intermediate regime with $2\,000 \geq Q \geq 1\,600$. For moderate field strengths ($1\,400 \geq Q \geq 600$) only flux separated solutions are found. The pattern for $Q = 1\,000$ is shown in Figure 22.5. Finally, when the imposed field is weak ($Q \leq 500$), magnetic flux is confined to a narrow network enclosing clusters of actively convecting and evolving plumes, as

Fig. 22.2. Compressible magnetoconvection with $\hat{R} = 10^5$, $\sigma = 1$, $\hat{\zeta} = 1.2$ and $Q = 3\,000$ in a box with $\lambda = 4$. The grey-scale image in the upper panel shows the variation of $|\mathbf{B}|^2$ across the top and bottom of the layer; dark (light) regions denote weak (strong) fields. Temperature fluctuations are indicated on the sides of the box and the arrows represent the tangential component of the velocity. The lower panel shows the temperature gradient $|\partial T/\partial z|$ at the upper and lower boundaries, with dark (light) regions denoting weak (strong) gradients; note that $|\partial T/\partial z| \propto T^4$ at the top. Rising and expanding plumes at the top boundary appear dark in the upper panel and light in the lower panel. (After Weiss et al. 2002.)

shown in Figure 22.6. Magnetic flux moves rapidly through this network, like a fluid, giving rise to intense but ephemeral fields at the corners. For $Q \ll 200$ convection becomes much more vigorous and magnetic flux is confined to isolated flux tubes, with intense magnetic fields, that are almost completely evacuated. The resulting computational difficulties provide an

Fig. 22.3. Time-dependent small-scale convection for $Q = 1\,600$. As Figure 22.2 but in a box with $\lambda = 8$.

effective lower bound to the values of Q that can be used in this series of numerical experiments.

22.4 Small-scale dynamo action

The model calculations described above leave open the question of what happens when $Q \ll 200$, or when no net field is imposed. Is turbulent convection capable of generating and maintaining a disordered magnetic field? Although the relevant regimes cannot yet be reached in our compressible runs, they are accessible if the Boussinesq approximation is adopted. Indeed, Cattaneo (1999) has shown that turbulent convection can act as a

Fig. 22.4. As Figure 22.3 but for a flux-separated solution, again with $Q = 1\,600$. There is now a region with vigorously convecting plumes, separated by a front from the magnetically dominated region.

small-scale dynamo if the magnetic Reynolds number, R_m, is sufficiently high. In this approximation the fluid is assumed to be incompressible and density fluctuations only enter through the buoyancy term in the equation of motion, so that both **u** and **B** are solenoidal. This simplification makes it possible to model much more vigorous convection, with $R = 5 \times 10^5$, $\sigma = 1$ and $\zeta = 0.2$, when $R_m \approx 1\,200$. The governing equations have been solved numerically, subject to standard idealized boundary conditions, in very wide boxes, with aspect ratios $\lambda = 10$ and $\lambda = 20$, requiring meshes with up to $1024^2 \times 96$ gridpoints.

Fig. 22.5. As Figure 22.4 but for $Q = 1\,000$ when the convection is more vigorous and the magnetic field is concentrated into a smaller fraction of the layer.

Cattaneo's (1999) results indicate that such a flow can act as a fast dynamo: small magnetic perturbations grow until the a disordered field is strong enough to limit dynamo action. In the final state, the magnetic energy is approximately one-fifth of the kinetic energy, significantly below the equipartition level. Figure 22.7 shows snapshots of the temperature and the magnetic field near the upper surface at two different times from a calculation with $\lambda = 20$ (Cattaneo, Lenz & Weiss 2001). It is immediately obvious that the magnetic pattern outlines a mesocellular scale that is several times larger than the cellular pattern of small-scale convection. The magnetic field provides a convenient diagnostic but mesoscale patterns can also be

Fig. 22.6. As Figure 22.4 but now for $Q = 200$ in a box with $\lambda = 4$. Magnetic flux is confined to narrow channels surrounding the clusters of vigorously convecting plumes. Flux moves rapidly along these channels and intense fields appear temporarily at the corners.

recognized in purely hydrodynamic calculations – and indeed in compressible models, where the up-down symmetry of Boussinesq convection has been broken. The mesocellular scale, which closely resembles mesogranules at the solar surface, seems to be an ubiquitous feature of highly nonlinear convection; it apparently results from collective interactions between the smaller-scale cells.

Small-scale dynamo action should clearly persist if a weak vertical magnetic field is imposed, at least in the limit as $Q \to 0$. The transition from a turbulent dynamo to magnetoconvection can be followed by gradually in-

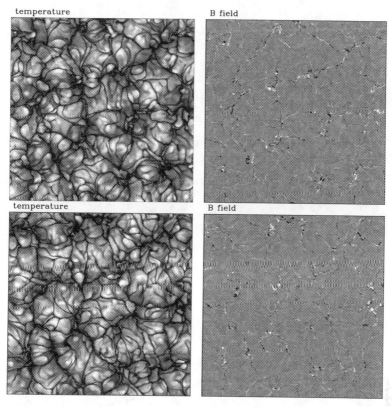

Fig. 22.7. Turbulent convection acting as a small-scale dynamo. Results for Boussinesq convection with $R = 5 \times 10^5$, $\sigma = 1$ and $\zeta = 0.2$ in an exceptionally wide box, with $\lambda = 20$. Left panels, the temperature near the upper surface; right panels, the magnetic field at the top of the layer, with black and white denoting oppositely directed fields. The upper and lower pairs correspond to different times. Note the contrast between the cellular pattern exhibited by the temperature and the mesoscale pattern of the magnetic field. (From Cattaneo et al. 2001.)

creasing Q (Emonet, Cattaneo & Weiss 2001; Emonet & Cattaneo 2001). Results obtained for the same values of R, σ and ζ but in a box with $\lambda = 10$ show that this transition is gradual. The six panels in Figure 22.8 show snapshots of the vertical field at the top of the convecting layer as Q is increased through the range $3\,125 \leq Q \leq 400\,000$. The lowest value of Q yields a pattern that is indistinguishable from that in Figure 22.7, though the magnetic and kinetic energies are now almost equal (but less than the kinetic energy in the original dynamo calculation for $Q = 0$). Case 2, with $Q = 12\,500$, still shows a mesoscale structure with fields of both signs but thereafter the field is in one direction only and forms a network that en-

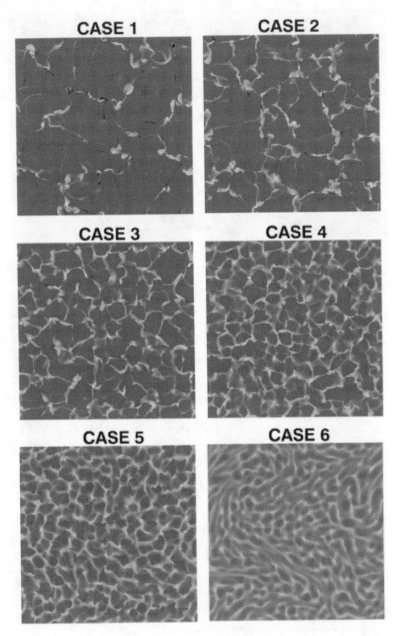

Fig. 22.8. The transition from small-scale dynamo action to magnetoconvection. Results for Boussinesq convection in a box with $\lambda = 10$ and the parameters of Figure 22.7, but now with an additional imposed magnetic field. Case 1: $Q = 3\,125$; Case 2: $Q = 12\,500$; Case 3: $Q = 25\,000$; Case 4: $Q = 50\,000$; Case 5: $Q = 100\,000$; Case 6: $Q = 200\,000$; Case 6: $Q = 400\,000$. The transition is continuous but there is a clearly visible distinction between the patterns in cases 2 and 3. (From Emonet et al. 2001.)

(a)

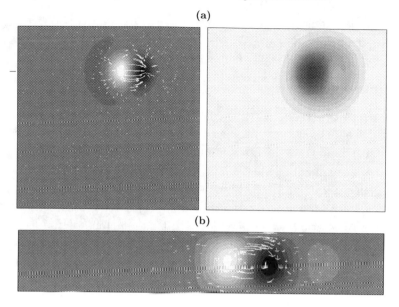

(b)

Fig. 22.9. A convecton in three-dimensional Boussinesq magnetoconvection. Results of a vertically truncated model calculation in a box with $\lambda = 6$ for $R = 5\,000$, $Q = 5\,500$, $\sigma = 1$ and $\zeta = 0.1$. (a) Horizontal sections near the top surface, showing (left) temperature, with arrows corresponding to the horizontal velocity and (right) the magnetic field strength. (b) Vertical section through the centre of the convecton, again showing temperature and velocity. The flow is periodic and reverses its direction. (From Blanchflower & Weiss 2002.)

closes the rising plumes, which spread out as they impinge upon the upper boundary. There is, of course, a complementary network surrounding sinking plumes at the lower boundary. Convection actually persists (though very weakly) up to $Q = 400\,000$, beyond the initial Hopf bifurcation, which appears to be subcritical. By that stage the magnetic energy is far greater than the kinetic energy of the feeble motion.

There is no reason why compressible convection should not also act as a turbulent dynamo, provided that the magnetic Reynolds number is sufficiently large. At present, however, it is not feasible to compute magnetohydrodynamic solutions in a compressible layer with $R_m \geq 1\,000$, so we have to rely on the Boussinesq results. So far only hints of flux separation have been reported. On the other hand, there are examples in both two and three dimensions of localized solutions ('convectons') where a single eddy is embedded in a strong magnetic field (Blanchflower 1999; Blanchflower & Weiss 2002). In these solutions magnetic flux is expelled from the convecton, so that local motion is relatively unimpeded, while flux conservation ensures that the ambient field is strong enough to suppress convection. Figure 22.9 shows the structure of a time-dependent solution obtained in a vertically truncated three-dimensional calculation (Blanchflower & Weiss 2002).

22.5 Conclusion

Small-scale features are only just resolved in the latest observations of magnetic structures at the solar photosphere, and they are beyond our reach in other stars; similarly, the most powerful supercomputers are barely able to reproduce behaviour at high Reynolds numbers ($R_m \geq 1\,000$). Nevertheless, it is possible to recognize a qualitative similarity between patterns of behaviour in numerical experiments and at the surface of the Sun. So far, though, we have not been able to analyse these effects properly.

On the smallest scale, intergranular magnetic fields may be locally intense but are constantly changing as magnetic flux moves through the network of cooler sinking fluid. That resembles the results in Figure 22.6. On a much larger scale, the umbrae of sunspots are relatively dark, but energy must still be supplied by convection. Small-scale, time-dependent motion, corresponding to the spatially modulated oscillations of Figure 22.3, is apparently concealed by a radiative blanket, penetrated by occasional plumes which appear as 'umbral dots'. Flux separation leads to magnetically dominated dark nuclei, which are separated from regions where larger umbral dots are prevalent (Weiss et al. 2002).

Magnetic flux is constantly emerging all over the Sun to form a 'magnetic carpet' (Title 2000). Ephemeral active regions break through the surface near the centres of supergranules and are split by granular convection into fragments with intense magnetic fields. These flux elements are carried by the supergranular motion into the network at cell boundaries, where they merge with opposing fields and are annihilated (Simon et al. 2001). Since the ephemeral regions are largely uncorrelated with the solar cycle they must be generated by small-scale dynamo action near the surface. Their properties indicate, however, that the turbulent dynamo lies below the level of photospheric granulation.

Flux separation is a fascinating effect in its own right. It involves a hysteretic transfer of stability from one chaotic attractor to another. Regarded as pattern formation, the two phases are separated by fronts (as in phase changes such as magnetization). The patterns themselves are essentially two-dimensional and it should therefore be possible to set up some model system that possesses similar properties. Work on that problem is in progress. This is yet another example of coherent structures in a complex system.

These days it has become fashionable to discuss complexity: for those of us who have been studying nonlinear problems it does perhaps come as a surprise to discover (rather like M. Jourdain in Moliére's *Le Bourgeois Gentilhomme*) that we have been talking about complexity for years.

Acknowledgements The results I have described were obtained in collaboration with Michael Proctor and Derek Brownjohn at Cambridge (using codes developed by Paul Matthews and Louis Tao) and with Fausto Cattaneo and Thierry Emonet at Chicago. We are grateful for support from PPARC.

References

Blanchflower, S. M., 1999, *Phys. Lett. A* **261**, 74.

Blanchflower, S. M. & Weiss, N. O., 2002, *Phys. Lett. A*, **294**, 297.

Cattaneo, F., Lenz, D. & Woioo, N., 2001, *ApJ* **503**, L91.

Cattaneo, F., 1999, *ApJ* **515**, L39.

Emonet, T. & Cattaneo, F., 2001, *ApJ* **560**, L197.

Emonet, T. Cattaneo, F., & Weiss, N. O., 2001, in Chossat, P., Armbruster, D. & Oprea, I., eds, *Dynamo and Dynamics, a Mathematical Challenge*, Kluwer, Dordrecht, p. 173.

Gough, D. O. & Tayler, R. J., 1966. *MNRAS* **133**, 85.

Rucklidge, A. M.. Weiss, N. O., Brownjohn, D. P., Matthews, P. C. & Proctor, M. R. E., 2000, *JFM* **419**, 283.

Simon, G. W., Title, A. M. & Weiss, N. O., 2001, *ApJ* **561**, 427.

Tao, L., Weiss, N. O., Brownjohn, D. P. & Proctor, M. R. E., 1998, *ApJ* **496**, L39.

Title, A. M., 2000. *Phil. Trans. R. Soc. Lond. A* **358**, 657.

Weiss, N. O., Proctor, M. R. E. & Brownjohn, D. P., 2002, *MNRAS*, **337**, 293.

23

Nonlinear magnetoconvection in the presence of a strong oblique field

KEITH JULIEN

Department of Applied Mathematics, University of Colorado, Boulder, CO 80309, USA

EDGAR KNOBLOCH & STEVEN M. TOBIAS

Department of Mathematics, University of Leeds, Leeds LS2 9JT, UK

Reduced partial differential equations valid for convection in a strong imposed magnetic field (vertical or oblique) are derived and discussed. These equations filter out fast, small-scale Alfvén waves, and are valid outside of passive horizontal boundary layers. In the regime in which the convective velocities are not strong enough to distort substantially the field, exact, fully nonlinear, single-mode solutions exist. These are determined from the reduced PDEs reformulated as a nonlinear eigenvalue problem whose solution also gives, for each Rayleigh number, the time-averaged Nusselt number and oscillation frequency together with the mean vertical temperature profile. In the oblique case a hysteretic transition between two distinct convection regimes is identified. Possible applications to sunspots are discussed.

23.1 Introduction

The study of convection in an imposed magnetic field is motivated primarily by astrophysical applications, particularly by the observed magnetic field dynamics in the solar convection zone (Hughes & Proctor 1988). Applications to sunspots (Thomas & Weiss 1992) have led several authors to investigate the suppression of convection by strong "vertical" or "horizontal" magnetic fields. However, the magnetic field in sunspots is neither vertical nor horizontal, and this has led to recent nonlinear investigation of convection in an oblique magnetic field (Matthews et al. 1992, Julien et al. 2000). Numerical simulations of magnetoconvection are unable to reach the parameter values, both in terms of field strengths and Reynolds number (Re), characteristic of convection in sunspots. Indeed, the former compounds the prohibitive temporal and spatial restrictions placed on high-Re simulations through the presence of high frequency Alfvén waves (if $Ma \ll 1$, where the Mach number $Ma := U/V_A$ and U, V_A are the flow and Alfvén speeds) and the

345

development of thin (magnetic) boundary layers (if $Rm \gg 1$, where the magnetic Reynolds number $Rm := UL/\eta$, L is the length scale and η is the ohmic diffusivity). Even with today's state-of-the-art computers these facts together with limitations on memory and computational speed place substantial constraints on the accessible parameter range. We develop here, from the primitive-variable Boussinesq equations, a reduced set of PDEs valid in the strong field limit ($Ma \ll 1$) where convective velocities are not large enough to distort the imposed field. These equations have the appealing property of filtering fast Alfvén waves and relaxing the need to resolve (magnetic or mechanical) boundary layers which can be determined, *a posteriori*, as a passive inner solution.

The reduced equations presented in (23.5) below admit exact, fully nonlinear, single-mode solutions for both two- and three-dimensional spatially periodic convection. The degree of nonlinearity is characterized by the distortion of the mean temperature profile, and the vertical structure can be followed from onset to high Re (or equivalently large Rayleigh number Ra) via a one-dimensional eigenvalue problem for the (time-averaged) Nusselt number. The derivation of this problem can be performed analytically, although the problem itself must be solved numerically. We find that in the strong field limit all competing steady patterns are degenerate in the sense that they transport the same amount of heat. This is so also for oscillatory patterns. In the overstable case two distinct modes of convection are uncovered. The first or "vertical field" mode is characterized by thin thermal boundary layers and a Nusselt number that increases rapidly with the applied Rayleigh number; this mode is typical of steady convection as well. The second or "horizontal field" mode is present in overstable convection only and has broad thermal boundary layers and a Nusselt number that remains small and approximately independent of the Rayleigh number. At large Rayleigh numbers this regime is characterized by a piecewise linear temperature profile with a small isothermal core. The "horizontal field" mode is favoured for substantial inclinations of the field and sufficiently small ohmic diffusivity. The transition between the two regimes is typically hysteretic and for fixed inclination and diffusivity may occur with increasing Re.

The dimensionless Boussinesq equations describing magnetoconvection in a plane horizontal layer are

$$D_t \mathbf{u} = -\nabla \pi + Ma^{-2}\mathbf{B} \cdot \nabla \mathbf{B} + \mathcal{R}T\hat{\mathbf{z}} + Re^{-1}\nabla^2 \mathbf{u} \tag{23.1}$$

$$D_t T = Pe^{-1}\nabla^2 T \tag{23.2}$$

$$D_t \mathbf{B} = \mathbf{B} \cdot \nabla \mathbf{u} + Rm^{-1}\nabla^2 \mathbf{B} , \tag{23.3}$$

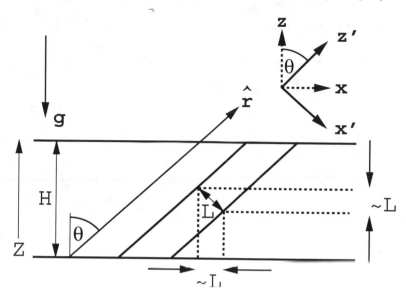

Fig. 23.1. A schematic diagram illustrating the appearance of a small length scale in the vertical due to magnetic alignment. The rotated coordinate system for the small scales is also shown.

with $D_t = \partial_t + \mathbf{u} \cdot \nabla$ and $\nabla \cdot \mathbf{u} = \nabla \cdot \mathbf{B} = 0$. Here $\mathbf{u} = (u, v, w)$ is the velocity field in Cartesian coordinates (x, y, z) with z vertically upwards, T denotes the temperature and π is the total (thermal and magnetic) pressure. The dimensionless magnetic field is assumed to be the superposition $\mathbf{B} = \hat{\mathbf{r}} + \mathbf{b}$ of an imposed oblique field of unit strength and a three-dimensional field $\mathbf{b}(x, y, z, t)$ due to the presence of convection. Here $\hat{\mathbf{r}} = (\sin \vartheta, 0, \cos \vartheta)$, where ϑ denotes the angle with respect to the vertical in the (x, z) plane (Fig. 23.1). The equations have been nondimensionalized with respect to characteristic horizontal length scale L, speed U, time scale L/U, magnetic field strength B_0, and temperature difference ΔT. The resulting dimensionless parameters are the Reynolds (Re), Péclet (Pe), magnetic Reynolds (Rm), buoyancy ($\mathcal{R}$), and the Alfvén speed Mach (Ma) numbers.

23.2 Reduced PDE description for $Ma \ll 1$

Given our primary focus on investigating magnetically constrained flows we set $Ma \ll 1$ for the purpose of asymptotic development. In the following we motivate both the geometrical framework and the subsequent mathematical assumptions that are used to derive the reduced equation set (23.5).

The schematic picture for magnetically constrained convective motions that arises from theory (Chandrasekhar 1961; Julien et al. 2000) and numerical simulations (Weiss et al. 1990, 1996) is one consisting of small aspect-ratio convection cells where $A := L/H \ll 1$ (Fig. 23.1). These cells align themselves with the imposed magnetic field $\hat{\mathbf{r}}$ as a consequence of the magnetic analogue of the Taylor-Proudman constraint (Taylor 1923, Julien et al. 2000). The tilted nature of the small aspect-ratio cells now implies the existence of two scales of motion in the vertical direction:

- a small scale $\tilde{z}$, controlled solely by the imposed field and to first-order unaffected by the presence of boundaries in the vertical. This scale is of the same order as the horizontal length scale L, and corresponds to the vertical cross-section of the cells (Fig. 23.1);
- a large scale Z of the order of the layer depth H on which convective motions are driven through buoyancy forcing (Fig. 23.1).

Mathematically, this picture suggests a dependence on four spatial variables $\mathbf{x} \equiv (x, y, \tilde{z}, Z)$ and a multiple scales expansion in the vertical with the corresponding derivatives in equations (23.1)–(23.3) interpreted as $\partial_z = \partial_{\tilde{z}} + A\partial_Z$. However, we find that because of the magnetic analogue of the Taylor-Proudman constraint, the solution cannot depend, at leading order, on the small scale along $\hat{\mathbf{r}}$. Therefore a more concise mathematical description is obtained by rotating the coordinate system for the small scales so that it is aligned with $\hat{\mathbf{r}}$, while maintaining the alignment of Z with gravity (Fig. 23.1). Hence $\mathbf{x}'^T = \mathcal{R}_\vartheta(x, y, \tilde{z})^T$, $\nabla' = \mathcal{R}_\vartheta^{-1}\nabla$, $\mathbf{u}' = \mathcal{R}_\vartheta\mathbf{u}$, and $\mathbf{b}' = \mathcal{R}_\vartheta\mathbf{b}$; $\mathcal{R}_\vartheta$ is the unitary rotation matrix in the (x, z) plane.

Note that the spatial dependence is now represented by the three variables (x', y') and Z (because no fields may exhibit z' dependence); hence (x', y', Z) comprises a non-orthogonal coordinate system. Note also that the characteristic scales are sufficiently small compared to the domain size to justify the use of periodic boundaries in (x', y').

We now summarize the derivation leading to the reduced equations (23.5) from the Boussinesq equations (23.1)–(23.3). An equivalent derivation based on an expansion in the Chandrasekhar number Q (for the vertical field case alone) is given in Julien et al. (1999). Using $Ma \ll 1$ as a small parameter, we write $\mathcal{R} = Ma^{-1}\mathcal{R}'$, and take Re, Pe and $Rm \approx \mathcal{O}(1)$. The aspect ratio is set, without loss of generality, as $A \equiv Ma$, with the fluid variables $\mathbf{u}' \approx \mathcal{O}(1)$, $\mathbf{b}' \approx \mathcal{O}(Ma)$, $\pi \approx \mathcal{O}(Ma^{-2})$. Finally

$$T = \overline{T}(Z) + Ma\,\theta(x', y', Z, t)\,, \tag{23.4}$$

where the overbar denotes the spatial average over small scales and time:

$\lim_{\tau \to \infty} \tau^{-1} \int_0^\tau dx' \, dy' \, dt$. The reduced equations are deduced from (23.1)–(23.3) using asymptotic expansions in Ma for the rescaled fluid variables, e.g.,

$$\mathbf{u}' = \mathbf{u}'_0 + Ma \, \mathbf{u}'_1 + Ma^2 \, \mathbf{u}'_2 + \cdots .$$

At leading order in the momentum, heat and induction equations one readily deduces

- a mean hydrostatic balance $\partial_Z \Pi_0 = \mathcal{R}' T_0$, where $\Pi_0(Z) \equiv \overline{\pi}_0$ and $T_0(Z) \equiv \overline{T}_0$ are slowly varying mean quantities.
- an invariance of the motion with respect to the small scales along $\widehat{\mathbf{r}} \equiv \widehat{\mathbf{z}}'$ leading to non-divergent motions in planes perpendicular to $\widehat{\mathbf{r}}$. We are therefore free to adopt the streamfunction representation $\mathbf{u}'_{0\perp} = \widehat{\mathbf{z}}' \times \nabla' \Psi_0$ and $\mathbf{b}'_{0\perp} = \widehat{\mathbf{z}}' \times \nabla' A_0$.
- the absence of any $\mathcal{O}(1)$ mean flows or mean fields. These quantities are generated at higher orders with $\overline{\mathbf{u}} \approx \mathcal{O}(Ma)$ and $\overline{\mathbf{b}} \approx \mathcal{O}(Ma^2)$.

The next order in Ma produces the reduced PDEs in the non-orthogonal coordinate system, valid in the strong field limit:

$$\left(D'_{0t} - Re^{-1}\nabla'^2_\perp \right) \nabla'^2_\perp \Psi_0 = \mathcal{R}' \sin \vartheta \; \partial_{y'}\theta_0 + \widehat{r}_z \partial_Z \nabla'^2_\perp A_0 + J_\perp(A_0, \nabla'^2_\perp A_0)$$

$$\left(D'_{0t} - Re^{-1}\nabla'^2_\perp \right) w'_0 = \mathcal{R}' \cos \vartheta \; \theta_0 + \widehat{r}_z \partial_Z b'_{03} + J_\perp(A_0, b'_{03})$$

$$\left(D'_{0t} - Rm^{-1}\nabla'^2_\perp \right) A_0 = \widehat{r}_z \partial_Z \Psi_0$$

$$\left(D'_{0t} - Rm^{-1}\nabla'^2_\perp \right) b'_{03} = \widehat{r}_z \partial_Z w'_0 + J_\perp(A_0, w'_0) \qquad (23.5)$$

$$\left(D'_{0t} - Pe^{-1}\nabla'^2_\perp \right) \theta_0 = - \left(\cos \vartheta \; w'_0 + \sin \vartheta \; \partial_{y'} \Psi_0 \right) \partial_Z \overline{T}_0$$

$$\partial_Z \overline{\left((\cos \vartheta \; w'_0 + \sin \vartheta \; \partial_{y'} \Psi_0)\theta_0 \right)} = Pe^{-1}\partial_{ZZ}\overline{T}_0 \,,$$

where $D'_{0t} = \partial_t + J_\perp[\Psi_0, \bullet]$ with Jacobian $J_\perp(f, g) := \partial_{x'} f \partial_{y'} g - \partial_{y'} f \partial_{x'} g$. Equations (23.5a–f) represent coupled equations for the magnetically aligned vorticity $\zeta = \nabla'^2_\perp \Psi_0$, velocity w'_0, current $j = \nabla'^2_\perp A_0$, and magnetic field b'_{03}. All these fields are sustained through buoyancy forcing whose distribution is governed by (23.5e). For $\vartheta = 0$, buoyancy forcing cannot sustain Ψ_0 and A_0 which necessarily decay without an additional external source. For this case alone the only surviving nonlinearity occurs in (23.5f) which describes the distortion of the mean temperature $\overline{T}_0$ via vertical convective flux $\overline{w_0 \theta_0} = \overline{(\cos \vartheta w'_0 + \sin \vartheta \partial_{y'} \Psi_0)\theta_0}$. The reduced equations are dynamically coupled on the large scale Z by vertical stretching due to the presence of $\widehat{\mathbf{r}}$.

23.2.1 Computational and Theoretical Advantages

The reduced equations (23.5) have several appealing features that are not present in the full Boussinesq equations (23.1)–(23.3). These include

- a relaxation of spatial resolution requirements. This occurs as a consequence of a reduction of vertical order of the equations with respect to the large-scale variable Z. It follows that the precise details of the vertical boundary conditions are not distinguished, and that any (mechanical or magnetic) boundary layers are passive and need not be resolved. This fact is already known from linear theory (Chandrasekhar 1961) and was established in the weakly nonlinear regime by Clune & Knobloch (1993). Julien et al. (2000) show that even in the strongly nonlinear regime the exact form of the mechanical boundary layer can be deduced from the interior (bulk) solution. Note that thermal boundary layers, if present, are retained through (23.5f).

- a relaxation of the timestepping/CFL criterion. It can be shown (in an analysis similar to Embid & Majda 1998) that the reduced equation set filters out the small scale fast Alfvén waves. These waves, which must be resolved in the Boussinesq equations, do not interact with the slow convective dynamics.

- existence of exact analytic solutions due to the simplified nonlinearities.

23.3 Exact Single-Mode Solutions

For single-mode solutions of the form $F(Z)\exp(i\omega t + i\mathbf{k}_\perp \cdot \mathbf{x}'_\perp)$ plus its complex conjugate, all nonlinearities in (23.5a-e) vanish identically with the exception of the convective flux term in (23.5f). The resulting equations can be reformulated into a single *nonlinear* complex eigenvalue problem for the vertical structure, namely

$$\mathrm{D}^2 W - \frac{1}{\hat{r}_z^2}\left(i\omega + Re^{-1}k_\perp^2\right)(i\omega + Rm^{-1}k_\perp^2)W \qquad (23.6)$$

$$+\frac{\mathcal{R}'KPe}{\hat{r}_z^2}\frac{(i\omega + Rm^{-1}k_\perp^2)(-i\omega + Pe^{-1}k_\perp^2)}{\omega^2 + Pe^{-2}k_\perp^4 + 2k_\perp^2|W|^2}\frac{\cos^2\vartheta k_x^2 + k_y^2}{k_\perp^2}W = 0 \ .$$

Here $W(Z) \equiv \cos\vartheta W'(Z) + ik_y \sin\vartheta \Psi(Z)$ is the complex amplitude of the vertical velocity, while $k_\perp \equiv |(k_x, k_y, 0)|$ is the wavenumber in the plane perpendicular to $\hat{\mathbf{r}}$. The constant K is given by

$$K = \left[\int_0^1 \frac{\omega^2 + Pe^{-2}k_\perp^4}{\omega^2 + Pe^{-2}k_\perp^4 + 2k_\perp^2|W|^2}dZ\right]^{-1} \qquad (23.7)$$

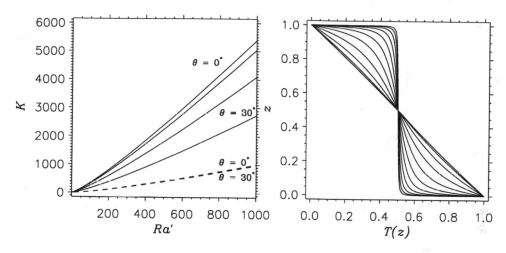

Fig. 23.2. Panel (a) (left) shows the (time averaged) Nusselt number K for steady (dashed line) and oscillatory (solid line) convection for $\vartheta = 0, 10°, 20°, 30°$ as a function of the scaled Rayleigh number Ra' when $\zeta = 0.1$ and $\sigma = 1.1$. Panel (b) (right) shows the development of an isothermal core in the mean temperature profile $\overline{T}(Z)$ with increasing Ra' when $\zeta = 0.1$, $\vartheta = 0$.

and for steady patterns is to be identified with the Nusselt number; for oscillatory patterns K represents the time-averaged Nusselt number.

Equation (23.6) is to be solved subject to impenetrable boundary conditions $W = 0$. Once this is done the mean temperature profile $\overline{T}(Z)$ can be found from the relation

$$D\overline{T}\left[1 + \frac{2k_\perp^2}{\omega^2 + Pe^{-2}k_\perp^4}|W|^2\right] = -KPe. \tag{23.8}$$

23.4 Results

The solutions of the problem (23.6-23.8) depend on the parameter set ($\mathcal{R}'$, Re, Pe, Rm), wavenumber $k_\perp$ and ϑ. For contact with previous work (Julien et al. 1999, 2000), we set $U = \kappa/L$ (the thermal diffusive velocity scale). It follows that $Pe = 1$, $Re^{-1} = \sigma$ (Prandtl number), $Rm^{-1} = \zeta$ (magnetic Prandtl number), and $A \equiv Ma = (\sigma\zeta Q)^{-1/4}$. We also define the scaled Rayleigh number $Ra' = \mathcal{R}'/\zeta$. We solve this problem on a discretized one-dimensional mesh using an iterative Newton-Raphson-Kantorovich scheme with $\mathcal{O}(10^{-10})$ accuracy in the L_2 norm of $W(Z)$ and the corresponding eigenvalues K and ω.

We present first the results for a vertical magnetic field ($\vartheta = 0$), and then discuss the oblique field ($\vartheta \neq 0$) in the so-called perpendicular case, i.e., when the roll axes are perpendicular ($k_y = 0$) to the plane containing $\mathbf{g}$ and $\hat{\mathbf{r}}$. All results are obtained with the wavenumber $k_\perp = 1$, $\sigma = 1.1$ and $\zeta = 0.1$.

In Fig. 23.2a we show the (time-averaged) Nusselt number K and frequency for both steady and overstable convection when $\vartheta = 0$ as a function of the scaled Rayleigh number Ra'. Observe that solutions can be obtained for highly supercritical Rayleigh numbers and that K increases monotonically with increasing Ra'. We also find that the frequency ω saturates. For both steady and oscillatory convection the temperature gradients are confined to thinner and thinner boundary layers at the top and bottom as Ra' increases; this process occurs more rapidly in the steady case. At the same time the bulk of the layer becomes more and more isothermal (see Fig. 23.2b). Midplane symmetry implies that these boundary layers are identical and that the isothermal interior has temperature $\overline{T} = 1/2$.

Calculations show that these results are not changed qualitatively when the magnetic field is tilted, provided that the tilt angle ϑ is not too large. However, with increasing tilt both steady and oscillatory convection become less efficient at transporting heat, and the Rayleigh number dependence of the Nusselt number becomes weaker (Fig. 23.2a). The increase in tilt angle leads to a larger Lorentz force, which in turn leads to a suppression of the heat transport. The resulting dependence on the tilt angle is much stronger in the oscillatory regime since ohmic diffusion now has only a finite time to reduce the Lorentz force due to field distortion before the flow reverses. In contrast in the steady case the Lorentz force exerts a much weaker effect and the reduction of the Nusselt number is largely due to a geometrical effect: the strong oblique magnetic field inclines the convection cells relative to the vertical allowing them more time to lose their upward buoyancy to adjacent descending fluid.

Figure 23.3a,b shows the corresponding results for the oscillatory mode when $\vartheta = 65°$. The figure reveals a remarkable behaviour: the Nusselt number K initially increases rapidly with Ra' as in the vertical magnetic field case, but then undergoes a hysteretic transition to a new state characterized by a small Nusselt number, and one that *decreases* slowly with increasing Ra'. As this state is followed to larger Rayleigh numbers we see that the mean temperature becomes almost piecewise linear (Fig. 23.3c), with a limited isothermal core. The extent of this core quickly saturates, in contrast to the case of a vertical field for which the isothermal core grows continuously with Ra' as the temperature gradients are compressed into ever thinner thermal boundary layers (as in Fig. 23.2b). Evidently, in this state increasing the heat input does not result in increased heat transport

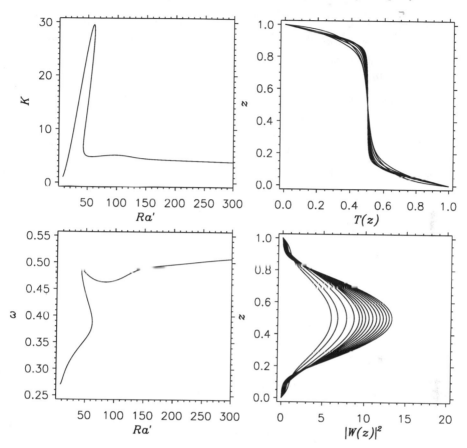

Fig. 23.3. (a) (top left) The (time-averaged) Nusselt number K for oscillatory perpendicular rolls as a function of the scaled Rayleigh number Ra' for $\vartheta = 65°$ and $\zeta = 0.1$, $\sigma = 1.1$. (b) (bottom left) The corresponding frequency ω. Note the hysteretic transition from the "vertical" convection mode to the "horizontal" convection mode with increasing Ra'. (c) (top right) Mean temperature profiles $\overline{T}(Z)$ and (d) (bottom right) the convection amplitude as measured by $|W(Z)|$ for increasing values of Ra' along the horizontal branch. Note the development of broad boundary layers of approximate thickness $1/2K$ and a small isothermal core. These properties are characteristic of the "horizontal" convection mode.

across the layer. Instead the added energy is all stored in the magnetic field perturbations (since the field strength is large this is achieved with small deformation of the field); moreover, the perturbation magnetic field suppresses the convective motion in the boundary layers near the top and bottom (see Fig. 23.3d) thereby reducing the transport of heat across the layer. In this regime (i.e., on the branch where the Nusselt number remains low as Ra' is increased) the system of perpendicular rolls therefore behaves much more like one with an imposed *horizontal* field (cf. Brownjohn et al. 1995). Similar results are obtained for parallel rolls ($k_x \equiv 0$), although the hysteretic transition is delayed to higher Ra'.

Note that the Rayleigh number must exceed a critical value before the "horizontal" convection mode sets in. This is because the flow in the interior must be strong enough to expel the magnetic field perturbation into the boundary layers at the top and bottom; this expulsion occurs primarily in the vertical direction because the vertical velocity in the interior is much larger than the horizontal velocity. In steady convection the resulting boundary layer thickness is determined by ohmic diffusion and is therefore narrow (because $\zeta \ll 1$). In contrast in an oscillatory flow the flow reversals prevent the formation of such narrow boundary layers and the boundary layer thickness is determined by the perturbation Lorentz force and not ohmic diffusion. A number of conclusions follow immediately from these considerations. First, the transition between the two regimes occurs at lower Rayleigh numbers when ϑ is larger. Indeed, for small values of ϑ the transition to the lower "horizontal branch" does not occur (for this value of ζ). This is so also if ζ is larger even though a " horizontal branch" may still be present. Moreover, since the ability of the magnetic field to suppress oscillatory convection increases with decreasing ζ the value of the Rayleigh number at which the transition from the "vertical field" regime to the "horizontal field" regime takes place is an increasing function of ζ. This argument also explains why the two convection regimes are only found in *oscillatory* convection.

It is tempting to speculate about the possible role of the fully nonlinear single-mode solutions discovered here for the structure of a sunspot. In the sunspot umbra the magnetic field is nearly vertical, while in the penumbra it is tilted. Since the critical Rayleigh number steadily decreases at small Ma (large Q) as the tilt angle increases the supercriticality, or effective Rayleigh number, increases with radial distance from the spot centre. Since the temperature difference across the layer, here supposed to be constant, determines the Rayleigh number convection will be less supercritical in the spot centre than farther out (this is consistent with the observed contrast in luminosity). Thus we identify our low Rayleigh number results with convection in the umbra, and our higher Rayleigh number results with convection in the penumbra. These regimes correspond in turn to our vertical and horizontal convection branches, i.e., the vertical convection mode is the umbral mode while the horizontal mode corresponds to the penumbral mode. If this interpretation of our results is correct it suggests a natural explanation for the abrupt change in the properties of the umbra and penumbra even when the effective Rayleigh number varies gradually across the spot. Furthermore, putting the question of pattern selection aside, observations of parallel (radial) rolls in the penumbra are consistent with our finding that these are the structures that transport heat most efficiently and are therefore the most luminous.

23.5 Conclusion

We have summarized the derivation of a new class of reduced PDEs (23.5) valid for magnetoconvection in the presence of a strong oblique field. These filter out the fast, small scale Alfvén waves and relax the need to resolve the horizontal mechanical and magnetic boundary layers. This reformulation promises to have significant applications in astrophysics and in particular for studies of convection in sunspots.

It is shown that the reduced set of equations admits exact fully nonlinear, single-mode solutions which are determined from the nonlinear eigenvalue problem (23.6). This is possible when the field strength is large and the distortion of the field by the flow remains small. When the imposed magnetic field is vertical the solutions of the nonlinear eigenvalue problem with $\omega = 0$ can be used to construct a variety of three-dimensional patterns, all of which have the same Nusselt number in the strong field regime. A similar degeneracy characterizes all oscillatory patterns in this regime. Weak selection among these patterns is due to subdominant terms not computed here. An inclined field breaks this degeneracy and the theory then describes two-dimensional structures only. Whenever these distort the magnetic field (i.e., $\mathbf{k_0} \times (\mathbf{g} \times \hat{\mathbf{r}}) \neq 0$) the behaviour of the system falls into two possible regimes. For small tilt angles the magnetic field plays a relatively minor role in inhibiting convection, and the Nusselt number is an increasing function of the Rayleigh number. If the tilt angle is increased past a threshold value (which depends on the value of ζ) a hysteretic transition may take place with increasing Rayleigh number from this "vertical field" regime to a "horizontal field" regime in which the field plays a major role in inhibiting the heat transport. The possible connection with observation of sunspots is discussed.

The work presented here can be extended to incorporate the effects of depth-dependent profiles of the diffusivities thereby breaking the Boussinesq midplane symmetry (Julien et al. 1999, 2000). In addition it is possible to explore the transition from steady to overstable motion with increasing depth. The procedure outlined above extends readily to the more realistic anelastic formulation of the basic equations of motion, as detailed in Julien et al. (2002).

Acknowledgments. This work was supported by NASA under SEC grant MAS-99026 (KJ), the Department of Energy under Grant No. DE-FG03-95ER-25251 (EK) and NASA under SPTP grant NAG5-2256 (SMT).

References

Brownjohn, D. P., Hurlburt, N. E., Proctor, M. R. E., & Weiss, N. O., 1995, *JFM*, **300**, 287

Chandrasekhar, S., 1961, *Hydrodynamic and Hydromagnetic Stability*, Oxford University Press

Clune, T., Knobloch, E., 1993, *Physica D*, **74**, 151

Embid, P. F., Majda, A. J., 1998, *GAFD*, **87**, 1

Hughes, D. W., Proctor, M. R. E., 1988, *Ann. Rev. Fluid Mech.*, **20**, 187

Julien, K., Knobloch, E. & Tobias, S. M., 1999, *Physica D*, **128**, 105

Julien, K., Knobloch, E. & Tobias, S. M., 2000, *JFM*, **410**, 285

Julien, K., Knobloch, E. & Tobias, S. M., 2002, in Ferriz-Mas, A. & Núñez-Jiménez, M., eds, *Advances in Nonlinear Dynamos*, Taylor and Francis, in press

Matthews, P. C., Hurlburt, N. E., Proctor, M. R. E. & Brownjohn, D. P., 1992, *JFM*, **240**, 559

Taylor, G. I., 1923, Proc. Roy. Soc. A, **104**, 213

Thomas, J. H. and Weiss, N. O., 1992, in Thomas, J. H. and Weiss, N. O., eds, *Sunspots: Theory and Observations*, Kluwer, p. 3

Weiss, N. O., Brownjohn, D. P., Hurlburt, N. E. & Proctor, M. R. E., 1990, *MNRAS*, **245**, 434

Weiss, N. O., Brownjohn, D. P., Matthews, P. C. & Proctor, M. R. E., 1996, *MNRAS*, **283**, 1153

24

Simulations of astrophysical fluids

MARCUS BRÜGGEN

Institute of Astronomy, Madingley Road, Cambridge CB3 0HA, UK, and International University Bremen, Campus Ring 1, 28759 Bremen, Germany

In this contribution I discuss how recent advances in numerical techniques and computational power can be applied to problems in astrophysical fluid mechanics. As a case in point some results of simulations of radio relics are presented which have provided strong support for a model that explains the origin of these peculiar objects. Radio relics are extended radio sources which do not appear to be associated with any radio galaxy. Here a model is presented which explains the origin of these relics in terms of old plasma that has been compressed by a shock wave. Having taken into account synchrotron, inverse Compton and adiabatic energy losses and gains, the relativistic electron population was evolved in time and synthetic radio maps were made which reproduce the observations remarkably well. Finally, some other examples are discussed where hydrodynamical simulations have proven very useful for astrophysical problems.

24.1 Introduction

With the advent of powerful computers and more accurate algorithms, simulations of astrophysical fluids have become increasingly useful. Most fields of astrophysics, such as solar physics, star formation, stellar evolution and cosmology have benefitted greatly from hydrodynamical simulations and hopes for further advances are high.

Essentially, there are two main approaches to the numerical solution of the equations of hydrodynamics: Finite-grid simulations and Smoothed Particle Hydrodynamics (SPH). In the former approach the equations are discretised on a computational mesh before they are solved. The latter method avoids the notion of a mesh and employs particles to track the fluid. Both methods have different strengths and weaknesses but here I will only be concerned with grid-based codes. These have proven especially useful for discontinuous

357

flows where shock-capturing advection schemes yield much greater accuracy than SPH codes. Moreover, it is generally easier to add more physics such as magnetic fields or radiation to the grid-based codes than to SPH codes.

One of the main challenges in the simulation of astrophysical fluids is to bridge the gap between the simulation of the macroscopic flow and different microphysical processes, such as nuclear reactions, X-ray and radio emission, just to name a few. The scales of the macroscopic flow and the microphysical processes are separated by many orders of magnitude, yet both 'worlds' are tightly interlinked. In this contribution I will show some results from simulations where we have attempted to bridge the gap between the large-scale flow and the different physical processes of radio emission from a relativistic plasma. These simulations have proven useful for the understanding of radio relics and radio galaxies. Despite the very general title, I can only give very few examples of the use of hydrodynamical simulations in astrophysics. In the course of this meeting we have heard of several applications in the field of stellar physics so that here I am going to give some examples from the realm of extragalactic astronomy.

Many problems in numerical hydrodynamics require a high resolution in order to describe the evolution of the system accurately enough. In turn, the use of large grids implies high demands in terms of both, computer memory and CPU time. One numerical technique which has been developed to increase the efficiency of these finite-difference simulations is the method of adaptive mesh refinement (AMR). Here the computational grid is continually adapted to the flow and resolution elements are only placed where they are needed. In the last section of this contribution I will show an example of such a simulation.

24.2 Radio relics

The jets of powerful radio galaxies inflate large cavities in the IGM that are filled with relativistic particles and magnetic fields. Synchrotron emission at radio frequencies reveals the presence of electrons with energies of the order of GeVs. These electrons have radiative lifetimes of the order of 100 Myr before radiative losses cause their radio emission to fade until they can no longer be observed. The remnants of radio galaxies and quasars are called 'fossil radio plasma' or a 'radio ghosts'.

For some years now, astronomers have observed faint, extended radio emission that cannot be linked to any nearby galaxy. Several of these mysterious objects have been observed and they have been called *radio relics*.

The so-called *cluster radio relics* are typically located near the periphery of the cluster; they often exhibit sharp emission edges and many of them show strong radio polarisation (Enßlin et al. 2001, Roettiger et al. 1999, Venturi et al. 1999, Enßlin & Gopal-Krishna 2001). It has also been noted that the location of some radio relics coincides with the location of shock waves that are produced by collisions between clusters of galaxies (Roettiger et al. 1998, 1995).

Enßlin & Brüggen (2002) modelled the radio relics in terms of blobs of fossil radio plasma that are being compressed by shock waves. These shock waves may have been produced by a merger between clusters of galaxies. The compression increases, both, the energy of the electrons and the strength of the magnetic fields. As a result, the luminosity of the fossil radio plasma may rise again to observable levels.

In order to compare this simple model with observations, we performed 3D magneto-hydrodynamical simulations using the ZEUS-3D code which was developed especially for problems in astrophysical hydrodynamics (Stone & Norman 1992a, b). The simulations were computed on a Cartesian grid with 200^3 equally spaced zones. The fluid flows in the x-direction with inflow boundary conditions at the lower boundary and outflow conditions at the outer boundary. The boundary conditions in the y- and z-directions were chosen to be reflecting. The simulation was set up such that a stationary shock formed at the centre of the computational domain.

In the pre-shock region a spherical bubble was set up, in which the density was lowered by a factor of 10 with respect to the environment. In turn, the temperature in the bubble was raised such that the bubble remained in pressure equilibrium with its surroundings. Within the bubble a tangled magnetic field was set up (for details see Enßlin & Brüggen 2002). The bubble was filled with around 10^4 uniformly distributed tracer particles that are advected with the flow. The simulations were performed on an SGI ORIGIN 3000 at the United Kingdom Astrophysical Fluids Facility (UKAFF).

Now, the radio maps are constructed using the passively advected tracer particles. Initially, each tracer particle is located inside the bubble and is associated with the same initial relativistic electron population. At the location of each tracer particle, the pressure, density and the magnetic field components are recorded. Then the electron spectrum for each tracer particle is evolved in time taking into account synchrotron, inverse Compton, and adiabatic energy losses and gains (for details see Enßlin & Brüggen 2002).

The pre-shock external gas density is set to about $5 \cdot 10^{-4}$ electrons/cm^3 and the temperature to 1-2 keV. The simulation box is assumed to have a size of 1 Mpc^3 and to be located at a distance of 100 Mpc from the observer.

The passage of a radio cocoon through a shock wave can be seen in Fig-

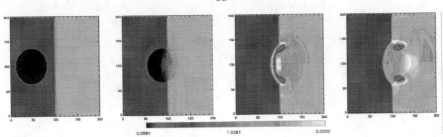

Fig. 24.1. Shock passage of a hot, magnetised bubble (a radio ghost) through a shock wave. The gas flows from the left to the right. Shown is the evolution of the density in a vertical cut through the centre of the box.

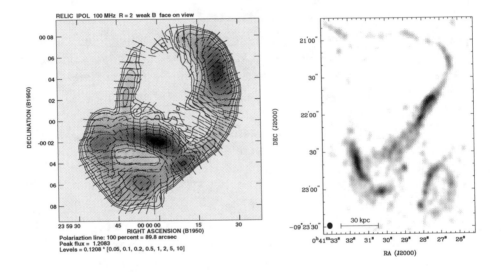

Fig. 24.2. Left: Synthetic radio map of a shocked radio ghost. The contours show the intensity and the dashes depict the polarisation (E-vector). Right: Observed radio relic in the cluster Abell 85 at 1.4 GHz.

ure 24.1. A possible evolutionary scenario of the radio morphology is as follows: In the early stages of the compression the torus has not formed and one would expect to find a sheet-like radio relic. The relic gets more and more edge brightened as the radio plasma accumulates in the torus. Finally, the thin sheet is disrupted and only a torus of radio plasma remains.

Thus we expect that the early, sheet-like radio relics have a 'younger', less steepened spectrum, while the later-stage, filamentary or toroidal relics have a much higher spectral age with a bent, steep radio spectrum.

These expectations are supported by observations of the filamentary relics in the cluster Abell 2256 that have a flat spectrum (Röttgering et al. 1994), whereas all the filamentary relics in the sample observed by Slee et al. (2001) exhibit steep, bent and therefore 'aged' spectra.

We have run a series of simulations varying, both the shock strength and the strength of the initial magnetic field. In the simulations with weak magnetic fields the final morphology is toroidal for the strong shock and shows two tori in the case of a weak shock wave. Such a double torus seems to be observed in the relic in Abell 85 (see Figure 24.2). The degree of polarisation is relatively high and the electric polarisation vectors tend to be perpendicular to the radio filaments which implies that the magnetic field lines follows the filaments.

24.2.1 Conclusion

We have presented 3-D MHD simulations of a hot, magnetised bubble that traverses a shock wave in a much colder and denser environment. This is assumed to be a fair model for a blob of radio plasma in the IGM which is passed by a merger shock wave. Synthetic radio maps of these simulations show filamentary and toroidal structures which are remarkably similar to the observed radio relics. Our simulations find polarisation patterns that indicate that the magnetic fields are mostly aligned with the filaments. This is also found in the observed radio relics.

We argue that the formation of filaments and tori is a generic feature of a hot bubble that is passed by a shock and whose internal sound speed exceeds the shock speed. We expect this to be a robust result that will also be found in more realistic simulations, e.g. with the proper equation of state, or more realistic initial magnetic field configurations.

We conclude that we have found strong evidence that cluster radio relics are formed when old radio plasma is compressed by a shock wave. On the basis of our simulations several predictions about the properties of radio relics can be made:

Our simulations indicate that the radius R of the bubble of radio plasma remains approximately constant during the passage of the shock. It was also found that the final structure consists of tori of minor radius r and major radius R. The compression factor of the radio plasma is thus proportional to $(R/r)^2$ and this ratio is a measure of the shock strength. Thus, the approximate compression factor of the radio plasma can be read off the radio map. Hence, if the pressure jump of the shock wave can be obtained from X-ray observations, the unknown adiabatic index of the radio plasma can be computed.

While the local radio polarisation reflects the complicated magnetic field structures, the total integrated polarisation of a relic reveals the 3-dimensional orientation of the shock wave. Since the compression aligns the fields with the shock plane, the sky-projected field distribution is aligned with the intersection of the shock plane and the plane of the sky. Thus, the direction of the total E-polarisation vector yields the sky-projected normal of the shock wave. Finally, the angle between the plane of the shock and the plane of the sky can be estimated from the fractional polarisation of the integrated flux.

Thus, the observations of radio relics may reveal, both, the unknown equation of state of the radio plasma and the geometry of the merger shock wave.

24.3 Radio galaxies

The gas permeating clusters of galaxies often shows a peak in the X-ray surface brightness which is interpreted as evidence for a cooling flow. In the picture of a cooling flow radiative cooling in the central regions of a cluster causes a slow subsonic inflow of gas which deposits mass at a rate of 10-1000 $M_\odot$/yr (see Fabian 1994 for a review). However, searches in all wavebands for the cooled material have revealed significantly less gas than predicted. The X-ray observatories Chandra and XMM have shown that the spectra from cooling flow regions show a remarkable lack of emission lines from gas with a temperature below about 1 keV (e.g. Peterson et al. 2001). At the same time it has been noted that cooling flow clusters often host strong radio sources at their centres. These radio sources inflate large bubbles of hot plasma that subsequently rise through the cluster atmosphere, thus mixing and stirring the cluster atmosphere. Together with Christian Kaiser, Torsten Enßlin and Eugene Churazov, I have performed highly resolved hydrodynamical simulations to show that rising bubbles can raise the cooling time in the inner cluster regions thereby reducing the deposition of cold gas.

A hydrodynamic simulation of the full problem is currently beyond our possibilities as the scales involved vary enormously: The velocities range from around $10\,\mathrm{km\,s^{-1}}$ in the outer cooling flow region to close to the speed of light in the radio jet. Likewise, the densities and energy densities vary by many orders of magnitude. Thus, one may have to break the problem up in order to make some progress.

Here we will focus on the later stages of the radio plasma where the bow shock of the initially overpressurised radio cocoon has vanished and the cocoon has come into approximate pressure equilibrium with its surroundings.

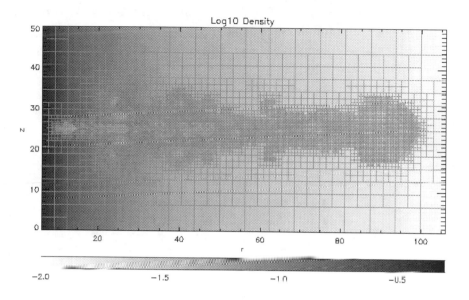

Fig. 24.3. Logarithmic contour plot of the density in one of our simulations of a radio galaxy. Hot gas which represents the radio plasma is injected near the bottom of the domain with zero initial velocity. Subsequently, buoyancy causes the gas to rise and Rayleigh-Taylor and Kelvin-Helmholtz instabilities create a mushroom-shaped plume. The lines drawn on top of the contour plot show the boundaries of the blocks (one block corresponds to 8×8 computational cells). One can see how the computational mesh is refined near the buoyant gas while the mesh remains quite coarse in places where the fluid is steady.

In a series of 3D simulations (Brüggen, Kaiser, Churazov & Enßlin 2002) hot matter was injected continuously into a small region off-set from the cluster centre. In agreement with previous analytic estimates we found that the bubbles evolve very differently depending on their luminosity. Using tracer particles we computed the efficiency of the bubbles to stir the intra-cluster medium and find that recurrent low-power sources are more effective in mixing the inner cluster region than rarer large outbursts. Moreover, we computed radio maps of the bubbles based on different assumptions about the magnetic field. In the radio band the bubbles closely resemble Fanaroff-Riley type I (FRI) sources (Fanaroff & Riley 1974). For the bubbles to be detectable for long enough to account for FRI sources, we found that reacceleration has to take place. The bubbles are generally difficult to detect, both, in the radio and in the X-ray band. Thus, another important result of these simulations is that it is possible to hide a significant amount of energy in the form of bubbles in clusters.

However, our simulations suffered from numerical diffusivity which made it impossible to make any firm conclusions about the evolution of the entropy in the cluster. Therefore, we went on to perform some 2D highly-resolved simulations using the FLASH code (Fryxell et al. 2000). This code uses the piece-wise parabolic method to solve the equations of hydrodynamics and is thus more accurate than the codes used in previous simulations. Moreover, the code employs an adaptive mesh that places resolution elements only where they are needed so that a huge effective resolution can be achieved. An example of such a simulation with an effective resolution of 1000×2000 is shown in Figure 24.3. The results of this work will appear in a forthcoming publication (Brüggen & Kaiser 2002).

I hope I was able to present some convincing examples for the use of hydrodynamical simulations in astrophysics. Even though these examples were chosen from the field of extragalactic astronomy, computer simulations have proven equally useful in other fields of astronomy such as stellar physics, where they are applied, for instance, to solar convection. A lot of work still needs to be done to include more physical processes, to overcome numerical limitations and to cover the vast range of scales. These are not tasks that are automatically resolved with the advent of more powerful computers. They require the development of novel and clever methods and algorithms.

Acknowledgements I am profoundly grateful to Douglas for his continuous support and wish to thank Mike, Sylvie and Jørgen for having organized such a tremendous meeting. This paper refers to work which has been done in collaboration with Torsten Enßlin, Christian Kaiser and Eugene Churazov. Some of the computations reported here were performed using the UK Astrophysical Fluids Facility (UKAFF). Some of the software used in this work was in part developed by the DOE supported ASCI/Alliances Center for Thermonuclear Flashes at the University of Chicago. This work was supported by the European Community Research and Training Network 'The Physics of the Intergalactic Medium'.

References

Brüggen, M. & Kaiser, C. R., 2002, *Nature* , **418**, 301.
Brüggen, M., Kaiser, C. R., Churazov, E. & Enßlin, T.A., 2002, *MNRAS*, **331**, 545
Enßlin, T. A. & Brüggen, M., 2002, *MNRAS*, **331**, 1011.
Enßlin T. A. & Gopal-Krishna, 2001, *A&A*, **366**, 26
Enßlin, T. A., Biermann, P. L., Klein, U. & Kohle, S., 1998, *A&A*, **332**, 395
Fabian, A. C., 1994, *ARAA*, **32**, 277
Fanaroff, B. L. & Riley, J. M., 1974, *MNRAS*, **167**, 31

Fryxell, B. et al. 2000, *ApJS*, **131**, 273

Peterson, J. R. et al., 2001, *A&A*, **365**, L104

Roettiger, K., Burns, J. O. & Pinkney, J., 1995, *ApJ*, **453**, 634

Roettiger, K., Burns, J. O. & Stone, J. M., 1999, *ApJ*, **518**, 603

Roettiger, K., Stone, J. M. & Mushotzky, R. F., 1998, *ApJ*, **493**, 62

Röttgering, H., Snellen, I., Miley, G., de Jong, J. P., Hanisch, R. J. & Perley, R., 1994, *ApJ*, **436**, 654

Slee, O. B., Roy, A. L., Murgia, M., Andernach, H. & Ehle, M., 2001, *AJ*, **122**, 1172

Stone, J. M. & Norman, M. L., 1992a, *ApJS*, **80**, 753

Stone, J. M. & Norman, M. L., 1992b, *ApJS*, **80**, 791

Venturi, T., Bardelli, S., Zambelli, G., Morganti, R. & Hunstead, R. W., 1999, in Böhringer, H., Feretti, L., Schücker, P., eds., *Diffuse Thermal and Relativistic Plasma in Galaxy Clusters*, MPE-Publ., Garching, p. 27

VI Dynamics

25

A magic electromagnetic field

DONALD LYNDEN-BELL

Institute of Astronomy, The Observatories,
Madingley Road, Cambridge CB3 0HA, UK
and Clare College, Cambridge, UK

An electromagnetic field of simple algebraic structure is simply derived. It turns out to be the $G = 0$ limit of the charged rotating Kerr-Newman metrics. These all have gyromagnetic ratio 2, the same as the Dirac electron. The charge and current distributions giving this high gyromagnetic ratio have charges of both signs rotating at close to the velocity of light.

It is conjectured that something similar may occur in the quantum electrodynamic charge distribution surrounding the point electron.

25.1 The electromagnetic field

Away from charges and currents, both the electrostatic potential, Φ, and the magnetostatic potential, χ, are harmonic. Thus $\Psi = \Phi + i\chi$ satisfies

$$\nabla^2 \Psi = 0 \ .$$

The solution obeying this equation everywhere – except the origin – and tending to zero at infinity is $\Psi = q/r$, but if we move the origin to $\mathbf{b}$ this solution becomes

$$\Psi = q \left/ \sqrt{(\mathbf{r} - \mathbf{b})^2} \right. \ .$$

This solution is harmonic whether q and $\mathbf{b}$ are real or complex.

To ensure no magnetic monopole, term q must be real, but we now consider the possibility that $\mathbf{b} = i\mathbf{a}$ where $\mathbf{a}$ is real so that $\mathbf{b}$ is pure imaginary. Then we shall have both an electric and a magnetic field with $\mathbf{F} = \mathbf{E} + i\mathbf{B} = -\boldsymbol{\nabla}\Psi$. Without loss of generality we may orient the z axis along $\mathbf{a}$ so that

$$\Psi = q \left/ \left(R^2 + (z - ia)^2\right)^{1/2} \right. \qquad \text{where} \qquad R^2 = x^2 + y^2 \ . \qquad (25.1)$$

This expression will be harmonic except at singularities and branch points.

The singularities lie at $R = a$ and $z = 0$. If we ask for no branch points at infinity then we may take the cut defined by the disk $z = 0$, $R \le a$, (but notice that we could take the cut around the sphere $r = a$, $z \ge 0$ say).

We may evaluate $-\nabla\Psi$ to obtain

$$\mathbf{F} = \mathbf{E} + i\mathbf{B} = q(\mathbf{r} - i\mathbf{a}) \Big/ \left[(\mathbf{r} - i\mathbf{a})^2\right]^{3/2} . \qquad (25.2)$$

The total charge is clearly q but the field also has a magnetic dipole moment. Indeed for $r > a$ we may use the Legendre polynomial expansion of Ψ

$$\Psi = \frac{q}{r} \sum_0^\infty \left(\frac{ia}{r}\right)^n P_n(\cos\theta) . \qquad (25.3)$$

Evidently all the P_{2n} have real coefficients and all the P_{2n+1} have imaginary coefficients so the magnetic potential is antisymmetrical about $z = 0$ while the electric potential is symmetrical. Evidently the magnetic moment is the coefficient of iP_1 which is qa while the electric quadrupole moment is qa^2, etc. The relativistic invariants of the field are contained in

$$F^2 = E^2 - B^2 + 2i\mathbf{E} \cdot \mathbf{B} = q^2 \Big/ \left[(\mathbf{r} - i\mathbf{a})^2\right]^2 .$$

Now $\left[(\mathbf{r} - i\mathbf{a})^2\right]^2$ is only imaginary if $(\mathbf{r} - i\mathbf{a})^2 = \pm\dfrac{1}{\sqrt{2}}(1 \pm i)|\mathbf{r} - i\mathbf{a}|^2$ which occurs when $\left(r^2 - a^2\right)/(2\mathbf{r} \cdot \mathbf{a}) = \pm 1$ as then the real and imaginary parts are equal in magnitude. This condition may be rewritten $(\mathbf{r} \pm \mathbf{a})^2 = 2a^2$ so $E^2 = B^2$ only on two spheres of radius $\sqrt{2}\,a$ centred on $(\mathbf{r} = \pm\mathbf{a})$. The circle in which they meet is the ring $z = 0$, $r = a$.

Figure 25.1 illustrates where $|\mathbf{B}| > |\mathbf{E}|$, etc. $\mathbf{E}$ and $\mathbf{B}$ are perpendicular when $(\mathbf{r} - i\mathbf{a})^2 = r^2 - a^2 - 2i\mathbf{a} \cdot \mathbf{r}$ is either purely real or purely imaginary; i.e., on the sphere $r = a$, and the plane $z = 0$. The Poynting vector is given by

$$\mathbf{F}^* \times \mathbf{F} = (\mathbf{E} - i\mathbf{B}) \times (\mathbf{E} + i\mathbf{B}) = 2i\mathbf{E} \times \mathbf{B} = 2iq^2\mathbf{a} \times \mathbf{r} \Big/ \left(r^2 + a^2\right)^3 ,$$

and the field energy density by $(8\pi)^{-1}\mathbf{F}^* \cdot \mathbf{F} = (8\pi)^{-1}\left(E^2 + B^2\right)$. The velocity of the Lorentz frame in which $\mathbf{E}$ and $\mathbf{B}$ are parallel is given by $\mathbf{v} = c\mathbf{V}$ where

$$\begin{aligned}
\mathbf{V}\big/\left(1 + V^2\right) &= \mathbf{E} \times \mathbf{B}\big/\left(E^2 + B^2\right) = \mathbf{a} \times \mathbf{r}\big/\left(a^2 + r^2\right) \\
&= \mathbf{F}^* \times \mathbf{F}\big/(2i\mathbf{F} \cdot \mathbf{F}^*) ;
\end{aligned}$$

squaring and solving for V we find

$$V = \left[a^2 + r^2 - \sqrt{(a^2 + r^2)^2 - 4a^2 R} \right] / (2aR)$$

$$= 2aR \Big/ \left[a^2 + r^2 + \sqrt{(a^2 + r^2)^2 - 4a^2 R^2} \right] = aR / (a^2 + \lambda) \ ,$$

where

$$\lambda = \tfrac{1}{2} \left[r^2 - a^2 + \sqrt{(r^2 - a^2)^2 + 4(\mathbf{a} \cdot \mathbf{r})^2} \right] \ ,$$

is defined with the positive root and μ is the same but for the negative root. λ and μ are spheroidal coordinates. Evidently $\Omega = V/R$ is constant on the confocal spheroids, $\lambda = $ constant which have a focal ring at the singularity ('This result is due to J. Gair.)

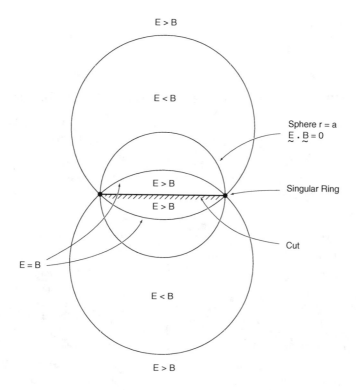

Fig. 25.1. Planar cut through the origin, orthogonal to the $z = 0$ plane, showing the delineation of regions of $E > B$ and $E < B$, for the potential given by eq. (28.1).

On the cut itself we have $R < a$ and $z = 0+$.

$$\mathbf{E} + i\mathbf{B} = q\left(\mathbf{R} - i\mathbf{a}\right) \Big/ i\left(a^2 - R^2\right)^{3/2} = -q\left(\mathbf{a} + i\mathbf{R}\right) \Big/ \left(a^2 - R^2\right)^{3/2} .$$

This gives an electric field vertically down into the disc and a magnetic field parallel to the disk surface for $R < a$ as though the disk has a Meissner effect. The corresponding charge density on the symmetry plane is

$$\sigma = -\left(q/2\pi\right) a \left(a^2 - R^2\right)^{-3/2} .$$

This charge density gives a divergent total charge but that divergence is cancelled by a ring of opposite charge on the edge which leaves the total charge not 'negative' but 'positive' $+q$. The total charge at axial distance less than R is $Q(< R) = -q\left[a(a^2 - R^2)^{-1/2} - 1\right]$, $R < a$. From the discontinuity in the $\mathbf{B}$ field across the cut we find $4\pi J_\phi = -2qR(a^2 - R^2)^{-3/2}$. This corresponds to the charge density given above rotating with angular velocity $\Omega = c/a$, reaching the velocity of light at the singularity. Again its effect is reversed by a ring current at the edge. The fields are illustrated in Figures 25.2 and 25.3.

25.2 The connection to Kerr's metric and the electron

A much more complicated but more intriguing derivation of the above results is to take the Kerr (1963) metric of a black hole of mass m and angular momentum mac. Then complexify it following Newman (1973) to get the Kerr-Newman metric of charge q, (Newman et al. 1965). Finally, take the limit with $G \to 0$ leaving the charge and the moment corresponding to 'a' but now in flat space. The resultant electromagnetic field is exactly that derived and discussed above, (Pekeris & Frankowski 1987). Carter (1968a) showed that all the Kerr-Newman metrics had the same gyromagnetic ratio as the Dirac electron. Does this mean that there is some relationship between the charge distribution of the Kerr-Newman metric and the charge distribution of the quantum electrodynamic field of a point electron?

Classical models of the electron had a problem over the gyromagnetic ratio. Even if all the charge were confined to a ring rotating at close to the velocity of light the magnetic moment generated gives a gyromagnetic ratio of one rather than the electron's value of 2.0023193044. It is of some interest to gain an understanding as to how the Kerr-Newman metric does it. The answer is that the charge distribution is not all of one sign. In fact a circular current dipole of two rings of opposite charge rotating uniformly about their common axis gives a net magnetic moment but no net charge.

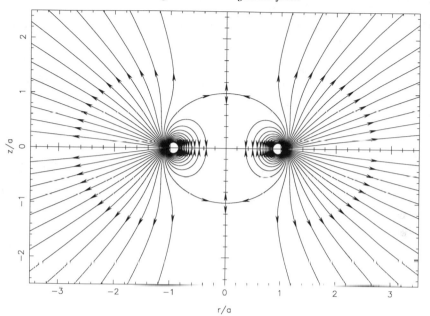

Fig. 25.2. A plot of electric field lines for the potential given by eq. (28.1).

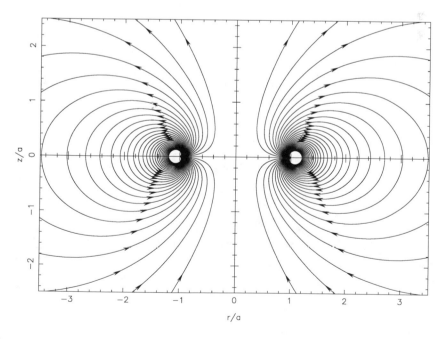

Fig. 25.3. A plot of magnetic field lines for the potential given by eq. (28.1).

The way our electromagnetic field gets its large magnetic dipole moment per unit net charge is that its much larger internal charges are of opposite signs but rotate together giving a magnetic dipole with relatively little net charge. We show elsewhere that this is a characteristic of relativistically rotating conductors!

25.3 Separability of motion in the field

Studies of separability of wave equations in the Kerr and Kerr-Newman metrics (Carter 1968b, Teukolsky 1972, 1973, Chandrasekhar 1976, Page 1976) have shown that Dirac's equation is separable in these metrics. This of course implies that it is still separable in their flat space limit as $G \to 0$. The criterion for the separability of Schrödinger's equation in a real potential in spheroidal coordinates is $\Phi = [\zeta(\lambda) - \eta(\mu)]/(\lambda - \mu)$ (Morse and Feshback 1953). Here λ and μ are spheroidal coordinates and ζ, η are arbitrary functions of their arguments.

The field that we derived so simply above is rewritten in spheroidal coordinates as follows: λ and μ are the roots for τ of the quadratic

$$\frac{x^2 + y^2}{a^2 + \tau} + \frac{z^2}{\tau} = 1 ,$$

where $x^2 + y^2 = R^2 = (\lambda + a^2)(\mu + a^2)/a^2$ and the metric is

$$ds^2 = dx^2 + dy^2 + dz^2 = \frac{\lambda - \mu}{4\lambda(\lambda + a^2)}d\lambda^2 + \frac{\lambda - \mu}{4\mu(\mu + a^2)}d\mu^2 + R^2 d\phi^2 .$$

To compare to Kerr's metric one uses the quasi-spherical form of spheroidal coordinates $\tilde{r} = \sqrt{\lambda}$, $\mu = -a^2 \cos^2 \vartheta$, $z = \tilde{r}\cos\vartheta$. Note however that $\tilde{r}$ is constant on spheroids and $\tilde{r} = 0$ is the disc $z = 0$, $R \leq a$. Also ϑ is not the θ of spherical polar coordinates but is constant in hyperboloids. Thus

$$ds^2 = \left(\tilde{r}^2 + a^2\cos^2\vartheta\right)/\left(\tilde{r}^2 + a^2\right)d\tilde{r}^2 + \left(\tilde{r}^2 + a^2\cos^2\vartheta\right)d\vartheta^2 +$$

$$+ \left(\tilde{r}^2 + a^2\right)\sin^2\vartheta d\phi^2 .$$

In spheroidal coordinates our potential $\Psi = q/\sqrt{(\mathbf{r} - i\mathbf{a})^2}$ takes the simple forms

$$\Psi = q \left/ \left(\sqrt{\lambda} - i\sqrt{-\mu}\right)\right. = q\frac{\sqrt{\lambda} + i\sqrt{-\mu}}{\lambda - \mu} = \frac{q}{\tilde{r} - ia\cos\vartheta} .$$

The second of these forms is exactly of the right type for separability of the Schrödinger equation but the similarity is partly misleading for Schrödinger's equation only separates in an electrostatic potential of that form. When the imaginary (magnetic) part is added Schrödinger's equation no longer separates although the Klein-Gordon equation now does separate (which it does not with only the electrostatic part). For a derivation and explanation of these results see Lynden-Bell (2000).

Systems with the same charge distribution but less magnetic field are given by taking $\psi = \alpha\Psi + (1-\alpha)\Psi^*$ for $\alpha < 1$. The magnetic fields are then multiplied by $2\alpha - 1$. These are weighted superpositions of discs rotating forwards and backwards so the net rotation is less fast and $\alpha = 1/2$ is static. These fields lose the magic of separability. For the other charge & current distributions with that property see Lynden-Bell (2000).

25.4 Eulogy

In closing, let me say that I still do not know the answer to the problem discussed in my joint paper with Douglas (Gough & Lynden-Bell 1968), i.e., "How *do* turbulent fluids with angular momentum like to rotate?" Nevertheless, I never expected to know the internal rotation of the Sun within my lifetime and I have immense admiration for Douglas – and the helioseismic fraternity – for having persisted in analysing solar pulsations until that became possible. Such is the real meat of good science.

References

Carter, B. 1968a, *Phys. Rev.* **174**, 1559
Carter, B. 1968b, *Commun. Math. Phys.* **10**, 280
Chandrasekhar, S. 1976, *Proc. R.Soc. London A* **349**, 571
Gough, D.O. & Lynden-Bell, D. 1968, *JFM*, **32**, 437
Kerr, R.P. 1963, *PRL*, **11**, 217
Lynden-Bell, D. 2000, *MNRAS*, **312**, 301
Morse, P.H. & Feshback, H. 1953, *Methods of Theoretical Physics*,
    McGraw Hill (NY)
Newman, E.T. 1973, *J. Math. Phys.* **14**, 102
Newman, E.T., Couch, E., Channapared, K., Exton, A, Prakesh, A., Torrance, R.
    1965, *J. Maths. Phys.* **6**, 918
Page, D.N. 1976, *Phys. Rev. D* **14**, 1509
Pekeris, C.L. & Frankowski, K. 1987, *Phys. Rev. A* **16**, 5118
Teukolsky, S.A. 1972). *PRL*, **29**, 16, 1114
Teukolsky, S.A. 1973, *ApJ*, **185**, 635

26

Continuum equations for stellar dynamics

EDWARD A. SPIEGEL

Department of Astronomy
Columbia University, New York, NY 10027, USA

JEAN-LUC THIFFEAULT

Department of Applied Physics and Applied Mathematics
Columbia University, New York, NY 10027, USA

The description of a stellar system as a continuous fluid represents a convenient first approximation to stellar dynamics, and its derivation from the kinetic theory is standard. The challenge lies in providing adequate closure approximations for the higher-order moments of the phase-space density function that appear in the fluid dynamical equations. Such closure approximations may be found using representations of the phase-space density as embodied in the kinetic theory. In the classic approach of Chapman and Enskog, one is led to the Navier–Stokes equations, which are known to be inaccurate when the mean free paths of particles are long, as they are in many stellar systems. To improve on the fluid description, we derive here a modified closure relation using a Fokker–Planck collision operator. To illustrate the nature of our approximation, we apply it to the study of gravitational instability. The instability proceeds in a qualitative manner as given by the Navier–Stokes equations but, in our description, the damped modes are considerably closer to marginality, especially at small scales.

26.1 A kinetic equation

If we have a system of N stars, with N very large, and wish to study its large-scale dynamics, we have to choose the level of detail we can profitably treat. Even if we could know the positions and velocities of all N stars for all times, we would be mainly interested in the global properties that are implied by this information. For just such reasons, many investigators prefer to find an approach that leads directly to a macroscopic description of the dynamics. As Ogorodnikov (1965) has put it, "In order to exhibit more clearly the kinematics of highly rarefied media, and of stellar systems in particular, it is useful to make a comparison with the motion of a fluid." However, traditional methods for deriving fluid equations are effective only for media in which the mean free paths of constituent particles are very short compared to all macroscopic scales of interest. This condition is not met in many stellar systems and plasmas and so we here describe an approach that is effective for deriving fluid equations for rarefied media such as stellar systems.

The first problem we must face is to decide what the kinetic description of a stellar dynamical system ought to be. Since an N-body description is not what we want to work with, even if we could, since that approach would have us computing the complicated trajectory of the stellar system through a phase space with large dimension. So we go straight to the description of the system in the six-dimensional phase space whose coordinates are the spatial coordinates (x^i) and the velocities (v^i) of the N stars, where $i = 1, 2, 3$.

A plot of the locations of each of the N stars in the six-dimensional phase space at some given time, would reveal a swarm of points whose detailed description would also be too complicated for us, at least in a first look at the problem. So instead, we concentrate on an ensemble mean of such a description and seek an equation for the density distribution of this mean. That equation, on which we base this work, is an evolution or continuity equation for the probability density in the six-dimensional phase space.

Since we treat the stars as points, the true density F in the six-dimensional phase space is a summation of delta functions at suitable locations. This density is advected by a six-dimensional phase velocity that Hamilton's equations tell us is solenoidal. Hence the total time derivative of the density is

$$\mathcal{D}F = 0 \, , \tag{26.1}$$

where the comoving derivative in phase space is defined as

$$\mathcal{D} := \partial_t + \mathbf{v} \cdot \nabla_\mathbf{x} + \mathbf{a} \cdot \nabla_\mathbf{v} \tag{26.2}$$

and the subscripts $\mathbf{x}$ and $\mathbf{v}$ on the gradient symbols indicate that they are gradients with respect to position and velocity respectively. As usual, $\mathbf{v} = \dot{\mathbf{x}}$ where the dot means total time derivative and the quantity $\mathbf{a}$ is the gravitational acceleration per unit mass; we assume that all the stars in the system have the same mass, m. The gravitational acceleration is given by the gradient of the gravitational potential per unit mass, which is a solution of Poisson's equation,

$$\Delta \Phi = 4\pi G \int F \, \mathrm{d}^3 v \, . \tag{26.3}$$

Let $f(\mathbf{x}, \mathbf{v}, t)$ be the ensemble mean of F. Then the total density will be $F = f + \hat{f}$ where $\hat{f}$ represents the fluctuations about the ensemble mean; $\hat{f}$

will involve the same summation of delta functions as does F plus a smooth background distribution with negative mass density arranged so that the ensemble average of $\hat{f}$ is zero. We similarly split the gravitational potential Φ into an ensemble mean part ϕ plus a fluctuating part $\hat{\phi}$. Then, if we take the ensemble average of (26.1), we obtain

$$\partial_t f + \mathbf{v} \cdot \nabla_{\mathbf{x}} f + (\nabla_{\mathbf{x}} \phi) \cdot \nabla_{\mathbf{v}} f = -\langle (\nabla_{\mathbf{x}} \hat{\phi}) \cdot \nabla_{\mathbf{v}} \hat{f} \rangle , \tag{26.4}$$

where

$$\Delta \phi = 4\pi G \int f \, \mathrm{d}^3 v . \tag{26.5}$$

The terms on the left hand side of (26.4) describe how the mean phase density, or distribution function, f, streams through the single-particle phase space. The right side represents the mean influence on the evolution of f exerted by the average of the fluctuation-interaction term. The latter represents the self-interactions of the system caused by fluctuating effects and may be thought of as representing the influence of collective modes (such as waves or quasiparticles) on the motions of the individual particles.

It is typical that the chance of close approaches of two stars in many stellar systems is small. Because of this, one commonly made approximation is to neglect the right side of (26.4) completely and so work with what is called the collisionless Boltzmann (or Vlasov) equation. More reasonably perhaps, one may try to give an expression for the way the self-interaction term affects the flow of the phase density through the phase space.

As in Boltzmann theory, we shall suppose that the right side of (26.4) may be expressed as a functional of f itself so that the kinetic equation is deterministic. That is, we assume that the kinetic equation takes the form

$$\partial_t f + \mathbf{v} \cdot \nabla_{\mathbf{x}} f + (\nabla_{\mathbf{x}} \phi) \cdot \nabla_{\mathbf{v}} f = \mathcal{C}[f] , \tag{26.6}$$

where $\mathcal{C}[\cdot]$ may be called a collision term in keeping with the terminology of kinetic theory even though it does not arise from direct binary collisions. This approach may be acceptable because it appears that the fluid description that we seek is not very sensitive to the details of the right side of (26.4). On the other hand, we must admit that this hope is founded on a very limited range of trial forms since the job of deriving the consequences of each form is laborious. Our aim here is to adopt one standard form for the collision term and use it to go on to a coarser description of the stellar system like that of fluid dynamics.

The parallel to the Boltzmann theory has been used to good effect in the study of plasma physics, as in the work of Rosenbluth, et al. (1957). As

Clemmow & Dougherty (1969) explain, those authors obtained their results by "expanding the Boltzmann collision operator under the approximation that all the deflections are small angle and cutting off the impact parameter at about the Debye length.... At first sight the success of that method is surprising, as any treatment dealing with binary collisions would seem to be discredited. The physical reason for the agreement is that, for the majority of particles, there is little difference between a succession of numerous small-angle collisions (regarded as instantaneous and occurring at random) and the stochastic deflections due to the presence of many nearby particles continually exerting weak forces. These two pictures of the dynamics are of course represented respectively by the Boltzmann and the Fokker–Planck" approaches.

Similar thoughts have been expressed in the context of stellar dynamics, most recently by Griv et al. (2001) and by kinetic theorists generally. As E. G. D. Cohen reports (1997), "when Academician Bogolubov and I discussed the nature of kinetic equations, he mentioned a discussion he had had with Professor A. Vlasov, where they had agreed that: *Yes, in first approximation the kinetic equations for gases with strong short-range forces (i.e. the Boltzmann equation) and for gases with long-range forces (i.e. the Vlasov equation) differ, but in higher approximations they will become more and more similar.* How right they were." A formal theory to buttress these remarks would be very comforting, but though we do not have one we shall adopt the point of view that the Fokker–Plank terms capture the essence of interactions of the stars in the system. Having thus supported our approach by the appeal to authority, we turn to the main purpose of this work, the derivation of continuum mechanics from the microscopic theory in a way that is not severely restricted to the case of short mean free paths.

26.2 The collision term

In the spirit of standard kinetic theories, we shall suppose that the effect of the collision term is to drive the system toward a local equilibrium, though the correct equilibrium of a stellar system is not known on purely theoretical grounds. The tendency to approach an equilibrium seems not even to require a collision term of the usual kind since the violent relaxation described by Lynden-Bell apparently can do the job. Nevertheless, we shall proceed in terms of the kinetic theory under the assumption that the spreading of the phase fluid through the phase space may be effected by a collision term. Furthermore, though the gravitational force has long range in physical space, we shall presume that this spreading takes the form of a diffusion of f

through velocity space, that is, by the agency of a Fokker–Planck form of the collision term. (This may not be completely unfounded since there seems to exist a form of gravitational shielding (Spiegel, 1998) that may support the idea of local behaviour.) In this spirit, we write

$$\mathcal{C}[f] = \frac{\partial}{\partial v^i}\left[\,A^i f + \tfrac{1}{2}\frac{\partial}{\partial v^j}\left(B^{ij}\,f\right)\right].\tag{26.7}$$

The coefficients A^i and B^{ij} are generally functions of $(\mathbf{x}, \mathbf{v}, t)$ and they may also be functionals of f. In this discussion, the choice of these coefficients in the Fokker–Planck description is adapted to the equilibrium that is expected or assumed. This equilibrium is a local one that satisfies the condition $\mathcal{C}[f_0] = 0$.

Our goal is to find equations that govern the dynamics of the macroscopic properties of the fluid embodied in the density, temperature and velocity. These are defined as:

$$\text{Mass density} \qquad \rho := \int mf\,\mathrm{d}^3v,\tag{26.8}$$

$$\text{Mean velocity} \qquad \mathbf{u} := \frac{1}{\rho}\int m\mathbf{v}f\,\mathrm{d}^3v,\tag{26.9}$$

$$\text{Temperature} \qquad T := \frac{m}{3R\rho}\int c^2 f\,\mathrm{d}^3v,\tag{26.10}$$

where the *peculiar velocity* is

$$\mathbf{c}(\mathbf{x}, \mathbf{v}, t) := \mathbf{v} - \mathbf{u}(\mathbf{x}, t)\,,\tag{26.11}$$

and $R = k/m$, k being Boltzmann's constant.

As to the nature of $\mathcal{C}[f]$, we shall design it so that it produces what may be the simplest plausible equilibrium, namely the Maxwell–Boltzmann distribution

$$f_0(\mathbf{x}, \mathbf{v}, t) = \frac{\rho}{m(2\pi RT)^{3/2}}\exp\left(-\frac{c^2}{2RT}\right).\tag{26.12}$$

Since T, ρ and $\mathbf{u}$ generally depend on $\mathbf{x}$ and t, this is a local equilibrium and we choose (Clemmow and Dougherty (1969))

$$A^i = -\tau^{-1}(v^i - u^i), \qquad B^{ij} = 2\tau^{-1}RT\,\delta^{ij}\,,\tag{26.13}$$

so that $\mathcal{C}[f_0] = 0$. We assume that the mean-free-time τ is a constant so that the Fokker–Planck operator is linear in f.

The collision term adopted here ensures the conservation of mass, momentum and energy in the system. This is reflected in the property

$$\int \psi^\alpha \, \mathcal{C}[f] \, \mathrm{d}^3 v = 0, \qquad \alpha = 0, \ldots, 4 \,, \tag{26.14}$$

where

$$\psi^\alpha = m \left(1, \mathbf{v}, \tfrac{1}{2} v^2\right) . \tag{26.15}$$

Thus we neglect the possible effects of dissipative processes and of evaporation of stars from the system.

The macroscopic quantities (26.8)–(26.10) enter the equilibrium distribution in (26.12) about which we are expanding. To ensure that the same macroscopic quantities that follow from f are those that determine f_0, we impose a consistency requirement known as the *matching conditions*,

$$\int \psi^\alpha \, f \, \mathrm{d}^3 v = \int \psi^\alpha \, f_0 \, \mathrm{d}^3 v, \qquad \alpha = 0, \ldots, 4 \,. \tag{26.16}$$

26.3 Fluid equations

When we multiply the kinetic equation (26.6) by the collisional invariants (26.15) and integrate over $\mathbf{v}$, the right-hand side does not contribute to the outcome, and we are left with

$$\partial_t \rho + \boldsymbol{\nabla} \cdot (\rho \, \mathbf{u}) = 0, \tag{26.17}$$

$$\partial_t \mathbf{u} + \mathbf{u} \cdot \boldsymbol{\nabla} \mathbf{u} = -\rho^{-1} \boldsymbol{\nabla} \cdot \mathbb{P} - \boldsymbol{\nabla} \phi, \tag{26.18}$$

$$\tfrac{3}{2} \rho R \left(\partial_t T + \mathbf{u} \cdot \boldsymbol{\nabla} T\right) = -\mathbb{P} : \boldsymbol{\nabla} \mathbf{u} - \boldsymbol{\nabla} \cdot \mathbf{q}, \tag{26.19}$$

where $\boldsymbol{\nabla}$ means $\nabla_{\mathbf{x}}$. Here the pressure tensor $\mathbb{P}$ and heat flux $\mathbf{q}$ are defined as

$$\mathbb{P} := \int m \mathbf{c} \mathbf{c} f \, \mathrm{d}^3 v, \qquad \mathbf{q} := \int \tfrac{1}{2} m c^2 \mathbf{c} f \, \mathrm{d}^3 v \,. \tag{26.20}$$

We see that the form of the macroscopic equations is just that of the usual fluid equations. This result is independent of the rarity of the medium. The usefulness of these equations depends entirely on how well we can prescribe the higher-order moments $\mathbb{P}$ and $\mathbf{q}$. A standard way to proceed is to solve (26.6) approximately for f. We shall follow this route also, but will deviate from the normally used prescription at a certain point.

We let $f = f_0 + \tau \, f_1 + \ldots$ and look first at order τ^0. We find that $\mathcal{C}[f_0] = 0$, and the solution f_0 is the Maxwell–Boltzmann equilibrium (26.12). From (26.20), we see that $\mathbb{P}_0 = p \, \mathbb{I}$ and $\mathbf{q}_0 = 0$, where the scalar pressure is given by $p := \rho R T$. If we stop at this order, Equations (26.17)–(26.19) are then the Euler equations for an ideal fluid.

At order τ^1, it is convenient to factor out the Maxwell–Boltzmann solution from f_1 and write the equation to be solved as

$$\widetilde{\mathcal{L}} \tilde{f}_1 = \mathcal{D} \ln f_0 , \qquad (26.21)$$

where $\tilde{f}_1 := f_1/f_0$ and

$$\widetilde{\mathcal{L}} \tilde{f} := -\mathbf{c} \cdot \nabla_{\mathbf{c}} \tilde{f} + R T \nabla_{\mathbf{c}} \cdot (\nabla_{\mathbf{c}} \tilde{f}) . \qquad (26.22)$$

The right-hand side of (26.21) may be written out as

$$\mathcal{D} \left[-\frac{c^2}{2RT} + \ln \frac{n}{(2\pi R T)^{3/2}} \right] = -\frac{1}{RT} \mathbf{c} \cdot \mathcal{D}\mathbf{c} + \left(\frac{c^2}{2RT} - \frac{3}{2} \right) \mathcal{D} \ln T + \mathcal{D} \ln \rho . \qquad (26.23)$$

We note that

$$\mathcal{D}\mathbf{c} = -\nabla\phi - \mathrm{D}\mathbf{u}/\mathrm{D}t - \mathbf{c} \cdot \nabla\mathbf{u}, \quad \text{where} \quad \mathrm{D}/\mathrm{D}t := \partial_t + \mathbf{u} \cdot \nabla . \qquad (26.24)$$

The operator $\widetilde{\mathcal{L}}$ maps polynomials in $\mathbf{c}$ to polynomials of the same degree so, given the form of (26.23), we may seek a solution $\tilde{f}_1$ to (26.21) as a cubic in $\mathbf{c}$. We write

$$\tilde{f}_1 = \mathfrak{a} + \mathfrak{b}_i \, c^i + + \mathfrak{c}_{ij} \, c^i c^j + \mathfrak{d}_{ijk} \, c^i c^j c^k , \qquad (26.25)$$

where $\mathfrak{a}$, $\mathfrak{b}$, $\mathfrak{c}$ and $\mathfrak{d}$ are functions of $\mathbf{x}$ and t, and symmetric in their indices; repeated indices are summed. Inserting (26.25) into the left-hand side of (26.21), we obtain

$$\widetilde{\mathcal{L}} \tilde{f}_1 = -3\mathfrak{d}_{ijk} \, c^i c^j c^k - 2\mathfrak{c}_{ij} \, c^i c^j + (6RT\mathfrak{d}_{ill} - \mathfrak{b}_i)c^i + 2RT\mathfrak{c}_{ll} . \qquad (26.26)$$

There is no $\mathfrak{a}$ term because it is annihilated by the Fokker–Planck collision operator. We now equate coefficients of $\mathbf{c}$ between (26.26) and (26.23), and find

$$\mathfrak{d}_{ijk} = -\frac{1}{18RT} \left(\delta_{ij} \nabla_{x^k} \ln T + \delta_{ik} \nabla_{x^j} \ln T + \delta_{kj} \nabla_{x^i} \ln T \right) , \qquad (26.27)$$

$$\mathfrak{c}_{ij} = -\frac{1}{4RT} \left[(\nabla_{x^j} u_i + \nabla_{x^i} u_j) + \frac{\mathrm{D} \ln T}{\mathrm{D}t} \delta_{ij} \right] , \qquad (26.28)$$

$$\mathfrak{b}_i = -\frac{1}{6} \nabla_{x^i} \ln T - \frac{1}{RT} \left(\nabla_{x^i}\phi + \frac{\mathrm{D}u_i}{\mathrm{D}t} \right) - \nabla_{x^i} \ln \rho . \qquad (26.29)$$

Since $\mathfrak{a}$ is still unspecified, we may use it to satisfy the matching condition (26.16) $\int f_1 \, \mathrm{d}^3 v = 0$, resulting from mass conservation. Only the terms even in $\mathbf{c}$ contribute, and we find

$$\int f_1\, \mathrm{d}^3 v = n(RT\, \mathfrak{c}_{ll} + \mathfrak{a}) = 0\,, \tag{26.30}$$

which allows us to solve for $\mathfrak{a}$ in terms of the trace of $\mathfrak{c}$.

The pressure tensor and heat flux (26.20) are then obtained from f_1 by performing straightforward Gaussian integrals, and we get

$$P_{ij} = p\, \delta_{ij} + 2pRT\mathfrak{c}_{ij}, \qquad Q_k = \tfrac{1}{2}pRT\,(5\mathfrak{b}_k + 21RT\, \mathfrak{d}_{kll})\,. \tag{26.31}$$

The $\mathfrak{a}$ term is absent from the pressure because we used the density matching condition (26.30). From (26.31) and (26.28), we find that the pressure tensor can be written

$$\mathbb{P} = p\,\mathbb{I} - 2\mu\,\mathbb{E} - \mu\left(\frac{\mathrm{D}\ln T}{\mathrm{D}t} + \tfrac{2}{3}\,\boldsymbol{\nabla}\cdot\mathbf{u}\right)\mathbb{I} \tag{26.32}$$

to first order in τ, where the viscosity $\mu := \tfrac{1}{2}p\tau$, and

$$E_{ij} := \tfrac{1}{2}\left(\nabla_{x^j} u_i + \nabla_{x^i} u_j - \tfrac{2}{3}\,\boldsymbol{\nabla}\cdot\mathbf{u}\,\delta_{ij}\right) \tag{26.33}$$

is the rate-of-strain tensor in traceless form.

From (26.31), (26.27), and (26.29), to first order in τ, the heat flux is

$$\mathbf{q} = -\eta\,\boldsymbol{\nabla}T - 3\eta\,T\left[\boldsymbol{\nabla}\ln p + \frac{1}{RT}\left(\frac{\mathrm{D}\mathbf{u}}{\mathrm{D}t} + \boldsymbol{\nabla}\phi\right)\right]\,, \tag{26.34}$$

where the thermal conductivity $\eta := (5/6)p\tau R$.

These results differ from those of the usual Navier–Stokes equations for which $\mathbb{P} = p\,\mathbb{I} - 2\mu\,\mathbb{E}$ and $\mathbf{q} = -\eta\,\boldsymbol{\nabla}T$. To get some understanding of the import of the additional terms found here we introduce the specific entropy

$$S = C_v\,\ln\!\left(p\,\rho^{-5/3}\right)\,, \tag{26.35}$$

where $C_v := 3R/2$ is the specific heat at constant volume. Since

$$\dot{S} := \frac{\mathrm{D}S}{\mathrm{D}t} = C_v\left[\frac{\mathrm{D}\ln T}{\mathrm{D}t} + \tfrac{2}{3}\,\boldsymbol{\nabla}\cdot\mathbf{u}\right]\,, \tag{26.36}$$

we find that

$$\mathbb{P} = p\left(1 - \frac{\tau}{2C_v}\,\dot{S}\right)\mathbb{I} - 2\mu\,\mathbb{E} + \mathrm{O}(\tau^2), \tag{26.37}$$

$$\mathbf{q} = -\eta\,\boldsymbol{\nabla}T + 3(\eta\,T/p)\boldsymbol{\nabla}\cdot\mathbb{T} + \mathrm{O}(\tau^2), \tag{26.38}$$

with $\mathbb{T} := (\mathbb{P} - p\,\mathbb{I})$. If we put these results into $\dot{S}$, we obtain

$$\dot{S} = -\frac{2C_v}{3p}\,[\mathbb{T} : \boldsymbol{\nabla}\mathbf{u} + \boldsymbol{\nabla}\cdot\mathbf{q}]\,. \tag{26.39}$$

Hence $\dot{S}$ can be seen to be $O(\tau)$ and so the additional terms in the pressure tensor are $O(\tau^2)$. A similar argument may be made for the new terms in the heat flux. Though these terms do not appear in the conventional fluid equations, they can be quite significant when the mean free paths are long.

If we eliminate the entropy using

$$\frac{1}{C_v}\,\dot{S} = \frac{\dot{p}}{p} - \frac{5}{3}\frac{\dot{\rho}}{\rho} \tag{26.40}$$

we encounter the combination $p(t) - \frac{1}{2}\tau\dot{p}(t)$ which are the first two terms of a Taylor series of p in $\tau/2$. We may then write

$$\mathbb{P} = p(t - \tfrac{1}{2}\tau)\,\mathbb{I} + \tfrac{10}{3}\,\mu\,\boldsymbol{\nabla}\cdot\mathbf{u}\,\mathbb{I} - 2\mu\,\mathbb{E} + O(\tau^2)\,. \tag{26.41}$$

We see that our procedure has taken account of the physical fact that the medium senses what particles were doing one collision time prior to the present time but is not yet aware of what they are doing at the present.

26.4 The Jeans instability

As an application of the equations of motion (26.17)–(26.19), we will use them together with the pressure tensor and heat flux derived in Section 26.3 to examine the Jeans criterion for gravitational instability. This instability, describing the gravitational collapse of a homogeneous medium, was first investigated by Jeans (1929). He found that perturbations above a critical wavelength (the Jeans length) were unstable to gravitational collapse, but that shorter wavelengths were unaffected due to the large-scale nature of the gravitational force. The Jeans length is the ratio of the adiabatic sound speed a_S to the gravitational frequency $\sqrt{4\pi G\rho}$. Pacholczyk & Stodólkiewicz (1959) and Kato & Kumar (1960) investigated the effect of viscosity and thermal conductivity on the instability and found that the collapse occurs above a critical wavelength given by the ratio of the isothermal sound speed a_T to the gravitational frequency. That critical wavelength is slightly larger than in the ideal case, which involves the adiabatic sound speed. Their interpretation for the increase in critical wavelength is that temperature gradients are adverse to the collapse, and a nonzero thermal conductivity allows the smoothing out of these gradients through very slow displacements of the medium; the mode is thus isothermal.

We will now investigate how the gravitational instability occurs in our set of equations. We take as our equilibrium a medium at rest and with uniform density ρ_0 and temperature T_0. We expand each fluid variable into an equilibrium piece and a small perturbation,

$$\rho = \rho_0(1 + \varphi), \quad T = T_0(1 + \theta), \quad p = \rho RT = p_0(1 + \varpi), \quad \varpi = \varphi + \theta \,.$$
$$(26.42)$$

The gravitational potential ϕ in (26.18) is obtained from the Poisson equation

$$\Delta\phi = 4\pi G\left(\rho - \rho_0\right), \tag{26.43}$$

where we choose $-\rho_0$ as a background "neutralising" density in order to have a proper uniform equilibrium about which to expand; the equilibrium velocity then vanishes. The density $-\rho_0$ is a repulsion term and may be regarded as a Newtonian analogue of Einstein's cosmological constant. To leave it out as Jeans did and jump straight to the linearised equation (26.47) below is expedient but questionable.

The linearised equations of motion (26.17)–(26.19) and Poisson equation (26.43) are

$$\partial_t\varphi + \boldsymbol{\nabla} \cdot \mathbf{u} = 0, \tag{26.44}$$
$$\rho_0\partial_t\mathbf{u} + \boldsymbol{\nabla} \cdot \mathbb{P} = -\rho_0\boldsymbol{\nabla}\phi, \tag{26.45}$$
$$\tfrac{3}{2}p_0\partial_t\theta + p_0\boldsymbol{\nabla} \cdot \mathbf{u} + \boldsymbol{\nabla} \cdot \mathbf{q} = 0, \tag{26.46}$$
$$\Delta\phi = 4\pi G\rho_0\,\varphi. \tag{26.47}$$

The pressure tensor $\mathbb{P}$ is given by (26.32) and the heat flux $\mathbf{q}$ by (26.34). We take the divergence of the velocity equation (26.45) and use the continuity equation (26.44) to eliminate $\boldsymbol{\nabla} \cdot \mathbf{u}$,

$$\rho_0\partial_t^2\varphi - \boldsymbol{\nabla}\boldsymbol{\nabla} : \mathbb{P} = 4\pi G\rho_0^2\,\varphi \,, \tag{26.48}$$

where we also used the Poisson equation (26.47) to eliminate ϕ. We then need to take two divergences of the linearised pressure tensor,

$$\boldsymbol{\nabla}\boldsymbol{\nabla} : \mathbb{P} = p_0\,\Delta(\varphi + \theta) + \mu\left(2\partial_t\Delta\varphi - \partial_t\Delta\theta\right). \tag{26.49}$$

We introduce the isothermal sound speed a_T, the kinematic viscosity ν, and the thermal diffusivity κ, through

$$a_T^2 := \frac{p_0}{\rho_0}, \qquad \nu := \frac{\mu}{\rho_0}, \qquad \kappa := \frac{\eta}{\rho_0\,C_p} = \frac{\eta}{\tfrac{5}{2}R\rho_0} \,. \tag{26.50}$$

Then, on inserting (26.49) into (26.48), we obtain

$$\partial_t^2\varphi = a_T^2\,\Delta(\varphi + \theta) + \nu\left(2\partial_t\Delta\varphi - \partial_t\Delta\theta\right) + \tfrac{5}{3}k_{\mathrm{J}}^2\,a_T^2\,\varphi \,, \tag{26.51}$$

where the Jeans wavenumber is given by

$$k_{\mathrm{J}}^2 := \frac{4\pi G\rho_0}{a_S^2} = \frac{3}{5}\,\frac{4\pi G\rho_0}{a_T^2} \,. \tag{26.52}$$

Next, we take the divergence of the linearised heat flux,

$$\boldsymbol{\nabla} \cdot \mathbf{q} = -\eta T_0 \, \Delta\theta - 3\eta T_0 a_T^{-2} \left[\partial_t \boldsymbol{\nabla} \cdot \mathbf{u} + \Delta\phi + a_T^2 \Delta(\varphi + \theta) \right] , \qquad (26.53)$$

and insert this into the temperature equation (26.46),

$$\partial_t \theta - \tfrac{2}{3}\partial_t \varphi - \tfrac{5}{3}\kappa \, a_T^{-2} \left[4a_T^2 \Delta\theta - 3 \left(\partial_t^2 \varphi - \tfrac{5}{3}k_J^2 \, a_T^2 \, \varphi - a_T^2 \Delta\varphi \right) \right] = 0 , \quad (26.54)$$

where again we eliminated $\boldsymbol{\nabla} \cdot \mathbf{u}$ using the continuity equation (26.44), and used the definition (26.50) of the thermal diffusivity κ.

Equations (26.51) and (26.54) are the equations required to derive a dispersion relation for the gravitational instability. It is convenient to use the viscous time μ/p_0 as unit of time and $a_T\mu/p_0$ as unit of length. Recycling the same symbols for the dimensionless quantities turns equations (26.51) and (26.54) into

$$\left(\partial_t^2 - \Delta - 2\partial_t \Delta - \tfrac{5}{3}k_J^2 \right) \varphi + (\partial_t - 1)\Delta\theta = 0, \qquad (26.55)$$

$$\left(3\partial_t - \tfrac{40}{3}\Delta \right) \theta + \left(-2\partial_t + 10 \left(\partial_t^2 - \tfrac{5}{3}k_J^2 - \Delta \right) \right) \varphi = 0. \qquad (26.56)$$

On letting $\varphi, \theta \sim \exp(ikx + \gamma t)$, we find the dispersion relation

$$\left(\tfrac{3}{5} + 2k^2 \right) \gamma^3 + \tfrac{22}{15} k^2 \, \gamma^2 + \left[\tfrac{22}{3} k^4 + \left(1 - \tfrac{10}{3} k_J^2 \right) k^2 - k_J^2 \right] \gamma$$
$$+ \tfrac{2}{3} \left(k^2 - \tfrac{5}{3} k_J^2 \right) k^2 = 0, \quad (26.57)$$

which may be compared to the expression obtained from Navier–Stokes by Kato & Kumar (1960),

$$\tfrac{3}{5}\gamma^3 + \tfrac{22}{15} k^2 \gamma^2 + \left(\tfrac{8}{9} k^4 + k^2 - k_J^2 \right) \gamma + \tfrac{2}{3} \left(k^2 - \tfrac{5}{3} k_J^2 \right) k^2 = 0 , \qquad (26.58)$$

with the Fokker–Planck values for the viscosity and thermal diffusivity inserted into their result. In each case the system is marginally stable with $\gamma = 0$ at $k^2 = (5/3)k_J^2$, and is damped for larger k. This is illustrated in Fig. 26.1. For $k = 0$, both dispersion relations predict a growth rate of $\gamma(k = 0) = \sqrt{5/3}\, k_J$ in dimensionless units, which with dimensions is $\sqrt{4\pi G\rho_0}$. This is consistent with the fact that dissipation is unimportant at large scales, so the growth rate at $k = 0$ involves only the gravitational time.

The asymptotic growth rate as $k \to \infty$ is $-3/4$ for Navier–Stokes and $-1/11$ for our system, independent of k. (Multiply by μ/p_0 to recover dimensions.) Thus, at large wavenumbers, the modes tend to be uniformly damped, both in our case and for Navier–Stokes. This is because at large k the fluid behaves like a Stokes flow, where we can ignore the inertial and gravitational terms completely, and the balance is between the Laplacian of the pressure and the viscosity, which have the same number of spatial

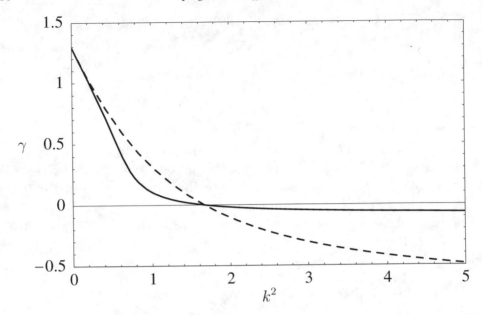

Fig. 26.1. Growth rate γ as a function of wavenumber k, for Navier–Stokes (dashed line) and the equations derived in Section 26.3 (solid). The growth rate of this mode is real and, in both cases, damping sets in above the isothermal Jeans wavenumber $(5/3)k_{\mathrm{J}}^2$ (here we have taken $k_{\mathrm{J}} = 1$ in dimensionless units).

derivatives; hence the lack of dependence on k. For our case (the equations of Section 26.3) there is a contribution from the new terms that shifts the damping rate considerably closer to marginality.

Coexisting with the real root associated with the instability, there is also a pair of unconditionally damped roots. The real part of these roots is plotted in Fig. 26.2. At $k = 0$ one of these roots is marginal (real part of growth rate equal to zero) and is an *isopycnal* mode (constant density). However, this mode is never destabilised and its growth rate immediately decreases as k increases away from zero.

For small k the two damped roots are distinct and real, but they come together at larger k and become a complex-conjugate pair with nonzero imaginary part (not plotted), indicating oscillatory behaviour. The Navier–Stokes case (dashed line) is seen to have heavily damped complex roots at large wavenumber.† But for our system of equation (solid line), the growth rate actually increases a little, looking as though it may be headed for a Hopf bifurcation (*i.e.*, overstability, where the real part of the growth

---

† In the Navier–Stokes system the complex roots come together again at very large k, but the growth rate continues to decrease with k.

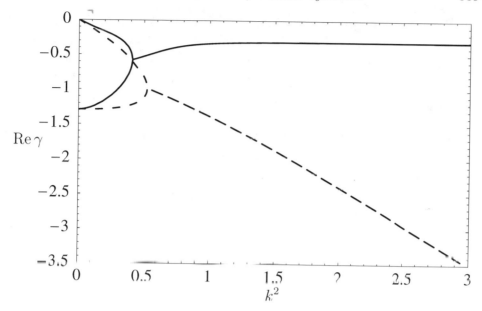

Fig. 26.2. Real part of the two complex roots γ as a function of wavenumber k, for Navier–Stokes (dashed line) and the equations derived in Section 26.3 (solid). Large wavenumbers are much more strongly damped for Navier–Stokes than in the system derived here.

rate becomes positive at nonzero imaginary part), before leveling off at an asymptotic value of the damping rate given by $-53/165 \simeq -0.3212$. Indeed, it can be shown that there is no Hopf bifurcation for any realisable parameter values in our equations, but the fact that the complex modes are somewhat "destabilised" by the new terms is intriguing (this is also true to a lesser extent for the real mode described above). This destabilisation has its source in the k^2 coefficient of the γ^3 term in the dispersion relation (26.57), which is not present in Navier–Stokes, but its physical significance is not yet apparent to us.

26.5 Conclusion

The basic approach in this as in other derivations of fluid equations from kinetic theory is to write the general moment equations (26.17)–(26.19). These are the fluid equations and, to complete them, we need closure relations for the pressure tensor and heat flux. This is an issue astronomers are familiar with from the study of radiative transfer. For the purpose, we could invent a phenomenological approximation as Eddington did in radiative transfer or we may pursue approximate solutions of the kinetic theory as Hilbert did by expanding in the collision time. The Hilbert expansion was developed by Chapman and Enskog in deriving the Navier–Stokes equations (Uhlenbeck & Ford, 1963) and we have pursued that line as well following earlier work (Chen, 2000; Chen *et al.*, 2000, 2001) based on the relaxation model of kinetic theory (Bhatnager et al., 1954; Welander, 1954). However, in that latter work, as here, we depart from the Chapman–Enskog approach in an essential way in not using results from lower orders to to simplify the results in the current order.

To express this idea in equations, let us consider what happens in general in such problems in the first order. Once we have expressed the one-particle distribution function as $f = f_0(1 + \tau\varphi)$ where τ is the (small) collision time, we are led to an equation for φ in the form

$$\mathcal{L}\varphi = \mathcal{D}f_0 + \mathcal{O}(\tau) , \qquad (26.59)$$

where $\mathcal{L}$ is the linearisation of the collision operator. In general, $\mathcal{L}$ is self-adjoint, as it is in the present study. Then $\mathcal{L}\psi^\alpha = 0$ implies $\int \psi^\alpha \mathcal{L}\varphi \, \mathrm{d}\mathbf{v} = 0$ and so we must have

$$\int \psi^\alpha \left[\mathcal{D}f_0\right] \mathrm{d}\mathbf{v} = \mathcal{O}(\tau) . \qquad (26.60)$$

Since ψ^α represents the collisionally invariant quantities, the fluid equations to the current order may serve as the solvability condition (26.60).

In the Chapman–Enskog procedure, this solvability condition is taken to be a lower order version of the fluid equations, here the Euler equation, and it is used to simplify the right side of (26.59). Then, the results are used in the general fluid equations. For both of these two conditions to be satisfied, we require τ to be very small indeed.

What we are doing here is to say that, to the first order, the fluid equations themselves are a realisation of condition (26.60) and that it is redundant to apply the same condition twice, once with $\mathrm{O}(\tau)$ retained and once with it omitted, as one does in the Chapman–Enskog method. Rather, we simply use the full condition (26.60) as a compatibility condition. It is for this reason that, in the first order theory, we allow ourselves to differ from the Chapman–Enskog results by terms $\mathrm{O}(\tau^2)$. In particular, we have for the trace of the pressure tensor

$$\mathrm{Tr}\,\mathbb{P} = 3\,p\left(1 - \frac{\tau}{2C_v}\dot{S}\right) + \mathrm{O}(\tau^2) . \qquad (26.61)$$

This result differs from the exact trace $(3p)$ by $O(\tau^2)$ and this, we suggest is allowed in a first order-theory.

We may add that in comparing the results of this approach to experiments on ultrasound we find that they do better than the usual Navier–Stokes version. Here we have an interesting example of a dictum of J. B. Keller: "Two theories may have the same accuracy but different domains of validity."

We regret that though, in honour of Douglas Gough's birthday, we have gone to second order in this approach (for only the relaxation model so far), we could not fit the derivations into the space we were allotted in this volume. So those results will have to be presented elsewhere. We are happy to report that, in that next order, the trace of our pressure tensor differs from the exact trace by $O(\tau^3)$. For now we must be content with mentioning that result and presenting our best wishes to Douglas on his birthday.

References

Bhatnager, P. L., Gross, E. P. & Krook, M., 1954, *Phys. Rev.*, **94**, 511

Chen, X., 2000, Columbia University Astronomy Department Ph.D. thesis

Chen, X., Rao, H. & Spiegel, E. A., 2000, *Phys. Lett. A*, **271**, 87

Chen, X., Rao, H. & Spiegel, E. A., 2001, *Phys. Rev. E*, **64**, 046308

Clemmow, P. C. & Dougherty, J. P., 1969, *Electrodynamics of Particles and Plasmas*, Addison-Wesley

Cohen, E. G. D., 1997, *Mathematical Models and Methods in Applied Sciences* **7**, 909

Griv, E., Gedalin, M. & Eichler, D., 2001, *ApJ*, **555**, L29

Jeans, J. H., 1929, *Astronomy and Cosmogony*, Cambridge University Press

Kato, S. & Kumar, S. S., 1960, *PASJ*, **12**, 290

Ogorodnikov, K. F., 1965, *Dynamics of stellar systems*, Pergamon

Pacholczyk, A. G. & Stodólkiewicz, J. S., 1959, *Bull. Acad. Polonaise des Sciences*, **7**, No. 11

Rosenbluth, M. N., MacDonald, W. M. & Judd, D. L., 1957, *Phys. Rev.*, **107**, 1

Spiegel, E. A., 1998, in Harvey, A., ed., *On Einstein's Path: Essays in Honor of Engelbert Schucking*, (Springer-Verlag) p. 465

Uhlenbeck, G. E. & Ford, G. W., 1963, *Lectures in Statistical Mechanics*, American Mathematical Society

Welander, P., 1954, *Ark. f. Fysik*, **7**, 507

27

Formation of planetary systems

DOUGLAS N. C. LIN

*Department of Astronomy & Astrophysics, University of California,
Santa Cruz, CA 95064, USA*

*The discovery of extrasolar planets and the determination of their orbital
properties have provided golden opportunities for new advancements in the
quest to understand the origin and evolution of planets and planetary systems. While their bizarre variety presents a challenge for the existing theories, their ubiquity suggests that planetary formation is a robust process.
Combining data obtained from solar system exploration, star formation studies and the searches for extra solar planets, we address some outstanding
issues concerning critical processes of grain condensation, planetesimal coagulation, and gas accretion. Some implications of these investigations are:
1) the amount of heavy elements available for planetary formation in protostellar disks is retained at a similar level as that empirically inferred for the
primordial solar nebula, through self regulated processes and 2) the critical
stages of planet formation, from grain condensation, planetesimal coagulation, to gas accretion, proceed on the timescale of a few million years.*

27.1 Observations

Ongoing searches of extra solar planets (ESPs) have led to their discovery
around ten per cent of the solar-type stars on various target lists (Marcy
& Butler 1998). The dynamical properties of many ESPs are very different
from those of planets in the solar system. The first ESP discovered, while
having a mass (M_p) similar to that of Jupiter (M_J), is located 100 times
closer to its host star 51 Peg than Jupiter is to the Sun (Mayor & Queloz
1995). The period (P) distribution of ESPs has a noticeable concentration
between 3-7 days. Nearly all of these short-period ESPs have negligible
orbital eccentricity e. A nearly logarithmic distribution of periods is found
for ESPs with $P > 7$ days. Those ESPs with periods greater than 21 days

have a nearly uniform distribution of eccentricities between 0 and ~ 0.7 (cf. $e \simeq 0.05$ for both Jupiter and Saturn).

The masses of ESPs projected along the line of sight (with an inclination angle i to their orbital axes), inferred from the radial velocity measurements, have a distribution which declines with $M_p \sin i$. Below an upper cutoff at $\sim 10 M_J$, many planets are found with $M_p \sin i < M_J$, despite the difficulties associated with their detection. The radii of ESPs are not known except for a $0.69 M_J$ planet around HD 209458, whose radius has been measured during its transit across the star to be 1.347 ± 0.060 R$_J$ where R$_J$ is Jupiter's radius (Brown et al. 2001). The low density of this planet indicates that it is composed mostly of hydrogen and helium. Two or more planets have been observed to orbit around v And, HD 168443 and HD 74156 (Butler et al. 1999). Signs of multiple planets are also found around a large fraction of all the stars with planets (Fischer et al. 2001). The planets around GJ 876 and HD 82943, are in mean motion resonances. The M_p and P distributions within these systems are different from those in the solar system.

The average metal abundance of planet-bearing stars is 0.17 ± 0.20 dex higher than that of the Sun (Gonzalez 1998, Santos et al. 2001). This association suggests that planets are either preferentially formed near metal-rich stars or their presence tends to enrich their host stars. The first scenario is consistent with the non-detection of any short-period planets in the metal-deficient globular cluster 47 Tuc. An observation in support of the second scenario is the discovery of ^{6}Li in a mature planet-bearing star HD 82943 (Israelian et al. 2001). Since, ^{6}Li burns at relatively low temperatures, it is expected to be depleted during HD 82943's pre main sequence evolution and its current content is probably added to the star after the temperature at the base of its convection zone has declined to the present value (Sandquist et al. 2002). Finally, the observed upper limit on the dispersion of iron and calcium abundance among 20 coeval G dwarfs in the Pleiades is less than 0.02 dex (Wilden et al., in preparation). This upper limit corresponds to a difference $\Delta M_z < 1 M_J$ in the total mass content of heavy elements if the interior of these stars is chemically homogeneous or less than $6 M_\oplus$ ($M_\oplus$ is the mass of the Earth) if the dispersion is limited to the convection zone. The discovery of any ESPs with masses greater than $1 M_J$ around these stars would imply that they are mostly composed of H and He.

In the solar system, the planets' coplanar orbits motivated Laplace to postulate that they were formed in a primordial solar nebula. In the past two decades, protostellar disks which resemble the hypothetical solar nebula have been directly observed and extensively studied. Surveys in the mid-infrared indicate that nearly all young stars have disks when they are formed

(Beckwith 1999). The radial extent of the resolved disks is in the range 10 – 1000 AU, and the corresponding inferred mass in dust is estimated to be 0.01 – 0.1 times the total heavy element content of their host stars (Beckwith et al. 1990). Using spectroscopic observations from the Infrared Space Observatory (ISO), between 1-10M_J of molecular hydrogen has been directly detected in some protostellar disks (Thi et al. 2001). The mass accretion rate from these disks to their host stars, is estimated from their UV radiation to be $\dot{M}_{\text{disk}} \sim 10^{-8}\,\text{M}_\odot\,\text{yr}^{-1}$ for stars with an age of 10^6 yr, and, in a rough correlation, decreases with age (Calvet et al. 2000). The observed spectral energy distributions can be fitted with models in which the disks' mid-plane temperature T_C, surface temperature T_S and surface density of dust particles Σ_p scale with radius R as $T_C \propto R^{-0.5}$, $T_S \propto R^{-0.6}$ and $\Sigma_p \propto R^{-1.5}$.

These disk properties are similar to those inferred from a "minimum mass solar nebula" (MMSN) model which is constructed under the assumption that planets were formed *in situ* by acquiring *all* the heavy elements near their orbits (Hayashi et al. 1985). From the current distribution of the planets in the solar system, the inferred surface density of metals (Σ_z) is proportional to $R^{-1.5}$. Augmenting the metal abundance (Z) with hydrogen and helium, the gas surface density is assumed to be $\Sigma_{\text{gas}} = \Sigma_p/Z$ so the mass within 40 AU is $\sim 20 M_J$. The MMSN model also assumes that the grains are heated by the solar radiation and the gas is heated by the hot grains through conduction and cooled through molecular-band radiation. The inferred $T_C \propto R^{-0.5}$ which is consistent with the distribution of condensation temperature deduced for the dominant material in the terrestrial planets and the satellites of Jupiter and Saturn (Lewis 1972).

The life span of disks with physical conditions similar to those in the MMSN is brief. Infrared radiation from the dust components fades on a timescale $\tau_{\text{disk}} < 3-10\,\text{Myr}$. The decline of the dust signatures could simply mean that the dust has coagulated into larger particles (sizes greater than μm), which radiate much less efficiently. Disks with a few Jupiter masses of H_2 are found around T Tauri stars with ages $\tau_* \sim 1-10$Myr but the dust to gas ratio in these disks is significantly larger than the solar value of Z (Thi et al. 2001). There is no evidence for the life time of the gaseous component to be any longer than that of the dust (Haisch et al. 2001). Thus there is an observed constraint on the formation time of gaseous planets: it is 10 Myr or less.

The above brief summary of observational data on ESPs and their nascent disks indicate that their formation is a robust and complex process. The traditional concept which regarded the solar system as prototype has been challenged by this diversity of planetary properties. Some of the important questions that are raised by these data can be stated as follows:

(i) Is the over abundance of heavy elements in their host stars the cause or consequence of the formation of ESPs?

(ii) What processes are the primary mechanisms for planet formation? What is the timescale and efficiency of planetary formation?

(iii) What conditions determine the range and distribution of planetary masses? What is the origin of planets' orbital diversity?

(iv) What physical effects regulate the mass distribution and dynamical properties of planetary systems?

Here, we adopt the conventional scenario that planets are formed in the protostellar disks through grain condensation, planetesimal coagulation, core formation and gas accretion. This scenario is motivated by the chondrule structure of the primitive meteorites and the cratered surfaces of asteroids, satellites and terrestrial planets. The inferred core structure of gaseous giant planets also provides supporting evidence for sequential growth.

27.2 Grain condensation and growth

In the sequential-growth scenario, the size distribution and surface density of dust particles in the disks are the important initial conditions of planet formation. For example, the MMSN model is based on the assumptions of homogeneous solar composition and efficient retention of heavy elements by the planets. These assumptions can only be satisfied if the heavy elements are thoroughly mixed as they condense into grains which then coagulate and grow into full-size planetary bodies without any significant migration.

The relatively low CO abundance observed in some protostellar disks suggests an efficient condensation of volatile gases into small grains (Thi et al. 2001). Thorough mixing of μm-size grains arises naturally since protostellar disks are expected to be intrinsically unstable to convection in the vertical direction (Lin & Papaloizou 1980). But drag by the turbulent eddies is less effective on the modest- and large-size particles. These particles settle towards and become more concentrated near the mid plane as they grow (Supulver & Lin 2001). While large particles follow the Keplerian speed (V_k) which is determined by the central stars' gravity, the mean motion of the gas is slightly modified by its pressure gradient in the radial direction. In most regions of the disk, the radial pressure gradient is negative, so the azimuthal velocity of the gas is generally slower than V_k. Through molecular collisions, gas drag force takes angular momentum from the particles and induces them to undergo orbital decay (Adachi et al. 1976). The particles' orbital evolution timescale (τ_p) is determined by their sizes (S_p), the surface density of gas (Σ_{gas}), and temperature T_C. For the MMSN model, $\tau_p > 1\,\text{Myr}$ for $S_p < 100\,\mu\text{m}$, $\tau_p \sim 10^4\,\text{yr}$ for $S_p \sim 1\,\text{cm}$, and $\tau_p > 1\,\text{Myr}$ for $S_p > 1\,\text{km}$ at $R \simeq 1\,\text{AU}$. Unless it can be bypassed, the rapid orbital decay of the sub-mm to m-size grains would prevent them from growing into planetesimals and limit the retention efficiency of heavy elements.

In order to identify some physical effects which may overcome this growth barrier, Goldreich & Ward (1973) proposed that, in a turbulence-free nebula, dust may sediment into a thin layer which eventually becomes gravitationally unstable and produces km-size fragments. But, in a turbulent nebula, gas drag by eddies induces small particles to acquire dispersion velocity (σ_p) which prevents their sedimentation into such a thin layer unless they have already acquired km sizes. Nevertheless, the mm-size or larger particles can sediment towards layers with scale heights $H_p(S_p)$ smaller than that of the gas H_{gas}. When the mean density of the dust exceeds that of the gas near the mid plane, the gas is dragged by the dust to attain the local Keplerian speed while the particles' orbital evolution timescale τ_p is lengthened (Weidenschilling & Cuzzi 1993). However, both qualitative estimates based on Richardson criterion and linear stability analysis suggest that the vertical shear induced by this flow pattern may be unstable (Geraud & Lin, in preparation), in which case mixing between the gas layers near and above the mid plane would lead to angular momentum exchange and the resumption of particles' rapid orbital decay.

Random motion of the particles also leads to collisions, fragmentation and coagulation. The growth timescale τ_{grow} of the particles is determined by the magnitude of Σ_p. The ratio of $\tau_{\text{grow}}/\tau_p$ is a function of $\Sigma_p/\Sigma_{\text{gas}}$. In regions of disk where the latter is of order the solar metallicity $Z_\odot$, τ_{grow} exceeds τ_p for particles with sizes between $10^2\,\text{m}$ and $1\,\text{mm}$, such that they would sediment into a layer with $H_p < H_{\text{gas}}$ and undergo radial migration before they can grow. In addition to its sub-Keplerian azimuthal motion, the disk gas also flows in the radial direction due to the viscous transport of angular momentum. For the α prescription of viscosity, the disk gas flows outwards near the mid plane and inwards in the regions surrounding the mid plane (Kley & Lin 1992). Over a limited size range (sub-mm to mm), particles may sediment and be swept outwards near the mid plane, causing them to accumulate (Takeuchi & Lin 2002). The combined effects of sedimentation and accumulation enhance the size-sorted particles' average density and reduce their growth timescale until their surface density Σ_p becomes comparable to the inferred Σ_z and τ_{grow} becomes less than τ_p. Thereafter, particles undergo *in situ* runaway growth through the size barrier.

This dust-accumulation scenario provides a natural explanation for the magnitude of Σ_z in the MMSN model as a threshold Σ_p for runaway grain growth. Since only those particles within a limited size range can be retained in the disk, they constitute the basic building block of much larger planetesimals. This inference is consistent with the structure of the most primitive meteorites which are mostly composed of sub-mm to mm size chondrules. In Section 27.5, we suggest that an upper limit to Σ_p may be set by the rapid emergence and the disruptive migration of giant planets. The implication of this conjecture is that planetary formation is a self regulated and robust process which occurs around stars with different metallicities.

27.3 Planetesimal dynamics

The process of planetesimal growth has been studied extensively using various approaches. An outstanding challenge to the theory of protoplanet formation is how cores of several Earth masses may emerge on a timescale comparable to or less than τ_{disk}. The outcome of the planetesimals' growth is determined by how their dynamical properties may be affected by scattering and collisions amongst themselves and by their interaction with the ambient gas.

The growth of planetesimals is complex because their speed relative to each other is affected by the differential Keplerian speed and their mutual gravitational interaction is perturbed by the Sun. These effects suppress the contribution of distant encounters regardless of their average eccentricities. Close encounters lead to scatterings which transfer energy, at a rate $\dot{E}_k$, from the Keplerian shear into the planetesimals' random epicyclic motion and increases in velocity dispersion σ_p (Safronov 1969). But some close encounters also result in inelastic physical collisions which, along with the gas drag, damp σ_p. In a collection of planetesimals with a range of M_ps, energy is transferred between different populations towards energy equipartition (Stewart & Wetherill 1988). In a dynamical equilibrium, σ_p is a decreasing function of the size S_p and it is less than the surface escape speed, V_p of the most abundant planetesimals (Aarseth et al. 1993).

Assuming all collisions are cohesive, the planetesimals' mass growth rate $\dot{M}_p$ is given by $\dot{M}_p = \Sigma_p \Omega_k A_p$. For modest-size planetesimals, the collisional cross section can be approximated with $A_p \simeq \pi S_p^2 (1+\Theta)$ where the Safronov number $\Theta = R_{\mathrm{Bondi}}/S_p$ and the Bondi radius $R_{\mathrm{Bondi}} = GM_p/\sigma_p^2$. In the cold limit where $\Theta \gg 1$, the growth timescale $\tau_{\mathrm{grow}} = M_p/\dot{M}_p \propto S_p^{-1}$, leading to an initial phase of runaway growth in which the largest planetesimals double their mass before the field planetesimals have grown (Palmer et al. 1993).

As the mass of the most massive planetesimals (MMPs) increases, so does the rate of their excitation of the field planetesimals' velocity dispersion σ_p. Nevertheless, the field planetesimals would remain dynamically cool if their size is sufficiently small (less than a few km) for the gas drag to be effective. Within this background, the Bondi radius of the MMPs increases with their mass M_p. When their mass becomes greater than $M_{\rm emb} \equiv 3(\sigma_d/V_k)^{3/2} M_*$, their Bondi radius exceeds their own Roche radius $R_{\rm Roche}$ and they become protoplanetary embryos. Thereafter, the embryos' Safronov number Θ attains an asymptotic value ($\sim R_{\rm Roche}/S_p$) because their gravitational perturbation on the field planetesimals at large distances is limited by the tidal effect of the central star (with a mass $M_* \gg M_p$). Regardless of their σ_p, the field planetesimals can only collide with the embryos if their semi-major axis lengths (a) differ by less than $R_f \simeq \lambda R_{\rm Roche}$ where the number $\lambda \sim (12)^{1/2}$ (Greenzweig & Lissauer 1990). (The annulus within a width of $+R_f$ from an embryo's a is commonly referred to as its feeding zone.) In this limit, the growth timescale $\tau_{\rm grow} \propto S_p$ such that the growth advantage of the more massive embryos cannot be maintained. The formation of the first embryos with $M_p > M_{\rm emb}$ is followed by the emergence of other siblings with comparable masses. This orderly mass evolution is referred to as *oligarchic growth* (Kokubo & Ida 1998).

27.4 The final assemblage of terrestrial planets

When their mass is marginally larger than $M_{\rm emb}$, several embryos may coexist in each other's feeding zone. As they interact and merge with each other, their mass increases until it reaches an "isolation mass" $\sim M_{\rm iso} \equiv (4M_{\rm disk}/3M_*)^{3/2} 3\lambda M_*$ where $M_{\rm disk} = \pi \Sigma_p a^2$ is the disk mass. Thereafter, embryos evolve into protoplanets with very little additional growth potential unless their feeding zones are replenished by the diffusion of field planetesimals, enlarged by their gas accretion, or relocated by their orbital evolution.

Energy continues to be transferred into the protoplanets' dispersive motion at a rate $\dot{E}_k$ which is determined by their mutual distant encounters. Due to the lack of an adequate supply of nearby residual low-mass field planetesimals in their feeding zones, the protoplanets' dispersive energy cannot be efficiently damped through dynamical friction. Consequently, the protoplanets' velocity dispersion σ_p would increase until their $\dot{E}_k$ is balanced by the rate of their own collisional energy dissipation. In this new energy equilibrium, the protoplanets' σ_p would become comparable to their escape speed V_p, their $\Theta \sim 1$, their A_p would be reduced to their geometric cross section, and their growth timescale $\tau_{\text{grow}} \propto S_p$ which would continue to promote the oligarchic growth.

At $a = 1\,\text{AU}$ in a MMSN, M_{iso} at 1 AU is comparable to the mass of the Moon and $\tau_{\text{grow}} > 10\,\text{Myr} > \tau_{\text{disk}}$ for a $1\,M_\oplus$ protoplanet. Thus, the final assemblage of Earth-mass planets through the oligarchic growth process would require multiple cohesive collisions between protoplanets with substantial masses and large relative speeds after the nebula gas was mostly depleted and the Sun has already evolved onto the main sequence. Collisions between protoplanets or embryos with comparable masses would lead to rapid and randomly oriented spins similar to those observed in the terrestrial planets (Dones & Tremaine 1993). Since their $\Theta \sim 1$, their $\sigma_p \sim V_p \sim 10\text{km s}^{-1}$ for protoplanets with $M_p \sim M_\oplus$. Outside 1 AU, such a magnitude in σ_p would result in $e \sim \sigma_p/V_k \sim 0.3$, larger than the present eccentricities of Venus, Earth and Mars but comparable to those of the long-period ESPs.

The gaseous giant planets must be formed prior to the depletion of the gas. If terrestrial planets' $\tau_{\text{grow}} > \tau_{\text{disk}}$, their formation would need to be preceded and perturbed by the emergence of the giant planets. For example, energy equipartition would induce the less massive protoplanets and residual embryos to attain large eccentricities such that they would become more vulnerable to be scattered into the Sun by the giant planets' ν_6 secular resonance (Chambers & Wetherill 2001). The depletion of the high-σ_p low-M_p protoplanets and embryos would stimulate a further drainage of the high-M_p protoplanets' dispersion energy and the damping of their eccentricities. But it would also lead to the bombardment of heavy elements onto the Sun after the mass of its convection zone has reduced to about $0.02M_\odot$.

We now consider some observable implications of the post-formation accretion of the residual planetesimals. If the Sun's convective envelope and radiative interior are chemically segregated, this would enrich the surface metallicity relative to the interior. It has been proposed (e.g., by Jeffery et al. 1997) that such a 'dirty solar model' can provide a resolution for the solar neutrino problem and possibly the faint-sun paradox because a lower internal Z would decrease its current neutrino production rate and was claimed to increase its past luminosity. Jeffery et al. estimated that the accretion of 65 $M_\oplus$ would provide a reduction in the detection rate of high-energy solar neutrinos by a factor of two. However, it was pointed out by Christensen-Dalsgaard & Gough (1998), based on detailed calculations by Christensen-Dalsgaard, Gough & Morgan (1979), that the resulting changes in solar structure would be entirely incompatible with helioseismic evidence; they also noted that accretion would have little effect on the luminosity history of the Sun. Furthermore, it should be kept in mind that recent data from the Sudbury Neutrino Observatory (see Shibahashi, this volume) demonstrate that the neutrino production rate in the solar core is consistent with standard solar models.

To estimate the observable effects of more modest accretion we note that the accretion of 1 $M_\oplus$ of residual planetesimals, assuming no mixing or settling beyond the convection zone, would enrich the Sun's observable metallicity ($Z_\odot$) in its surface layer by $\sim 0.006 Z_\odot$. The corresponding reduction in the interior Z, relative to standard solar models, would probably be undetectable from the helioseismic results. However, the post-formation accretion of all the terrestrial planets, with a total mass $M_t \sim 3 M_\oplus$, would change Z at the level of $0.02 Z_\odot$ which might be marginally detectable with helioseismology.

There are other observations which provide some constraints on the amount of the post-formation accretion of heavy elements. For example, if the heavy-element contaminant is confined to their convection zone, the observed upper limit in the Z dispersion (less than 5%) among the G dwarfs in the Pleiades cluster (see Section 27.1) would imply $\Delta M_z \sim 6 M_\oplus$. Similarly, if the presence of ^{6}Li on the surface of the planet-bearing star HD 82943 is due to the post-formation pollution of its convection zone, its line strength would correspond to $\Delta M_z \sim 13 M_\oplus$. (Any mixing between the stellar convective and radiative regions would require the post-formation accretion of more solid material.) Although $\Delta M_z > M_t$, it is comparable to the inferred solid core mass ($M_{core} \simeq 10 M_\oplus$) of giant planets (Wuchterl et al. 2000) and it may be attained either through the accretion of a large population of residual embryos or a gaseous giant planet (Sandquist et al. 2002). With a chemically inhomogeneous interior, the seismology of HD 82943 is likely to be significantly different from that of the Sun.

The main theoretical paradox associated with the oligarchic growth is the requirement for the outer giant planets to form before the inner terrestrial planets. In conventional models of giant planet formation, gas accretion is preceded by the formation of a core with a mass $M_{core} \sim 10 M_\odot$ (see Section 27.5). The relatively low T_C in the outer part of the nebula is favorable for the condensation of volatile grains which provides a three-fold enhancement in Σ_p. The width of the feeding zone for a given M_p also increases with a. Both effects cause M_{iso} to increase with a. Nonetheless, $M_{iso} < M_{core}$ and $\tau_{grow} > \tau_{disk}$ (which make giant planet formation difficult) throughout the nebula unless Σ_p is several times that of the MMSN value (Lissauer 1993). But large Σ_p is inconsistent with the chemical homogeneity among the Pleiades stars and the observed mass distribution of dust disks (Section 27.1). Enhanced Σ_p is also likely be accompanied by large Σ_{gas} which would lead to the formation of massive protoplanets and promote orbital migration (Section 27.5).

The isolation bottleneck (and the above paradoxical requirement for low-e terrestrial planets) may be bypassed if the feeding zone is continually replenished (Bryden et al. 2000a) either by the diffusion of field planetesimals or the relocation of the protoplanets' orbit. Although gas drag induces small (km-size) planetesimals to undergo orbital decay, they would be captured onto the mean motion resonances of and avoid collisions with the protoplanets if their migration timescale across the feeding zones is larger than the libration timescales of the protoplanets' mean motion resonance. Scattering and collisions also induce the protoplanets to undergo wandering Brownian motion (Ida et al. 2000) which may lead to an adequate replenishment of the field planetesimals in their feeding zone. This effect would sustain the runaway growth until the protoplanets have depleted most of the heavy-element content in the inner MMSN or their mass becomes a few $M_\oplus$ in the outer MMSN. The main attraction of this extended runaway growth is that Earth-mass protoplanets may be able to emerge from the largest embryos with small eccentricities on a timescale less than 1 Myr at 1 AU. The remaining uncertainty is whether embryos can attain low σ_p despite perturbation by nearby siblings.

27.5 Giant planet formation through gas accretion

Since giant planets are primarily composed of gas, they must be formed in a gaseous environment with $\tau_{grow} < \tau_{disk}$. After its mass becomes larger than a few lunar masses, the surface escape speed of a protoplanetary core (V_p) exceeds the gas sound speed (c_s) in the MMSN. Initially, low-mass cores are surrounded by thin gaseous envelopes which are heated at their base by the release of gravitational energy from the contracting gas and the impinging planetesimals. The gas sedimentation rate $\dot{M}_{gas}$ onto the core is limited by the efficiency of heat transfer through the envelope and $\dot{M}_{gas}$ for low-mass cores is generally smaller than their Bondi accretion rate $\dot{M}_{Bondi}$ from the disk so that a quasi hydrostatic envelope is established. For $M_p < 10 M_\oplus$, $\dot{M}_{gas}$ is also smaller than the solid particle accretion rate $\dot{M}_p$ so that the gas would remain a minor fraction of the protoplanet's total mass until the planetesimals in its feeding zone is depleted (Pollack et al. 1996). Although gas accretion also enlarges the protoplanet's feeding zone and induces it to acquire additional planetesimals, its growth time τ_{grow} exceeds τ_{disk} in the MMSN model.

Several processes can lead to significant reductions in the present estimate of τ_{grow}. In the conventional analysis, the core's orbital migration is neglected such that $\tau_{\mathrm{grow}} < \tau_{\mathrm{disk}}$ requires Σ_p to be several times that of the MMSN model. But such a large Σ_p is inconsistent with the the small ΔZ among the Pleiades stars (Section 27.1). Alternatively, the feeding zones may be replenished if they are continually relocated by the protoplanets' orbital migration (Section 27.4). In addition to the Brownian motion induced by planetesimal scatterings, the orbits of protoplanets may also be affected by tidal interaction with their nascent disks even when their $M_p < 10 M_\oplus$ (Ward 1986).

In the conventional numerical models, the structure of the protoplanets is assumed to be spherically symmetric. The main radiation transfer bottleneck is associated with a radiative zone in the protoplanets' envelope (Pollack et al. 1996). But the envelope extends to the protoplanets' R_{Roche} where under the combined influence of the disk's differential rotation and the central star's tidal torque, gas acquires a large amount of spin angular momentum. Within R_{Roche}, the stellar perturbation induces shock dissipation which leads to efficient angular momentum transport. Preliminary 3-D radiative hydrodynamical simulations also show the onset of large circulation pattern in the protoplanetary envelope which may provide a much more efficient heat transport than that in the spherically symmetric models.

When the protoplanet's mass M_p exceeds $10 M_\oplus$, its envelope can no longer be supported by its own pressure (Mizuno 1980). The collapse of the envelope also leads to runaway gas accretion with $\dot{M}_{\mathrm{gas}} \gg \dot{M}_p$. This critical mass is comparable to M_{core} in the gaseous planets of the solar system. Although the critical mass does not depend on the protoplanets' a, the emergence of the gaseous planets probably occurs mainly beyond a few AU because the particle density interior to that may be inadequate to promote the formation of $10 M_\oplus$ cores.

For $M_p < M_J$ and $a \sim$ a few AU, the protoplanet's Bondi radius R_{Bondi} is smaller than both its Roche radius R_{Roche} and the scale height H_{gas} of the gas, so that its growth time $\tau_{\mathrm{grow}} \sim 10^3 (M_J/M_p)$ yr. As M_p approaches to its asymptotic values, τ_{grow} must become large compared with τ_{disk}. The termination of the protoplanet's rapid gas accretion requires the depletion of gas near its orbit on a timescale shorter than τ_{grow}. For $M_p \sim M_J$, the protoplanet's $R_{\mathrm{Bondi}} \sim R_{\mathrm{Roche}} \sim H_{\mathrm{gas}}$ and its tidal torque begins to perturb the disk flow near its orbit. Protoplanets excite density waves which carry negative/positive angular momentum flux at their inner/outer Lindblad resonances (Goldreich & Tremaine 1980). This flux of deficit/excess angular momentum is deposited into the gas as the waves dissipate during their propagation inward/outward from the resonances (Papaloizou & Lin 1984). If the rate of tidal angular momentum transport exceeds that of due to turbulent viscosity (ν), a gap would form in the vicinity of the protoplanet's orbit (Lin & Papaloizou 1986a). Gap formation quenches gas supply onto the protoplanet, reducing the sedimentation rate $\dot{M}_{\mathrm{gas}}$.

The gap formation criteria are estimated to be $M_p \simeq (40\nu/\Omega a^2)M_*$ and $R_{\mathrm{Roche}} > H_{\mathrm{gas}}$. Numerical simulations (Bryden et al. 1999, Kley 1999) confirm that gaps do begin to open when the above criteria are satisfied. But they also show that accretion onto the planet can continue through the gap to increase M_p. Nevertheless, when the gap formation criteria are satisfied, Σ_{gas} near the protoplanet's orbit is reduced by more than 3 order of magnitude from that elsewhere in the disk. Consequently, $\tau_{\mathrm{grow}} > \tau_{\mathrm{disk}}$ and M_p cannot be substantially increased thereafter. This termination process provides an explanation for the apparent upper limit in the mass distribution of ESPs. Many ESPs have masses comparable to or less than that of Saturn. According to the gap formation criteria, small asymptotic masses would be attainable in cold disk regions where $H_{\mathrm{gas}}/R < 0.05$ and viscosity parameter $\alpha < 10^{-3}$. Gas accretion would also be quenched if the disk is globally depleted on a timescale shorter than τ_{grow}. However, gas depletion over the entire disk also modifies the global potential which can induce secular resonance between multiple planets around a common host star (Nagasawa et al. 2002).

After the gap formation, protoplanets continue to interact tidally with the disk which provides a conduit for angular momentum transfer from the interior to the exterior regions of the disk (Goldreich & Tremaine 1980). Protoplanets, with M_p much less than the disk mass, migrate along with the viscous evolution of nearby disk fluid elements (Lin & Papaloizou 1986b). On a global viscous evolutionary timescale (τ_ν), protoplanets formed at small disk radii would migrated inward as the disk gas interior to their orbit loses angular momentum and is accreted. Short-period ESPs may have formed beyond several AUs and migrated to the vicinity of their host stars (Lin et al. 1996). The inward migration of protoplanets would be halted near their host stars if they enter into disk cavities induced by the stellar magnetosphere or if they receive angular momentum through their tidal interaction with their rapidly spinning host stars. Over time, the drainage of the host stars' angular momentum would lead to their spin down, causing a reversal in the tidal angular momentum transfer flux and the resumption of the protoplanets' inward migration.

Both Σ_{gas} and $\dot{M}_{\text{disk}}$ are observed to decrease with the stellar age (τ_*). Thus both τ_{grow} and τ_ν also decrease with τ_*. In some massive disks, such as those which have undergone recent FU Ori outbursts, both Σ_{gas} and Σ_p are much larger than those in the MMSN model. During the epoch when both τ_{grow} and τ_ν are shorter than τ_*, several protoplanets may form and migrate into their host stars. Although most residual particles interior to their orbits are captured onto and swept clean by the migrating protoplanets' mean motion resonances, gas and particles in the external disk expand in the protoplanets' wake and repopulate the entire disk with a lower Σ_p. Thus the magnitude of Σ_p may be self regulated with an upper limit which is set by the condition $\tau_{\text{grow}} \sim \tau_{\text{disk}}$. In this case, the ESPs are the last survivors of a series of protoplanetary formation, migration and disruption. The consumption of protoplanets is not expected to leave traceable effects on the structure of their young host stars which have extended convection zones. But they may regulate the spin velocity of their host stars and cause large variations in $\dot{M}_{\text{disk}}$ similar to the FU Ori outbursts.

The planet-star interaction cannot halt the protoplanets' migration at intermediate or large distance from the host stars, and the intermediate-period ESPs (with periods of weeks and months) may have stopped their migration as a consequence of timely depletion of the disk. The large observed eccentricities e of many ESPs is in contrast with the classical theoretical expectation of nearly circular orbits. During their formation, the isolated protoplanets' eccentricities are damped and excited as a result of their interaction with their nascent disk at their corotation and Lindblad resonances respectively. For protoplanets with $M_p \sim M_J$, the corotation resonances lead to e damping faster than the excitation effects of the Lindblad resonances (Goldreich & Tremaine 1980). But protoplanets with $M_p > 10 M_J$ open relatively wide gaps which contain the protoplanets' corotation resonances. Consequently, their eccentricities are excited by their Lindblad resonances with little damping (Artymowicz 1993, Papaloizou et al. 2001). Appreciable eccentricities are also observed among ESPs with $M \sin i \sim M_J$. For these ESPs, gravitational scattering by other planets (Rasio & Ford 1996, Lin & Ida 1997) also leads to e excitation. The early onset of dynamical instability among ESPs may also lead to the disruption of their nascent disks and the termination of their migration.

27.6 Formation of multiple planet systems

ESPs have been found around $\sim 10\%$ of the nearby stars on various search target lists. But signs of multiple planets are found around more than half of all the stars with planets (Fischer et al. 2001). Around a host star, the formation of a gap around one emerging planet leads to a positive pressure gradient which induces the disk gas to attain an azimuthal velocity which is greater than V_k. In this region, the tail-wind hydrodynamic drag exerted by the gas on the small solid particles induces their orbits to expand. This outward drift is in contrast to other regions of the disk where the gas has a sub-Keplerian speed which induces the small solid particles to spiral inwards. This barrier causes solid particles to accumulate just beyond the outer edge of the gap. The enhanced Σ_p provides a favorable location for the formation of an additional protoplanetary core with an orbital radius approximately twice that of the original protoplanet (Bryden et al. 2000b).

A system of three planets is found around a nearby G dwarf star, v And. Their 4 days, 8 months and 2 years periods suggest that the second and third planets probably formed after the first planet has already attained to its present location but before the two of them have migrated significantly, i.e. $\tau_{\rm grow}$ must be comparable to τ_ν. After their formation, each planet clears a gap which is centered on its orbital radius a. The residual gas between the planets forms rings. For planets with orbits separated by at least several $H_{\rm gas}$, the rings between them may be preserved in a manner analogous to the shepherded rings of Saturn and Uranus. In such a configuration, angular momentum is transferred from the inner disk to the inner planet, the ring, the outer planet and then to the outer disk. The presence of a ring between any pair of protoplanets prevents them from evolving close to each other and achieving orbital resonance. Nevertheless, these planets interact with each other along their migration paths. The gravitational perturbation between the planets causes their orbits to precess. Between successive apocenter passages, angular momentum is transferred between the planets by an amount which is determined by the differential longitude of periastron. But the orbital energy of each planet is essentially conserved so that the angular momentum exchange leads to a modulation in the e's of both planets. The gravitation potential of a MMSN disk also induces a precession of the protoplanets' orbit, generally at a higher rate than that due to their secular interaction. Consequently, the efficiency of e excitation may be limited.

In contrast, the density waves excited by a pair of closely separated protoplanets propagate throughout the ring, and non-local dissipation of these waves leads to gas leakage from the ring edges into the gaps. After the ring is depleted, the separation between the planets tends to decline as a result of angular momentum exchange between them and the surrounding inner and outer disk. For a disk with moderate viscosity, the timescale for the planets to approach each other is less than τ_{disk}. As these planets approach low-order mean motion resonances, they exchange both angular momentum and energy which results in a modulation of the planets' e and a. The amount of energy exchange in a system is determined by how close to exact resonance it is (cf. Murray & Dermott 2000). Energy exchange between the resonant planets is oscillatory whereas that between planets and disks is monotonic and uni-directional. Thus, the two resonant planets continue to migrate independently until they are sufficiently close to resonances that the rate of energy transfer through the planets' resonant interaction become comparable to that due to planet-disk interaction. Thereafter, the two planets are locked in resonance and migrate together.

Two planets locked in a 2:1 mean motion resonance have been discovered around GJ 876. Other resonant planets have also been reported. These systems provide the strongest evidence for orbital migration (Lee & Peale 2001) since they are unlikely to have formed in such special configuration. In those cases where the kinematic configuration is well determined, the resonant capture condition can be inferred. For example, in order for the planets around GJ 876 to be captured into their present resonant configuration, their viscous evolutionary timescale τ_ν must be of order 10^5 yr, which is consistent with that expected for a Jupiter-mass planet in the MMSN or a protostellar disks with $\dot{M}_{\mathrm{disk}} \sim 10^{-8} M_\odot$ yr^{-1}.

After the resonant capture, the outer planet continues to lose angular momentum and energy to the disk exterior to its orbit. Although the inner planet is too far removed from the outer disk region, it, nonetheless, loses angular momentum and energy through its resonant interaction with the outer planet. But the rates of energy and angular momentum losses are constrained by an adiabatic invariant such that they lead to both decreases in a and increases in e for both planets. This trend for eccentricities to increase is also found in the tidally driven outward expansion of the Galilean satellites (Peale et al. 1979, Lin & Papaloizou 1979) and the resonant capture of Kuiper Belt Objects by Neptune (Malhotra 1996, Ida et al. 2000). Provided a decreases at a modest pace, the rate of growth of e due to resonant migration would exceed the damping rate of e due to the planet-disk interaction through corotation resonances. Without a more effective damping process, the eccentricity of the migrating resonant planets would increase indefinitely until their orbits become unstable. But with a modest e, the radial excursion of the outer planet would enable it to venture outside the gap where non-linear e dissipation occurs.

The increase in the migrating resonant planets' e's enlarges the width of higher order mean motion resonances, as well as increases the magnitude of the planets' disturbing function and the energy exchange rate associated with the resonant interaction. For each resonance, the exchange of energy and angular momentum between the resonant planets causes their a and e to modulate on the appropriate resonant and secular libration timescales. For critical sets of values of the orbital parameters, pairs of resonances overlap triggering the onset of dynamical instability. The subsequent evolution of unstable systems can lead to e excitation, disk clearing and planet ejection. These processes together account for the observed eccentricity and period distributions among the ESPs as well as the origin of freely floating planets.

Acknowledgements I thank P. Bodenheimer, G. Bryden, D. Fischer, P. Geraud, P. Gu, S. Ida, W. Kley, R. Mardling, M. Nagasawa, E. Sandquist and T. Takeuchi for useful conversation, and NASA and NSF for support.

References

Aarseth, S. J., Lin, D. N. C. & Palmer, P. L., 1993, *ApJ*, **403**, 351

Adachi, I., Hayashi, C. & Nakazawa, K., 1976, *Prog, Theor. Phys.*, **56**, 1756

Artymowicz, P., 1993, *ApJ*, **419**, 166

Beckwith, S. V. W., Sargent, A. I., Chini, R. S. & Gusten, R., 1990, *AJ*, **99**, 924

Beckwith, S. V. W., 1999, in Lada, C. J. & Kylafis, N. D., eds, *The Origins of Stars and Planetary Systems*, Kluwer Academic Publishers: Dordrecht, p. 579

Brown, T. M., Charbonneau, D., Gilliland, R. L., Noyes, R. W. & Burrows, A., 2001, *ApJ*, **552**, 699

Bryden, G., Chen, X. M., Lin, D. N. C., Nelson, R. P. & Papaloizou, J. C. B., 1999, *ApJ*, **514**, 344

Bryden, G., Lin, D. N. C. & Ida, S., 2000a, *ApJ*, **544**, 481

Bryden, G., Rozyczka, M., Lin, D. N. C. & Bodenheimer, P., 2000b, *ApJ*, **540**, 1091

Butler, R. P., Marcy, G. W., Fischer, D. A., Brown, T. M., Contos, A. R., Karzennik, S. G., Nisenson, P. & Noyes, R. W., 1999, *ApJ*, **526**, 916

Calvet, N., Hartmann, L. & Strom, S. E., 2000, in Mannings, V., Boss, A. P. & Russell, S. S., eds, *Protostars and Planets IV*, Univ. of Arizona Press: Tucson, p. 377

Chambers, J. & Wetherill, G., 2001, *Meteorites, Planet. Sci.*, **36**, 381

Christensen-Dalsgaard, J. & Gough, D. O., 1998, *The Observatory*, **118**, 25

Christensen-Dalsgaard, J., Gough, D. O. & Morgan, J. G., 1979, *A&A*, **73**, 121

Dones, L. & Tremaine, S., 1993, *Icarus*, **103**, 67

Fischer, D. A., Marcy, G. W., Butler, R. P., Vogt, S. S., Frink, S. & Apps, K., 2001, *ApJ*, **551**, 1107

Goldreich, P. & Tremaine, S., 1980, *ApJ*, **241**, 425

Goldreich, P. & Ward, W. R., 1973, *ApJ*, **183**, 1051

Gonzalez, G., 1998, *A&A*, **334**, 221

Greenzweig, Y. & Lissauer, J., 1990, *Icarus*, **87**, 40

Haisch, K. E. J., Lada, E. A. & Lada, C. J., 2001, *ApJ*, **553**, L153

Hayashi, C. Nakazawa, K. & Nakagawa, Y. 1985, in Black, D. C. & Matthews, M.S., eds, *Protostars and Planets II*, Univ. of Arizona Press: Tucson, p. 1100

Ida, S., Bryden, G., Lin, D. N. C. & Tanaka, H., 2000, *ApJ*, **534**, 428

Israelian, G., Santos, N. C., Mayor, M. & Rebolo, R., 2001, *Nature*, **411**, 163

Jeffery, C. S., Bailey, M. E. & Chambers, J. E., 1997, *The Observatory*, **117**, 224

Kley, W., 1999, *MNRAS*, **303**, 696

Kley, W. & Lin, D. N. C., 1992, *ApJ*, **397**, 600

Kokubo, E. & Ida, S., 1998, *Icarus*, **131**, 171

Lee, M. H. & Peale, S. J., 2001, *ApJ*, **567**, 596

Lewis, J. S., 1972, *Icarus*, **16**, 241

Lin, D. N. C., Bodenheimer, P. & Richardson, D.C., 1996, *Nature*, **380**, 606

Lin, D. N. C. & Ida, S., 1997, *ApJ*, **477**, 781

Lin, D. N. C. & Papaloizou, J., 1979, *MNRAS*, **188**, 191

Lin, D. N. C. & Papaloizou, J. C. B., 1980, *MNRAS*, **191**, 37

Lin, D. N. C. & Papaloizou, J. C. B., 1986a, *ApJ*, **307**, 395

Lin, D. N. C. & Papaloizou, J. C. B., 1986b, *ApJ*, **309**, 846

Lissauer, J. J., 1993. *ARAA*, **31**, 129

Malhotra, R., 1996, *BAAS DPS Meeting*, **28**, 1082

Marcy, G. & Butler, R. P., 1998, *ARAA* **36**, 57

Mayor, M. & Queloz, D., 1995, *Nature*, **378**, 355

Mizuno, H., 1980, *Prog. Theor. Phys.* **64**, 544

Murray, C. D. & Dermott, S. F., 2000, *Solar System Dynamics*, Cambridge University Press: Cambridge

Nagasawa, M., Lin, D. N. C. & Ida, S., 2002, *ApJ*, submitted

Palmer, P. L., Lin, D. N. C. & Aarseth, S. J., 1993, *ApJ*, **403**, 336

Papaloizou, J. C. B. & Lin, D. N. C., 1984, *ApJ*, **285**, 818

Papaloizou, J. C. B., Nelson, R. P. & Masset, F., 2001, *A&A*, **366**, 263

Peale, S. J., Cassen, P. & Reynolds, R. T., 1979, *Science*, **203**, 892

Pollack, J. B., Hubickyj, O., Bodenheimer, P., Lissauer, J. J., et al., 1996, *Icarus*, **124**, 62

Rasio, F. A. & Ford, E. B., 1996, *Science*, **274**, 954

Safronov V., 1969, *Evolution of Protoplanetary Clouds and Formation of the Earth and Planets*, Nauka Press: Moscow (English transl. 1972)

Sandquist, E. L., Dokter, J. J., Lin, D. N. C. & Mardling, R. A., 2002, *ApJ*, **572**, 1012.

Santos, N. C., Israelian, G. & Mayor, M., 2001, *A&A*, **373**, 1019

Stewart, G. R. & Wetherill, G. W., 1988, *Icarus*, **79**, 542

Supulver, K. & Lin, D. N. C., 2001, *Icarus*, **146**, 525

Takeuchi, T. & Lin, D. N. C., 2002, *ApJ*, submitted

Thi, W. F., van Dishoeck, E. F., Blake, G. A., et al., 2001, *ApJ*, **561**, 1074

Ward, W. R., 1986, *Icarus*, **67**, 164

Weidenschilling, S. J. & Cuzzi, J. N., 1993, in Levy, E. H. & Lunine, J. I., eds, *Protostars and Planets III*, Univ. of Arizona Press: Tucson, p. 1031

Wuchterl, G., Guillot, T. & Lissauer, J.J., 2000, in Mannings, V., Boss, A. P. & Russell, S. S., eds, *Protostars and Planets IV*, Univ. of Arizona Press: Tucson, p. 1081

28

The solar-cycle global warming as inferred from sky brightness variation

WASABURO UNNO

Dept. Astronomy, Univ. of Tokyo, Bunkyo-ku, Tokyo 113-0033; and
Senjikan Future Study Institute, 4-15-12 Kichijoji, 180-0003, Japan

HIROMOTO SHIBAHASHI

Dept. Astronomy, Univ. of Tokyo, Bunkyo-ku, Tokyo 113-0033, Japan

In succession to our paper dedicated to Ed Spiegel, we proceed to establish a proportionality relation between the solar-cycle variation of the sky-brightness and that of the global warming. The increase of the optical depth appearing in the sky brightness may cause the solar-cycle global warming of a few degrees from the minimum to the maximum.
We wish to dedicate this paper to Douglas, in celebration of his 60th birthday anniversary.

28.1 Introduction

Solar magnetism not only controls the solar activity but also influences significantly the structure of the convection zone (Gough, 2001). On the other hand, the influence of solar activity on terrestrial meteorology such as found in tree rings, etc., has long been the subject of discussion (Eddy, 1976) but without finding the definitive causal relation explaining the physics involved. Recently, however, Sakurai (2002) analysed data of the sky background brightness observed with the Norikura coronagraph over 47 years (1951-1997) and found a clear 11.8-year periodicity as well as the marked annual variation, both exceeding the 95 per cent confidence level.

The annual variation is apparently meteorological, e.g., the famous Chinese yellow soil particles (rising up to 100 thousand feet high! – old Chinese sayings). The solar-cycle variation is also considered to be caused by increased aerosol formation (Sakurai, 2002); but if the solar activity changes the chemistry in the upper atmosphere; the observed time lag of 2 to 4 years of the sky-brightness variation relative to sunspot maximum is somewhat enigmatic.

In our preceding paper (dedicated to Ed Spiegel; Unno et al., 2002), we have proposed a model of the diffusion wave propagation of aerosol seed particles to explain the solar-cycle variation of the sky brightness and its phase delay.

The sky brightness is about 75 in units of $10^{-6}I_{\odot}$ on the average, where $I_{\odot}$ means a spectral intensity of the solar disk center at around 5300 Å. Random monthly fluctuations of about 30 to 40 in the same units are superposed on the annual and the solar-cycle components; the latter are some 15 % and 10 % in amplitude, respectively, in units of the average sky brightness.

The variations in sky brightness imply an optical-depth variation which would affect the global warming through the greenhouse effect. The present study attempts to coordinate the sky brightness and the greenhouse effect by solving the radiative transfer and to estimate the solar-cycle global warming from the sky-brightness variation.

28.2 Radiative transfer in the earth atmosphere

The equation of radiative transfer in the terrestrial atmosphere is given by

$$\mu\frac{\mathrm{d}I_{\nu}}{\mathrm{d}t} = (\kappa_{\nu} + \sigma_{\nu})I_{\nu} - \kappa_{\nu}B_{\nu} - \sigma_{\nu}J_{\nu} - (\kappa_{\nu} + \sigma_{\nu})S^{\mathrm{direct}}_{\odot\nu}, \qquad (28.1)$$

where isotropic scattering is assumed for simplicity. Here I_{ν}, J_{ν} and B_{ν} are the (monochromatic) specific intensity, mean specific intensity and Planck function respectively; μ is the direction cosine of the radiation, κ_{ν} and σ_{ν} denote the absorption and scattering coefficients, $\mathrm{d}t \equiv -\rho\mathrm{d}z$ (t is measured inward), and the last term denotes the contribution from the intensity of the direct solar radiation.

This equation describes both the thermal radiation field dominated by the infrared radiation and the scattered radiation field dominated by the visible solar radiation. Integrating equation (28.1) over both the infrared and visible frequency ranges, we obtain

$$\mu\frac{\mathrm{d}I}{\mathrm{d}\tau} = I - \frac{\kappa}{\kappa + \sigma}B - \frac{\sigma}{\kappa + \sigma}J - \frac{\kappa_V}{\kappa + \sigma}S_0 e^{-\tau^*_V} \qquad (28.2)$$

and

$$\mu\frac{\mathrm{d}I_V}{\mathrm{d}\tau_V} = I_V - \frac{\sigma_V}{\kappa_V + \sigma_V}S_{\odot}e^{-\tau_V/\mu_{\odot}}, \qquad (28.3)$$

respectively, where

$$\mathrm{d}\tau = (\kappa + \sigma)\mathrm{d}t,$$

$$\mathrm{d}\tau^*_V = \kappa_V\mathrm{d}t,$$

$$\mathrm{d}\tau_V = (\kappa_V + \sigma_V)\mathrm{d}t,$$

$$S_0 = \left(\frac{1}{16\pi}\right)\left(\frac{R_\odot}{a}\right)^2 \sigma T_\odot^4,$$

and

$$S_\odot = \left(\frac{\mu_\odot}{4\pi}\right)\left(\frac{R_\odot}{a}\right)^2 \sigma T_\odot^4 = 4\mu_\odot S_0.$$

Here σ denotes the Stefan-Boltzmann constant, $T_\odot$ ($=5780$ K) the effective temperature of the sun, and $R_\odot/a$ ($=2.3$ light-sec$/500$ light-sec $= 4.61 \times 10^{-3}$) is the solar radius in AU; $4\pi S_0$ is the solar energy flux per unit area averaged over the entire earth surface, while $4\pi S_\odot$ is the solar constant (1.37 kW/m^2); $\mu_\odot$ is the cosine of the angle of the sun from the zenith.

28.3 Radiative equilibrium model

For simplicity, we discuss here the radiative model to calculate the greenhouse effect in the earth atmosphere.

Equation (28.2) averaged over the whole solid angle results in

$$\frac{\mathrm{d}H}{\mathrm{d}\tau} = \frac{\kappa}{\kappa+\sigma}(J-B) - \frac{\kappa_V}{\kappa+\sigma}S_0 e^{-\tau_V^*},$$

where $J \equiv \frac{1}{2}\int_{-1}^{1} I\mathrm{d}\mu$ and $H \equiv \frac{1}{2}\int_{-1}^{1}\mu I\mathrm{d}\mu$. Assuming radiative equilibrium, so

$$\int_0^\infty \kappa_\nu(J_\nu - B_\nu)\mathrm{d}\nu = 0,$$

and grey absorption ($\kappa_\nu = \kappa$), we have $J = B$ and $H = S_0 e^{-\tau_V^*}$. As will be discussed in a subsequent section, the radiative model seems to be not so bad, perhaps because of the large heat capacity of the ground and oceans. Multiplying equation (28.2) by μ and averaging over the entire solid angle, we obtain $\mathrm{d}J/\mathrm{d}\tau = 3H = 3S_0 e^{-\tau_V^*}$ by using the Eddington approximation

$$K \equiv \frac{1}{2}\int_{-1}^{1}\mu^2 I\mathrm{d}\mu = \frac{1}{3}J.$$

Integrating this equation with respect to τ, and using the boundary condition $J(0) = 2H(0)$ at $\tau = 0$, we obtain

$$\begin{aligned}
J = B = \pi^{-1}\sigma T^4 &= 2S_0 + 3[(\kappa+\sigma)/\kappa_V]S_0(1 - e^{-\tau_V^*}),\\
&\simeq S_0(3\tau + 2) \qquad \text{for small } \tau_V^*.
\end{aligned}$$

Hence,

$$T^4 = \frac{3}{16}\left(\tau + \frac{2}{3}\right)\left(\frac{R_\odot}{a}\right)^2 T_\odot^4. \qquad (28.4)$$

Thus the temperature in the uppermost atmosphere $T(0)$ is given by

$$T(0) = 2^{-3/4} \left(\frac{R_\odot}{a} \right)^{1/2} T_\odot.$$

For comparison, the standard earth temperature $T_\oplus$ is defined by the black-body temperature with which the diluted solar radiation received by the area $\pi R_\oplus^2$ is emitted from the area $4\pi R_\oplus^2$, so that

$$T_\oplus = 2^{-1/2} \left(\frac{R_\odot}{a} \right)^{1/2} T_\odot = 2^{1/4} T(0). \tag{28.5}$$

The terrestrial atmosphere is not in radiative equilibrium. There are atmospheric and oceanic circulations on various scales. However, the oceans and the ground act as large heat reservoirs, and on average the thermal equilibrium condition may well be satisfied. The situation can be checked roughly by comparing the temperature stratification calculated in this way. We see from the model – equations (28.4) and (28.5) – that

$$T^4 = \frac{3}{4} \left(\frac{2}{3} + \tau \right) T_\oplus^4,$$

and

$$T_\oplus = 277.5 \,\mathrm{K}.$$

The τ-dependence is the greenhouse effect:

$$(\Delta T)_{\mathrm{GH}} = \frac{3}{16} \tau T_\oplus.$$

Now take this as the basis of the parameter fitting by putting $\tau = \tau_0 + \tau_{\mathrm{GH}}$. Here, τ_0 indicates the height in the atmosphere, and τ_{GH} the greenhouse effect. The average ground temperature T_{ground} is considered to be equal to $T_\oplus$ without the greenhouse effect ($\tau_{\mathrm{GH}} = 0$), and is taken from observation to be $T_{\mathrm{ground}} = 15°\mathrm{C} = 288 \,\mathrm{K}$ with τ_{GH} included. Then, $\tau_0 = 2/3$ for the ground level and the average greenhouse effect is given by $\tau_{\mathrm{GH}} = 0.88$. At the ground level, the increment of the greenhouse effect in the solar cycle is given by

$$\delta(\Delta T_{\mathrm{ground}})_{\mathrm{GH}} = 10.5 \left(\frac{\delta \tau_{\mathrm{GH}}}{\tau_{\mathrm{GH}}} \right)_{\mathrm{solar\ cycle}} \quad \mathrm{K}.$$

The last factor should be estimated from the sky brightness data.

The increase of τ by aerosol formation will give rise to the global warming. We now estimate the increase of opacity from the sky-brightness variation.

28.4 Sky brightness

Equation (28.3) can be easily solved with the help of the operator calculus; writing $D_{\tau_V} \equiv \mathrm{d}/\mathrm{d}\tau_V$, we have

$$
\begin{aligned}
I_V &= \frac{1}{\mu D_{\tau_V} - 1}\left[-\frac{\sigma_V}{\kappa_V + \sigma_V}S_\odot e^{-\tau_V/\mu_\odot}\right] \\
&= e^{\tau_V/\mu}\left(\frac{1}{\mu}\right)\int_0^{\tau_V}\mathrm{d}\tau_V\left[-\frac{\sigma_V}{\kappa_V + \sigma_V}S_\odot e^{-\tau_V(\mu_\odot^{-1}+\mu^{-1})}\right] \\
&= \frac{1}{\mu(\mu_\odot^{-1}+\mu^{-1})}\frac{\sigma_V}{\kappa_V + \sigma_V}S_\odot\left[e^{-\tau_V/\mu_\odot}+e^{\tau_V/\mu}\right].
\end{aligned}
$$

For sky brightness near the sun, $\mu = -\mu_\odot(1-\epsilon)$, and to the first order in ϵ, we obtain

$$
I_V = s_V I_\odot, \tag{28.6}
$$

where

$$
s_V \equiv \left(\frac{\sigma_V}{\kappa_V + \sigma_V}\right)\tau_V \tag{28.7}
$$

and $I_\odot$ denotes the apparent solar brightness:

$$
I_\odot = \frac{1}{4\pi}\left(\frac{R_\odot}{a}\right)^2 \sigma T_\odot^4 e^{-\tau_V/\mu_\odot}.
$$

28.5 Solar-cycle global warming

Sakurai's analysis gives that the sky brightness $I_V/I_\odot$ is 75×10^{-6}, and its solar-cycle variation is about $10\,\%$ of it. These values are supposed to be s_V and $(\delta s_V)_{\text{solar cycle}}$, or $s_V = 7.5 \times 10^{-5}$ and $(\delta s_V)_{\text{solar cycle}}/s_V \sim 0.1$. The corresponding change in the infrared optical depth τ will be given by $(\delta\tau)_{\text{solar cycle}}/\tau = k(\delta s_V)_{\text{solar cycle}}/s_V \sim 0.1k$. Here we have introduced the factor k to absorb possible differences in the properties of sources for κ between the (unknown) aerosols which are responsible for the solar-cycle variation in τ and the other sources (such as H_2O and CO_2) responsible for the steady infrared radiation. If the optical properties are the same for aerosols as other sources, $k = 1$. We estimate the solar-cycle global warming to be

$$
\delta[(\Delta T)_{\text{GH}}]_{\text{solar cycle}} \simeq \left[\frac{(\delta\tau_V)_{\text{solar cycle}}}{\tau_V}\right](\Delta T)_{\text{GH}} \sim 0.1k \times 10.5\,\text{K}. \tag{28.8}
$$

At present, the value of k is not known precisely. There will be a phase delay of 2 to 4 years for the solar-cycle global warming compared to the sunspot cycle.

28.6 Summary

One of the terrestrial phenomena associated with solar activity is the night sky light or the permanent aurora which is activated mostly by high energy UV photons. Therefore, there would be practically no time delay from the solar activity. The polar aurora, on the other hand, shows a complicated behaviour in the frequency of occurrence, probably because of the presence of coronal hole streamers which sweep the earth orbit more favorably when the active region lies on a certain heliographic latitude (see, e.g., Bone 1996). The situation seems to be different between the sky brightness and the aurora, to which UV photons and high speed plasma, respectively, could be the primary cause of activation.

In the preceding paper (Unno et al., 2002), the origin of the correlation between sky brightness and sunspot activity has been considered to be the modulation of aerosol contents in the upper atmosphere caused by the solar activity, as suggested by Sakurai (2002). In this working hypothesis, aerosol formation is considered to propagate as damping diffusion waves, and the phase delay is interpreted mainly to be due to the time required for propagation to the effective depth of aerosol formation and partly to the phase delay inherent to the diffusion wave propagation. This interpretation leads to the consequence that the optical-depth variation appearing in the sky brightness causes the global warming through the greenhouse effect. By solving the radiative transfer, we expect the solar-cycle global warming of a few degrees from the minimum to the maximum.

References

Bone, N., 1996, *The Aurora*, 2nd ed. (John Wiley & Sons), p. 116.
Eddy, J. A., 1976, *Science*, **192**, 1189.
Gough, D. O., 2001, *Nature*, **410**, 313.
Sakurai, T., 2002, *Earth, Planets and Space*, **54**, 153.
Unno, W., Shibahashi, H., & Yuasa, M., 2002, *PASJ*, submitted.